FABRICATE

FABRICATE
RETHINKING DESIGN AND CONSTRUCTION

ACHIM MENGES / BOB SHEIL / RUAIRI GLYNN / MARILENA SKAVARA

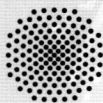
University of Stuttgart

ICD Institute for Computational Design and Construction

UCL

The Bartlett School of Architecture

CONTENTS

INTRODUCTION

- 8 FOREWORD
- 10 ACKNOWLEDGEMENTS
- 12 INTRODUCTION
- 16 THE TRIUMPH OF THE TURNIP
  Anthony Hauck, Michael Bergin, Phil Bernstein

1 RETHINKING PRODUCTION FUTURES

- 24 COOPERATIVE FABRICATION OF SPATIAL METAL STRUCTURES
  Stefana Parascho, Augusto Gandia, Ammar Mirjan, Fabio Gramazio and Matthias Kohler
- 30 INFINITE VARIATIONS, RADICAL STRATEGIES
  Martin Self and Emmanuel Vercruysse
- 36 AUTOMATED DESIGN-TO-FABRICATION FOR ARCHITECTURAL ENVELOPES: A STADIUM SKIN CASE STUDY
  James Warton, Heath May and Rodovan Kovacevic
- 44 ROBOTIC WOOD TECTONICS
  Philip F. Yuan and Hua Chai
- 50 RAPID ASSEMBLY WITH BENDING-STABILISED STRUCTURES
  Joseph M. Gattas, Kim Baber, Yousef Al-Qaryouti and Ting-Uei Lee
- 58 A PREFABRICATED DINING PAVILION: USING STRUCTURAL SKELETONS, DEVELOPABLE OFFSET MESHES, KERF-CUT AND BENT SHEET MATERIALS
  Henry Louth, David Reeves, Benjamin Koren, Shajay Bhooshan and Patrik Schumacher
- 68 OPEN CAGE-SHELL DESIGN AND FABRICATION (HEALING PAVILION)
  Benjamin Ball and Gaston Nogues
- 74 MAGGIE'S AT THE ROBERT PARFETT BUILDING, MANCHESTER
  Richard Maddock, Xavier de Kestelier, Roger Ridsdill Smith and Darron Haylock

2 RETHINKING MATERIALISATION

- 84 INFUNDIBULIFORMS: KINETIC SYSTEMS, ADDITIVE MANUFACTURING FOR CABLE NETS AND TENSILE SURFACE CONTROL
  Wes Mcgee, Kathy Velikov, Geoffrey Thün and Dan Tish
- 92 ROBOTIC INTEGRAL ATTACHMENT
  Christopher Robeller, Volker Helm, Andreas Thoma, Fabio Gramazio, Matthias Kohler and Yves Weinand
- 98 LACE WALL: EXTENDING DESIGN INTUITION THROUGH MACHINE LEARNING
  Martin Tamke, Mateusz Zwierzycki, Anders Holden Deleuran, Yuliya Sinke Baranovskaya, Ida Friis Tinning and Mette Ramsgaard Thomsen
- 106 ROBOTIC FABRICATION OF STONE ASSEMBLY DETAILS
  Inés Ariza, Brandon Clifford, James B. Durham, Wes McGee, Caitlin T. Mueller and T. Shan Sutherland
- 114 ADAPTIVE ROBOTIC FABRICATION FOR CONDITIONS OF MATERIAL INCONSISTENCY: INCREASING THE GEOMETRIC ACCURACY OF INCREMENTALLY FORMED METAL PANELS
  Paul Nicholas, Mateusz Zwierzycki, Esben Clausen Nørgaard, David Stasiuk, Christopher Hutchinson and Mette Thomsen
- 122 DIGITAL FABRICATION OF NON-STANDARD SOUND-DIFFUSING PANELS IN THE LARGE HALL OF THE ELBPHILHARMONIE
  Benjamin S. Koren and Tobias Müller
- 130 QUALIFYING FRP COMPOSITES FOR HIGH-RISE BUILDING FACADES
  William Kreysler
- 138 THE 2016 SERPENTINE PAVILION: A CASE STUDY IN LARGE-SCALE GFRP STRUCTURAL DESIGN AND ASSEMBLY
  James Kingman, Jeg Dudley and Ricardo Baptista

Q&A

148 Q&A: BIOGRAPHIES

150 Q&A 1
JENNY SABIN AND
MARIO CARPO

158 Q&A 2
MONICA PONCE DE LEON,
VIRGINIA SAN FRATELLO AND
RONALD RAEL

166 Q&A 3
CARL BASS, BOB SHEIL AND
ACHIM MENGES

174 Q&A 4
ANTOINE PICON AND
BOB SHEIL

3 RETHINKING ADDITIVE STRATEGIES

178 DISCRETE COMPUTATION FOR ADDITIVE MANUFACTURING
Gilles Retsin, Manuel Jiménez García and Vicente Soler

184 CILLLIA: METHOD OF 3D PRINTING MICRO-PILLAR STRUCTURES ON SURFACES
Jifei Ou, Gershon Dublon, Chin-Yi Cheng, Karl Willis and Hiroshi Ishii

190 FUSED FILAMENT FABRICATION FOR MULTI-KINEMATIC-STATE CLIMATE-RESPONSIVE APERTURE
David Correa and Achim Menges

196 3D METAL PRINTING AS STRUCTURE FOR ARCHITECTURAL AND SCULPTURAL PROJECTS
Paul Kassabian, Graham Cranston, Juhun Lee, Ralph Helmick, Sarah Rodrigo

202 MOBILE ROBOTIC FABRICATION SYSTEM FOR FILAMENT STRUCTURES
Maria Yablonina, Marshall Prado, Ehsan Baharlou, Tobias Schwinn and Achim Menges

210 THE SMART TAKES FROM THE STRONG: 3D PRINTING STAY-IN-PLACE FORMWORK FOR CONCRETE SLAB CONSTRUCTION
Mania Aghaei Meibodi, Mathias Bernhard, Andrei Jipa and Benjamin Dillenburger

218 PROCESS CHAIN FOR THE ROBOTIC CONTROLLED PRODUCTION OF NON-STANDARD, DOUBLE-CURVED, FIBRE-REINFORCED CONCRETE PANELS WITH AN ADAPTIVE MOULD
Hendrik Lindemann, Jörg Petri, Stefan Neudecker and Harald Kloft

224 ELYTRA FILAMENT PAVILION: ROBOTIC FILAMENT WINDING FOR STRUCTURAL COMPOSITE BUILDING SYSTEMS
Marshall Prado, Moritz Dörstelmann, James Solly, Achim Menges and Jan Knippers

4 RETHINKING CONSTRUCTIONAL LOGICS

234 SENSORIAL PLAYSCAPE: ADVANCED STRUCTURAL, MATERIAL AND RESPONSIVE CAPACITIES OF TEXTILE HYBRID ARCHITECTURES AS THERAPEUTIC ENVIRONMENTS FOR SOCIAL PLAY
Sean Ahlquist

242 BENDING-ACTIVE PLATES: PLANNING AND CONSTRUCTION
Simon Schleicher, Riccardo La Magna and Jan Knippers

250 PRECAST CONCRETE SHELLS: A STRUCTURAL CHALLENGE
Stefan Peters, Andreas Trummer, Gernot Parmann and Felix Amtsberg

258 FROM LAMINATION TO ASSEMBLY: MODELLING THE SEINE MUSICALE
Hanno Stehling, Fabian Scheurer, Jean Roulier, Hélori Geglo and Mathias Hofmann

264 SCALING ARCHITECTURAL ROBOTICS: CONSTRUCTION OF THE KIRK KAPITAL HEADQUARTERS
Asbjørn Søndergaard and Jelle Feringa

272 MPAVILION 2015
AL_A

280 MULTI-PERFORMATIVE SKINS
Edoardo Tibuzzi and Deyan Marzev

286 THE ARMADILLO VAULT: BALANCING COMPUTATION AND TRADITIONAL CRAFT
Philippe Block, Tom Van Mele, Matthias Rippmann and David Escobedo

END

295 EDITORS: BIOGRAPHIES

294 CONTRIBUTORS: BIOGRAPHIES

304 SPONSORS / COLOPHON

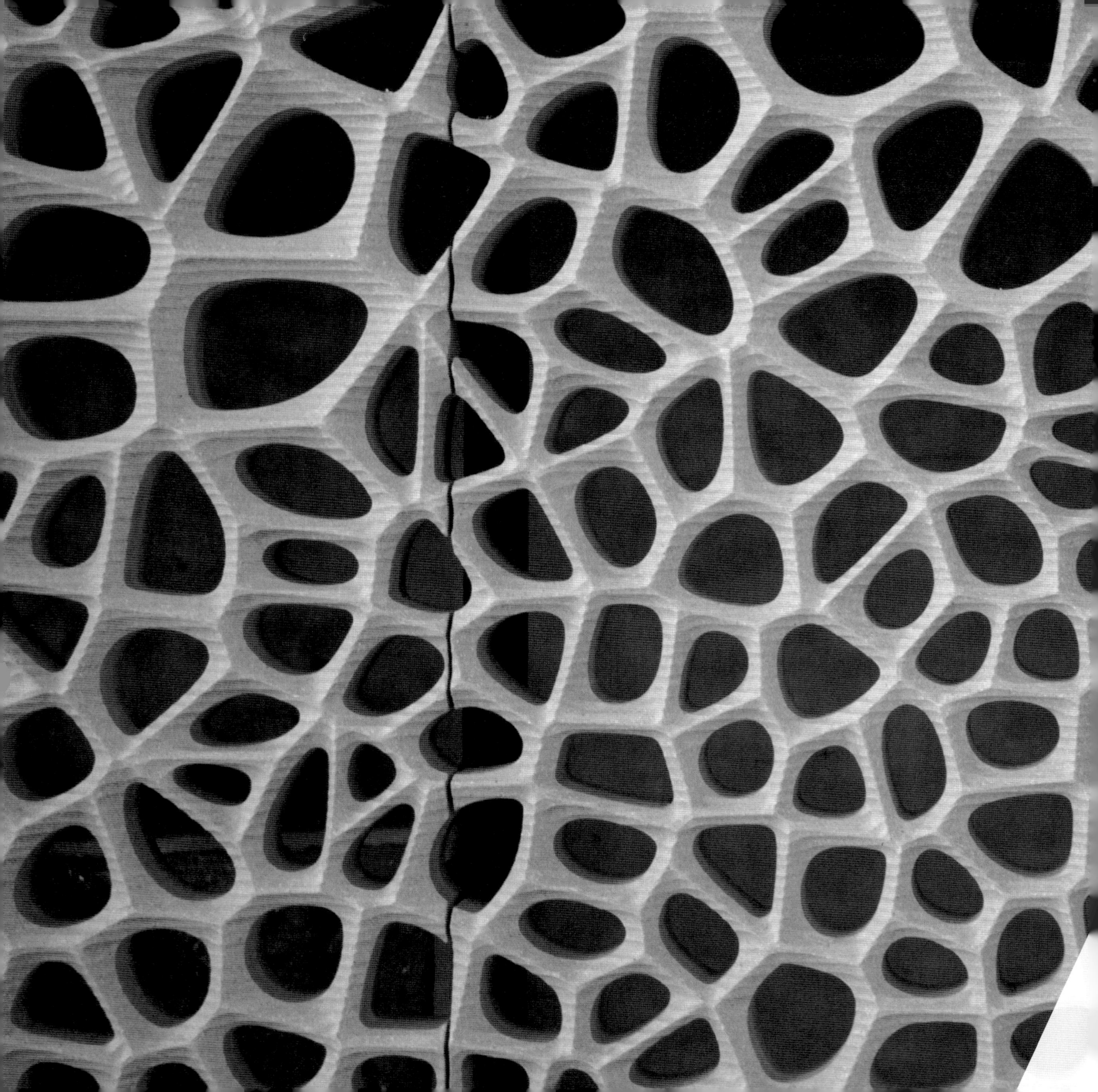

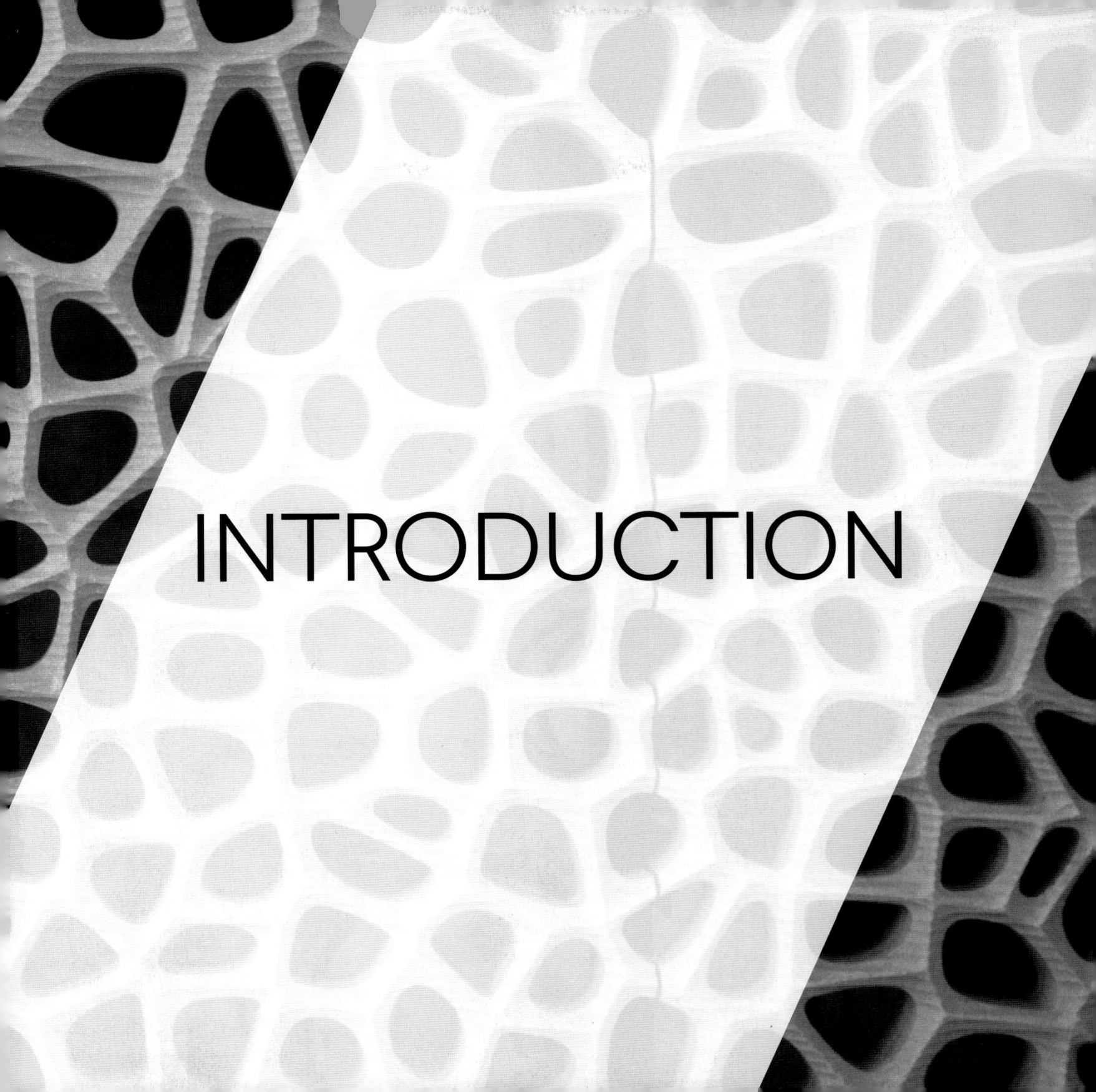

INTRODUCTION

FOREWORD
MARK BURRY & JANE BURRY

So, here we are with the third iteration of the built environment's thinking and making triennial celebration: FABRICATE 2017. What began as a polar comparison between the 2011 and 2014 events has developed into a series through which to take stock of the fast-changing fabrication design landscape. Rather than simply revealing greater sophistication and quicker processes six years after the inaugural event, we believe that the contents of this volume attest to a seismic shift in both professional outreach and construction application, and to a striking expansion in the field of players. Surveying this landscape, practices, industries and design education institutions have even more to draw upon that will boost their confidence as regards the required time commitment and budget to push research and education in fabrication further. Shifts in perspective concomitant with greater investment will be rewarded by the cultural and practical benefits of adventurous, high quality, responsive architecture produced with correspondingly reduced costs and minimised environmental impact. Looking beyond the immediate glamour of FABRICATE, these benefits are the ultimate endgame.

In their introduction to FABRICATE 2014, the editors rued that the preponderance of experimentation at that time was largely constrained to a relatively small number of progressive schools, insulated from mainstream reality and relatively out of step with the reality of wider practice and industry take-up. While pavilions have been popping up around the globe – London's startling annual contribution at the Serpentine being one of the better examples – many adventurous structures have been constrained to inhabiting the forecourts of the educational facilities in which they are spawned, perhaps speaking more loudly to the converted than to the desired new audience.

Traditionally, as we all know, there is a reticence to take risks in the construction industry, and a hesitation to shift ground, even when presented with the exciting possibilities manifestly demonstrated by an eye-catching and palpably successful built example. Even the great innovator Gaudí, while admitting that design by its very nature is experimental and innovative, asked his closest confidants why one would risk employing novel materials and construction techniques instead of expertly applying traditional craft and practice.

Yet we are in the twenty-first century now, with a broader set of urgent and inescapable imperatives: greater awareness of environmental responsibility, not least energy use and its impact on global warming; the need to take ethical account of where materials come from; which resources to eschew for their risk of being depleted; where building waste ultimately ends up; and what happens to buildings after they are finished with and demolished.

While pavilions have been crucial prototypical conversation starters, the editors are reassured to find that this year's take has moved to more built examples, some even at a heroic scale. It is rewarding, too, to see more projects that speak of generic and high impact (as opposed to project-specific) transformations to the way we design and make.

FABRICATE surely wants to pull us towards a post-digital design-making maturity where design intention, computational abstraction and automated fabrication and assembly are positioned more as the norm than as the exception within a shifting set of design priorities. Thus we look for exception not just through stand-out, game-changing examples but through a palpable shift in criteria towards enhanced building performance, in addition to the usual preoccupations about appearance. Work addressing this expanded field of challenges can be readily discerned through surveying this book's contents.

Whereas we might have been collectively drawn to more singular iconic formalist adventures in recent years, there seems to be greater emphasis on process in FABRICATE 2017's published projects, not necessarily pointing solely to a concrete outcome but to fresh approaches to transforming ideas into fabricated, tangible outcomes. We can see this by looking at the acknowledgements in many of the exhibited projects – most notably the growing range of disciplines contributing to this fast-evolving dialogue. These transdisciplinary teams include not just the designers, computation and robotics experts and builders, but also – and increasingly – materials scientists and engineers, industrial designers, process and systems specialists, diverse manufacturers and informed end-user participants, to name just a few.

The projects here have been selected from an almost overwhelmingly large pool of candidates, and together are the conspicuous vector for the questions we might embrace over the coming years. What is the role of schools of architecture and design in all this? Realistically, how can schools participate fully in the face of burgeoning student numbers (in many countries) that make the necessary access to hands-on experimentation with expensive machinery difficult to achieve? And what radical changes in syllabus will be required to ensure that students and researchers are appropriately acclimatised so that they can participate meaningfully in the increasingly diverse design and build teams, beyond mere speculative engagement?

These questions are not necessarily in the purview of FABRICATE 2017 in terms of providing answers, but the event and this published record will help fuel the argument for further change in the design and construction industries' priorities. They will especially stimulate greater confidence to make the most of transdisciplinary opportunities. These are opportunities that learning and research institutions such as universities are uniquely equipped to provide, yet so rarely seem to be able to fulfil beyond the rhetoric. Turning from the imminent present, the thinking/making community have John Ruskin as their friend, for it was he who so eloquently called on thinkers and makers to make themselves consciously aware of each other's contributions:

And yet more, in each several profession, no master should be too proud to do its hardest work. The painter should grind his own colours; the architect work in the mason's yard with his men; the master-manufacturer be himself a more skilful operative than any man in his mills; and the distinction between one man and another be only in experience and skill, and the authority and wealth which these must naturally and justly obtain.[1]

Any sceptic who wonders why design schools invest in robots and 5-axis routers over a century and a half later should be clear that such technology is not about assimilating the expertise of others or about dabbling dilettantism. Rather, these contemporary tools are the vital horizon expanders for disciplines otherwise cauterised by their own sense of historical continuity. In this fast-paced (r)evolution, FABRICATE 2017 is the latest in a unique series of conferences that not only demonstrate the rapidly shifting ground of the endgame but also, more crucially, illuminate fascinating alternative pathways to boosting ongoing professional relevance.

1. John Ruskin, 'The Nature of Gothic', *The Stones of Venice*, Book II.VI.XXI (1853).

ACKNOWLEDGEMENTS

As the editors of FABRICATE 2017, we have many people to thank. In the first instance, with over 250 submissions from 45 countries, we wish to thank everyone who responded to the call for works, an achievement requiring perseverance and commitment on all sides. We also thank all authors and collaborators on every selected project, and everyone who has kindly agreed to present. Each submission took time and involved others in addition to those who authored the content, so we wish to thank all the teams behind every submission – the assistants, the copy editors, the photographers and the IT teams who ensured that our networks didn't go down hours before the deadline. We also wish to thank the administrators who ensured that every submission was received, catalogued and made available on stable platforms so that the business of processing them ran smoothly, no matter where those who needed to access them were at any given time. We are indebted to more than 30[1] distinguished peer reviewers who fulfilled their duties with impeccable professionalism and care, especially as many did so in the holiday period of late summer. You know who you are, and we thank you.

Each edition of FABRICATE has adopted different organisational models. In 2011, both the event and the book were handled by a small local team at The Bartlett School of Architecture, UCL, led by the project's founders, Ruairi Glynn and Bob Sheil. FABRICATE 2014 was managed by a team at ETH Zurich, led by chairs Fabio Gramazio and Matthias Kohler, with Silke Langenberg and consultancy from Marilena Skavara. Arrangements for FABRICATE 2017 have been approached as a collaboration between the Institute for Computational Design, University of Stuttgart, and The Bartlett School of Architecture, UCL, with ICD taking the lead on the conference.

In Stuttgart, Achim would like to express his gratitude to the entire FABRICATE team at the Institute for Computational Design and Construction. First and foremost, thank you to Nicola Burggraf for her extraordinary effort in heading the administration team and for her tireless engagement in preparing FABRICATE in Stuttgart since January 2016. Britta Kurka and Scottie McDaniel have also contributed extensively to various organisational matters. Without this team, the conference would not have been possible. Thank you! In addition, thank you to all the other ICD researchers and students who helped with the wide range of aspects that needed to be taken care of for such a major event. We are lacking the space here to mention all contributions in detail but, again, FABRICATE would not have been possible without the tremendous effort and passion of this fantastic group of people – very well done! Thanks, too, to the University of Stuttgart for providing an excellent context for our research activities, which included making it possible for us to host such a remarkable event. Finally, I would like to acknowledge the significant support of the DFG Deutsche Forschungsgemeinschaft in recognising the scientific importance of the conference and for granting the additional means to make it happen.

In London, Ruairi and Bob wish to start by thanking Marilena Skavara, who has been involved in every event and production since 2011 as an utterly pivotal figure in the entire enterprise. Now co-editor, Marilena's key role is fully recognised, as is the impact of her strategic contribution and judgment. So now, as an editorial team of three, we each wish to thank the following people, who have helped us to assemble this wonderful publication. Eli Lee, our project editor, has approached it with unlimited reserves of patience and good humour, and her assistant, Aleema Gray, transcribed the interviews between our keynotes with forensic skill and concentration. We offer sincere thanks to Dan Lockwood and Patrick Morrissey, our meticulous proofreader and designer respectively, both of whom have executed their tasks with elegance, patience and beauty. Thank you to Lara Speicher, Publishing Manager at UCL Press, and her team, including Chris Penfold, Jaimee Biggins and Alison Major – we have hugely enjoyed working on our second project with you within six months. It's also deeply satisfying and enjoyable to be back in partnership with Riverside Architectural Press, led by the inspiring Philip Beesley and his general manager Salvador Miranda. Thank you to James Curwen, Luis Rego and Thomas Abbs for their generous assistance. Finally, we wish to thank those responsible for the tactile experience that you, as reader, are enjoying now. To Tom Maes and Hugh Jolly of Albe de Coker printers: we salute your passion for craft and detail, and deeply appreciate your kindness and tolerance – qualities that are vital in the making of well-made things.

Finally, from Stuttgart and London, we extend our sincere thanks to all our sponsors, as it is their support that enables FABRICATE to be disseminated widely. Thank you to our Diamond sponsor Autodesk, our Platinum sponsors Arup, KUKA, Ostseestaal and BigRep and our Gold sponsors FARO, Design-to-Production and Trimble.

Achim Menges, Bob Sheil, Ruairi Glynn and Marilena Skavara

1. FABRICATE 2017 peer reviewers: Francis Aish, Ehsan Baharlou, Martin Bechthold, Mirco Becker, Daniel Bosia, Mark Burry, Brandon Clifford, Xavier De Kastellier, Moritz Doerstelmann, David Gerber, Ruairi Glynn, Volker Helm, Axel Kilian, Nathan King, Branko Kolarevic, Toni Kotnik, Oliver Krieg, Julian Lienhard, Achim Menges, Philippe Morel, Caitlin Mueller, Brady Peters, Marshall Prado, Mette Ramsgaard, Matthias Rippmann, Christopher Robeller, Stanislav Roudavski, Jenny Sabin, Bob Sheil, Simon Schleicher, Tobias Schwinn, Asbjørn Søndergaard, Robert Stuart-Smith, Oliver Tessmann, Skylar Tibbits, Lauren Vasey, Tobias Wallisser, Jan Willmann and Dylan Wood.

INTRODUCTION
BOB SHEIL & ACHIM MENGES

Welcome to FABRICATE 2017: 'Rethinking Design and Construction'. This is the third volume in a triennial series of conference publications that began with 'Making Digital Architecture' in 2011 at The Bartlett School of Architecture, University College London. From these origins in one of the world's leading cities for design excellence, in 2014 FABRICATE moved to ETH in Zurich, a pioneering science and technology university, where its theme was 'Negotiating Design and Making'. In 2017, we are at the Institute for Computational Design and Construction, University of Stuttgart, a world-renowned research lab in design for construction, located in Europe's innovative and forward-thinking industrial heartland.

Each FABRICATE conference and book evolves from an open call for 'works in progress', with a submission deadline ten months prior to the conference. The call is designed to attract submissions from industry and practice as well as academia, and asks for an abstract on the trajectory of the work, including where it will be by the time the conference takes place. Selected projects are then invited to resubmit full papers for a second round. The conference theme emerges during this phase, while papers are also categorised into notional sub-themes.

Thus FABRICATE is itself a work in progress *about* works in progress, an event and a publication that converge at a point and place in time where what is being made is both still evolving and ready for sharing. This approach extends to how authors are encouraged to further translate their work as speakers. They are encouraged to go off-script, question and reinvent their medium, reveal what lies between the lines and add any new ideas that have come into play. Writing, after all, is no different to drawing or making – they are all forms of representation that we rely on to make sense of the world. So not only are the evidence and documentation of design and making critical to FABRICATE, but talking, showing and rethinking are, too.

In the beginning, the idea of a conference that explored the currents between technology, design and industry emerged from the need to understand the ever-changing shape of the world around us. In the six years since the first event, we have received over 800 submissions from more than 40 institutions across 30 countries. From this pool, we have selected 96 papers for publication and 48 for presentation, alongside 12 highly distinguished keynote lectures. A team of eight Conference Chairs, 12 editors, 12 Panel Chairs and 90 peer reviewers have been intimately involved throughout. FABRICATE is now widely regarded as the leading international forum in which centres of excellence in architecture, design, engineering and manufacturing can engage, collaborate and create. It has become a unique public platform for open debate on how these disciplines exchange and evolve their design and making expertise.

At its heart, FABRICATE is about *doing*: the where, who, what, why and how of *doing*. While it was not initially envisaged as a series, that it would become one was perhaps inevitable. From 'Making Digital Architecture' (2011) to 'Negotiating Design and Making' (2014), FABRICATE has set the agenda during an extraordinary period for the built environment – one which has empowered the designer with new tools and capabilities, challenged the orthodoxy of standardised processes and witnessed inspiring collaborations in which expert representation has met expert realisation, and vice versa.

Back in 2011, it might have seemed that The Bartlett School of Architecture, at University College London, was not an obvious institution for such a conference. At that time – and for the preceding decade – the School was more renowned for its culture of experimentation through 'hands-on' drawing, narrative speculation and cultural theory*, as well as for its influential work in digital theory and interaction. But in 2001,

* Each remains a vital strand in the School's operations today.

1

2

1. New Computational Construction Laboratory at the University of Stuttgart.

2. The new Integrative Technologies & Architectural Design Research MSc Programme at the Institute for Computational Design and Construction at the University of Stuttgart.

one of the earliest books on the subject, *Computer Aided Manufacture in Architecture – The Pursuit of Novelty* (Architectural Press) was authored by Nick Callicott, one of the School's workshop-based design tutors. The book was in part influenced by *The Idea of Building: Thought and Action in the Design and Production of Buildings* (Taylor and Francis, 1992) by Stephen Groak, Callicott's key mentor at UCL and later Head of Research and Development at ARUP. Groak and other key figures at The Bartlett, such as Stephen Gage, Peter Cook, Christine Hawley and Alan Penn, understood that the School's reputation for design excellence would not come from the studios or seminar rooms alone. Instead, it would rely on a well-equipped and strategically staffed workshop.

Among the craft and making experts The Bartlett hired was furniture designer Bim Burton, son of the late Richard Burton, who famously worked with Frei Otto and John Makepeace at Hooke Park. In the 1990s, experimentation through drawing was the dominant paradigm. By 2001, experimentation in digital forms of making was growing in scope, and by 2011, as a result of a £1m investment in digital technologies, The Bartlett had developed a complementary strand of research and teaching activity entirely focused on making. Furthermore, the School is now opening expanded facilities for new postgraduate programmes, including an MArch in Design for Performance and Interaction, an MArch in Design for Manufacture and a pioneering MEng in Engineering and Architectural Design.

The inaugural FABRICATE event in 2011, 'Making Digital Architecture' (chaired by Ruairi Glynn and Bob Sheil), presented two primary spheres, academic and practice. The extraordinary range of projects issuing from both were further categorised into the following themes: 'Physical Processes', 'Material Systems', 'Machines and The Bespoke' and 'Representation and Manufacture'. By FABRICATE 2014, 'Negotiating Design and Making', at ETH Zurich, the event's themes had evolved into 'Challenging The Thresholds', 'Material Exuberance', 'Forming Machines' and 'Living Assemblies'. Chaired by Fabio Gramazio and Matthias Kohler, pioneers of robotics in architecture, the event was a survey of ground-breaking approaches to fabrication from the most innovative research labs in the world. This conference indicated a clear shift in the field, with many leading institutions recognising the possibility of a far richer relationship between the drawn and the made. The next step was to engage with industry – and where else to go but Stuttgart?

3 & 4. Construction underway for The Bartlett School of Architecture's new fabrication facilities at Here East, Hackney Wick, London. Opening September 2017.
Image: Paul Smoothy (taken January 2017).

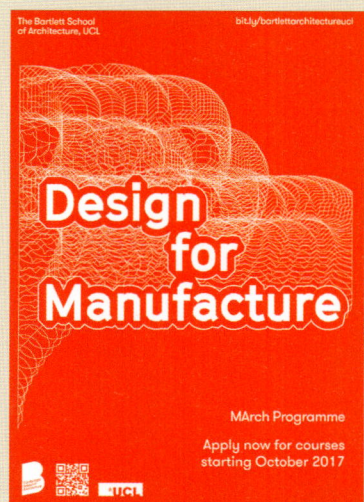 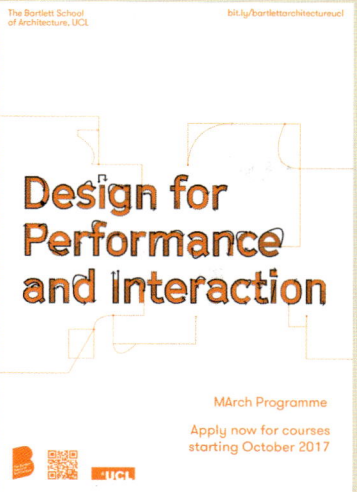 

5. Three of several new programmes being launched by The Bartlett School of Architecture, UCL, in September 2017.

FABRICATE 2017 recognises how much has changed since 2011. Once the final selection of papers was made, it was clear that the chosen projects were of a significantly larger scale in terms of both size and reach. The categories 'Production', 'Materialisation', 'Additive Strategies' and 'Construction' made immediate sense, as did the conference title, 'Rethinking Design and Construction'.

One key point that has emerged is that we can no longer talk in general terms about 'digital architecture'. Such a generalisation no longer seems to do justice to the multifaceted cultures of computational design and digital fabrication, their finely differentiated approaches and their diverse physical manifestations – which are explored both in the book and at the conference. Equally, we are witnessing a rapidly blurring boundary between computational design and digital fabrication. The clear line that once existed between these two domains has become increasingly questioned by cyber-physical productions systems and challenged by new forms of man-machine collaboration (designer and robot collaborations, in most cases), which form the basis of a significant number of submissions.

In Germany, this leap forward in the way we design, engineer and produce is often referred to as 'Industry 4.0', a term that originates from the high-tech strategy of the German government and indicates the significance of these developments as catalysts in a fourth industrial revolution. FABRICATE 2017 suggests that Industry 4.0 will have – indeed, is already having – a profound impact on the way the future built environment is conceived, designed and materialised. Stuttgart is situated in the heartland of advanced manufacturing, and the south of Germany is home to a large number of 'micro-leaders' and 'hidden champions', small-to-medium scale businesses that are leading in their respective technological fields. The University of Stuttgart is renowned for creatively engaging with these advanced industries and bringing to such engagements the rigour and insight of its skilled and specialist faculty and students. The most recent manifestation of this spirit is the ICD's new Computational Construction Laboratory (Fig. 1). It is no coincidence that the opening of this lab coincides with this year's conference, as it is this very spirit that we consider the ICD's modest contribution to the continuing success of FABRICATE.

'Rethinking Design and Construction' therefore constitutes both a critical assessment and a provocation. On the one hand, it reflects the work of a range of extraordinary thinkers who are challenging old approaches to design and making through ingenious ways of mastering digital technologies and lateral thinking. On the other, it serves as a rallying cry to use computational technologies as vehicles for creative exploration and for making ground-breaking and bold collaborations that would otherwise not happen in academia or industry by themselves.

THE TRIUMPH OF THE TURNIP

ANTHONY HAUCK / MICHAEL BERGIN / PHIL BERNSTEIN
Autodesk

Surrounded by dusty workmen, a man stands in a Florence piazza holding a turnip (Fig. 1). He raises his voice over the sounds of construction echoing down the nearby streets and alleys and carves a shape from the vegetable with a small knife. He glances up after every few words to confirm that he's understood. When his carving is done, he lectures further, pointing out various details of his sculpture, explaining, answering questions, rotating the carving into the sunlight. When he's satisfied, when the workmen nod, the master builder eats the turnip. There's no need to preserve it. One detail of Brunelleschi's Dome of Santa Maria del Fiore will be the only record of this conversation (Fig. 2).

In Rome, a man labours over a drawing, recording his design intent in detail sufficient enough to ensure its accurate realisation by the workmen of his day. As far as he is concerned, the building he draws is the real building, the realisation of his imagined edifice. Whatever occupied space might be built from these instructions is a copy of the perfect original, rooted in refined geometric mathematics, defined on the several sheets and models that summarise its creator's thoughts (Fig. 3).

Brunelleschi and Alberti stand on either side of a historical shift between the Renaissance master builder and the modern architectural profession. In his treatise of 1452, *De Re Aedificatoria,* Leon Battista Alberti broke with the tradition of the master builder, as exemplified by Brunelleschi, in suggesting that originating ideas about buildings was a separate and privileged predecessor to the act of actual construction. He declared:

> "...certainly it is enough if you give honest Advice, and correct Draughts such as to apply themselves to you. If afterwards you undertake to supervise and compleat the Work, you will find it very difficult to avoid being made answerable for all the Faults and Mistakes committed either by the Ignorance or Negligence of other Men: Upon which Account you must take care to have the Assistance of honest, diligent and severe Overseers to look after the Workmen under you."[1]

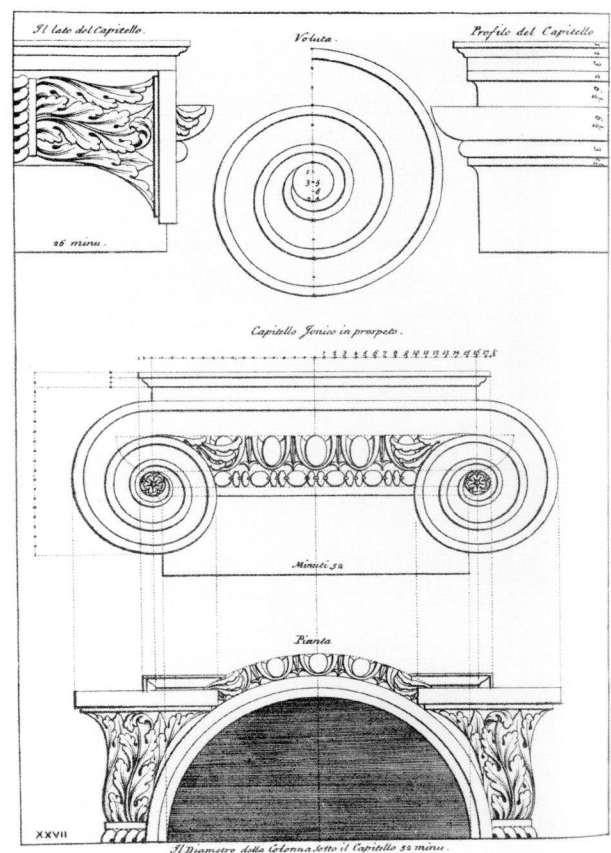

firms was a testament to a fee-for-service business model attended by a commensurate need to minimise labour costs as a means of maximising profits. Few questions were raised at the time concerning the composition of instruments of service dictated by contractual obligations and regulatory environments, and the initial penetration of computation into practice was limited to recording decisions arrived at by other methods. In many cases, efficiencies gained in more quickly producing and revising drawing sets were applied to design revisions, extending a project's exploratory dimensions more deeply, with decisions formerly confined to conceptual, schematic and design development stages allowed to spill into the construction document phase. This gain in flexibility was largely unforeseen by a profession adopting technology in the hope of production cost reductions, but the additional time afforded to design decisions was a quiet hint that computation might have more to offer than the mere transposition of previously physical activities to the digital realm.

In one sense, building engineers were quicker to transition to the emerging paradigm of design and drawing production that was finally termed Building Information Modelling (BIM). Accustomed to specifying manufactured components in the form of standard steel shapes, structural engineers were quick to adopt computational capabilities to model forces and select appropriate steel within the analytical environment. The rapid and sometimes instantaneous feedback of these computational structural design environments prefigured further advances of this type emerging today, conveying the promise of improved decisions about buildings when the professional can easily access relevant information and encoded expertise to inform building choices. Taking advantage of manufacturing standardisation and digital artefacts as proxies for fabricated objects, engineers in the early days of BIM were able to more closely unite design decisions to characteristics of materials and manufacturing conventions through the medium of computation (Fig. 4).

1. Rutabaga or swede (Swedish turnip) or turnip or yellow turnip (Brassica napobrassica), vintage engraved illustration. *Dictionary of Words and Things*, Larive and Fleury, 1895. Image: Morphart Creation/Shutterstock.

2. Brunelleschi's Dome, Santa Maria del Fiore, Florence, Italy. Image: Marcus Obal.

3. Architectural drawing for an Ionic capital by Italian Renaissance architect, Leon Battista Alberti (1404-1472) from his *Ten Books on Architecture*, ca. 1480. Image: Everett Historical/Shutterstock.

4. Office layout, Autodesk AutoCAD 2.18, 1985. Image: Autodesk.

This is tantamount to inventing the modern architectural profession with its notion of design drawings and limited liability. Alberti made one of the first declarations of a representational strategy in construction, while simultaneously dividing intention from realisation – design from construction – in a separation of responsibilities that has persisted for more than half a millennium into the building industry of today, resulting in a process of delivery and modification of contemporary buildings that is widely regarded by both owners and building professionals as being inefficient, risky, expensive and often an incomplete or inadequate realisation of the project's original intent. Instruments of service contain errors, are misunderstood or otherwise imperfectly convey design intent by filtering design representations through drawing conventions that date back centuries.

The first additions of computation to the architectural profession did little to advance building quality or delivery, focused as they were on raising the efficiency of producing conventional drawing sets. The rapid and thorough adoption of CAD software by architectural

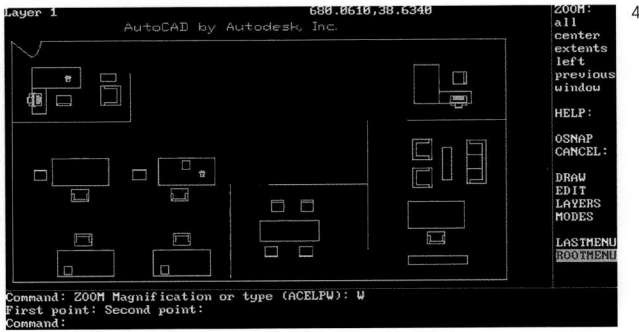

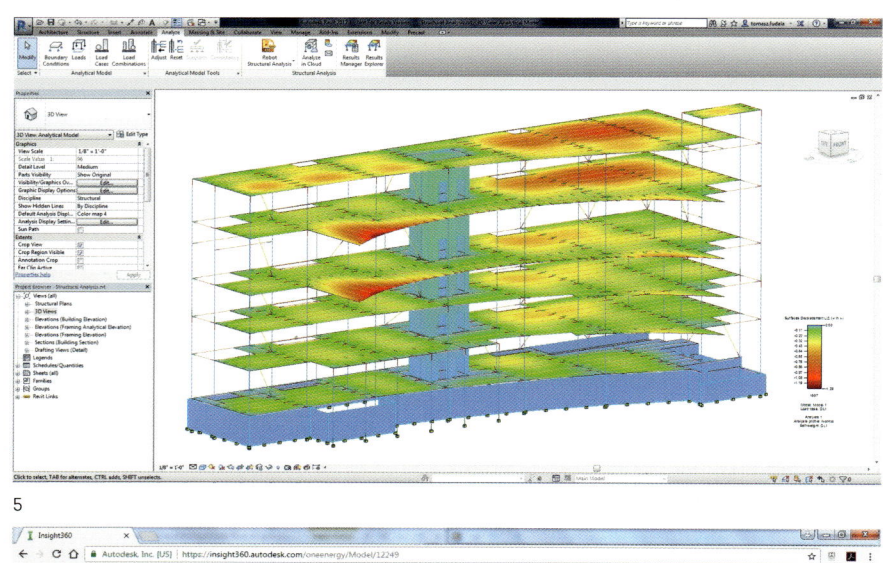

5

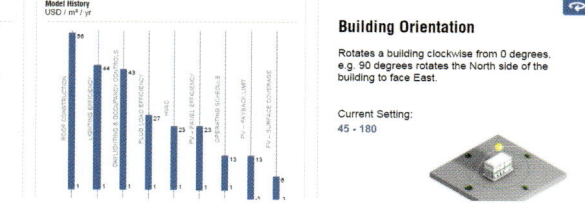

6

Corresponding activities in the architectural professions of the time remained divorced from computational environments and largely the responsibility of the specification group either within a large architectural firm or as an outsourced service to smaller firms, with most architects concentrating on the new possibilities for design understanding afforded by three-dimensional BIM and increasingly advanced computational rendering. For the first time, architects with less than decades of experience could easily understand the experiential aspects of space and light of their design choices before they were instantiated by construction. While the relatively rapid adoption of BIM by the architectural profession can once again be attributed to a quest for higher production efficiency, some architects quickly understood the possibilities in increasingly detailed digital representations of buildings as the medium of improved understanding and decisions (Fig. 5).

Like CAD before it, BIM arose from a fundamentally piecemeal digital representation of finished objects. CAD software conceives of drawing sets as complex arrangements of lines, arcs and circles. Building information models are digital assemblies of generic or specific manufactured objects, owing much more to their roots in software designed to support fabrication and construction metaphors than to the design process. As in physical construction, buildings in BIM are emergent phenomena resulting from the positioning of components in precise relationships. As the BIM assembly becomes more elaborate, it begins to exhibit emergent behaviours that cannot be predicted by an anecdotal understanding of the individual objects employed in its composition. Enhanced BIM environments reveal and predict such behaviours through increasingly sophisticated analytical tools calculating building properties such as energy consumption, quantity and spatial distribution of daylight and the modelled movement of inhabitants through space. The increasing use of virtual reality systems by architects simulates inhabitation of the project from its earliest stages of design, a technology that affords even the most unsophisticated of clients a visceral understanding of basic architectural choices while they remain open to discussion and adjustment (Fig. 6).

The ready availability of such predictive information has implicitly expanded the scope of architectural practice following Alberti's exhortation to offer "honest advice and correct Draughts" to include counsel that cannot be complete without an understanding of materials, fabrication techniques and the interplay of building systems. It is in this expansion of architectural knowledge facilitated by computational and data delivery technologies that we see the possibility of uniting the philosophies implicitly and explicitly shared by Brunelleschi and Alberti. Where the former's regard of his verbal instructions and vegetable instruments of service as perishable media to convey intent has given way to the indefinite preservation of digital artefacts of design and construction planning, his sense of differentiation in understanding construction means and methods has regained relevance to the architectural profession today. It was Brunelleschi's accurate assertion that he could complete the Santa Maria del Fiore Dome without the need for supporting scaffolding that won him the commission. However, Alberti's "correct Draughts" remains the standard of care of architects to their clients even today, but the scope of architectural "Draughts" has become nearly as extensive as virtual construction of the intended building.

The facilitation of building representation by digital environments has served to further blur Alberti's fundamental division between design intent and construction means and methods, already under attack by the modern economic pressures that compel a faster speed of project delivery. The former stately progression of conceptual design, schematic design, design development, construction documentation, fabrication, construction, operation and renovation has given way to an increasingly optimised process of overlapping phases dependent on the delivery of complete trade packages that, in effect, become existing conditions to be accounted for in subsequent deliverables for the same project. In a building market intolerant of sites fallow of anticipated revenue, design differentiation has begun to arise in the most advanced firms not only from a clear experiential vision of inhabited space, but also from a growing knowledge of the possibilities inherent in new materials and the growing sophistication of building systems production.

Advances in the available insular, optical and structural properties of glass alone over the preceding century have afforded options in architectural design that were previously impractical if not impossible, realising facades and interiors that not only fulfil design intent but also meet stringent performance goals rooted in environmental and human behavioural sciences. A lack of understanding of material properties and construction methodologies not only limits architectural business opportunities, but also limits the architect to building choices available only through advanced fabrication and construction methods.

As BIM environments and their analytical elaborations and generative design successors gain computational capabilities and information access through resources available through cloud connectivity, the architectural profession has an opportunity to assert a role explicitly ceded by Alberti and implicitly occupied by Brunelleschi: that of the master builder. At first glance impractical in an age of proliferating, specialised and necessarily complicated building trades, the enhanced capabilities of digital environments, with their rapid evaluation of modelled building performance characteristics and delivery of highly relevant information critical to improved building decisions, offer architects a means to confidently reassert primacy in the process of conceiving and realising buildings.

With the explosive growth of computational power in the second decade of the twenty-first century, the profession is entering a third era, beyond CAD and BIM, of potentially transformative digital capabilities in design and construction. Highly responsive computer processes of physical representation and simulation coupled with digital processes of fabrication, including material science, additive and subtractive manufacturing and robotic construction, are poised to change the essential landscape in which buildings are designed, built and operated (Fig. 7).

Projecting this evolution forward, the third era in design computation takes advantage of the best qualities of both Brunelleschi and Alberti's positions regarding the instrument of representation. Robust simulation and nearly unlimited computing power will

5. Structural Simulation, Autodesk Revit, 2016. Image: Autodesk.

6. Insight 360 Energy Simulation, Autodesk, 2016. Image: Autodesk.

7. The exponential growth of available computing capacity affords the possibility of high confidence in predicting the outcomes of building design and construction decisions. Source: www.singularity.com.

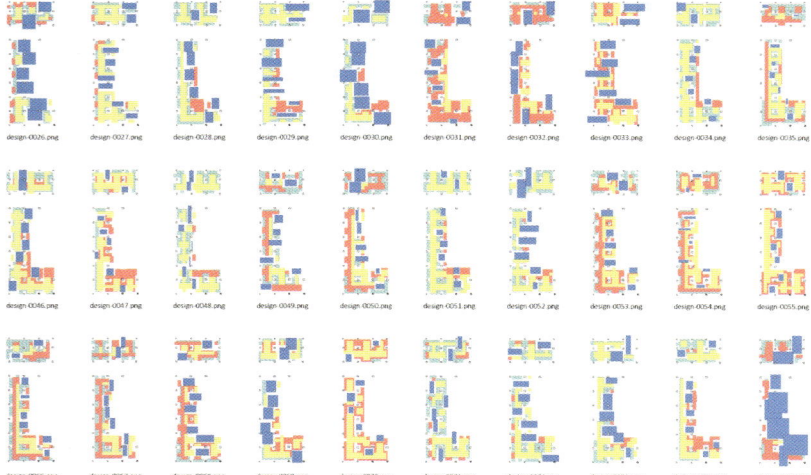

8. A generative design system employed in the design of Autodesk's office in Toronto produced many alternative solutions, each respecting the goals and constraints specified by the design team.

combine with machine intelligence and generative design to deliver a further unification of intent and realisation. Reality capture, digital fabrication and immersive design environments will provide a functionally identical model of digital and physical space. As design tools evolve, Brunelleschi's turnip and its associated conversation will become forever persistent in structured databases. The connectivity of this data will provide opportunities for machine learning, pattern recognition and design synthesis. Generative design systems will support the explicit modelling of knowledge from a variety of domain experts such that when the design requirements change new instructions in the form of drawings or models that describe the author's intent will be automatically generated (Fig. 8).

This new class of design systems will allow all stakeholders in a building project to represent their intent at the level of detail that best corresponds with functional properties in models used for design, construction and building management. In the event of a budget change at a late stage of the project, future design environments will provide consulting professionals with recommended alterations to support the new requirement. Effects of selected updates or native interventions will propagate to the redesign of multiple elements in the project. Instead of manual redesign and drawing changes (CAD) or editing parametric models (BIM), design synthesis algorithms will pursue declared goals while respecting project constraints. The intent of the design team will be preserved or compromised through various strategies for generating solutions employed by the design system. As the accuracy and speed of simulations increase, a wealth of building performance data will become available and complex trade-offs between alternative approaches will be intuitively revealed. Compensation for design and construction services, as well as the standard of care for professional building services, will become associated with the ability to offer the guaranteed level of building performance ensured by these tools.

To realise a vision, the master builder must synthesise many competing objectives relative to changing external conditions. The archetypal master builder understands how design decisions reinforce intent through a sophisticated understanding of aesthetic, performance, constructability, cost and schedule objectives. In the next era of design practice, any stakeholder will be able to understand the propagating effects of a change and offer feedback that directly influences design decisions. Ease of design changes will ensure that any compromise of intent is comprehensively evaluated before construction takes place. The ability to quickly and effectively balance the needs of participants in the design and construction process will embolden the master builder to deliver on their vision (Fig. 9).

Digital artefacts of design and construction are increasingly employed as operational avatars for the functioning building, joined to a wide spectrum of physical sensors to convey gross and subtle operational behaviours into digital representations where options for elaboration and modification can be readily explored at minimal cost. In the coming era of widely available statistical performance information as furnished by a highly connected built environment, the knowledge and experience once sequestered in fragmented form across many design and construction experts, owners and facilities managers will be consolidated and available to inform all design, fabrication and construction decisions. Reality capture technology will provide a 'mirrored' representation between the digital artefact and the developing physical manifestation during the construction process and throughout the lifecycle of the building. Design and construction firms that embrace and extend the possibilities of digital enrichment will lead future building projects. Firms that fail to grasp the gains offered by the coming era of connectivity will find themselves becoming irrelevant in an approaching time of exacting standards applied to desired building performance with the ready means to confirm predicted project behaviours (Fig. 10).

9

10

The vast increase in computational and informational capacities afforded to building professionals will also affect material and product supply chains, with fabricators differentiated during procurement not merely on bidding price, but also on the flexibility of deliverables and their ingenuity in engineering. In an era where a sophisticated computational and informational infrastructure will widely afford capabilities once confined to a few specialists, creativity in meeting and guaranteeing building performance outcomes will become a key distinction for both building professionals and material suppliers. As building performance becomes measurable and understandable through physical connectivity and digital representation, the standard of care for all building professionals and manufacturing entities will become higher and commonly verifiable, leading to a reintegration of design and making with attendant dramatic improvements in both the systems of project delivery and the practice of architecture and the built environment.

Notes

1. 1755 London printing (pdf), p.693, available at http:// http://archimedes.mpiwg-berlin.mpg.de/docuserver/images/archimedes/alber_archi_003_en_1785/downloads/alber_archi_003_en_1785.text.pdf (accessed 28 December 2016).

9. Robotics increasingly inform procedures on construction sites, and will be directly driven from design deliverables as the construction site transforms into a bespoke factory for a building project. Image: Autodesk.

10. Information derived directly from digital design environments has already begun to shape subsequent fabrication processes, sometimes with little further human intervention. Image: Autodesk.

COOPERATIVE FABRICATION OF SPATIAL METAL STRUCTURES

STEFANA PARASCHO / AUGUSTO GANDIA / AMMAR MIRJAN / FABIO GRAMAZIO / MATTHIAS KOHLER
Gramazio Kohler Research, ETH Zurich

Machines have the ability to manipulate material cooperatively, enabling them to materialise structures that could not otherwise be realised individually. Operating with more than one (mechanical) arm allows for the exploitation of assembly processes by performing material manipulations on a shared fabrication task. The work presented here is an investigation of such cooperative robotic construction, wherein two industrial robots assemble a spatial metal structure consisting of discrete steel tubes. The developed construction method relies on the alternate positioning of building members into triangulated configurations, where one robot temporarily stabilises the assembly while the other places a tube and vice versa. The intricate geometric dependencies of this structural system, as well as the fact that the machines limit each other's operational range, led to the exploration of robotic simulation and path planning strategies as an integral part of the design process. The experimental results of realising a space frame structure at an architectural scale (Fig. 2) validate this approach.

Space frame structures have been traditionally constrained to regular systems using standardised elements and joints, or have required the development of complex prefabricated joints for the construction of differentiated structures (Chilton, 2000). The implementation of an industrial robot into the building process offers a new approach for the construction of non-regular spatial structures, since a 6-axis robotic arm can precisely move, position, orient and hold a building element in space, something a human cannot accomplish without a reference system and support structure (Gramazio et al., 2014).

In the work presented here, the digital fabrication of space frame structures is further explored with a cooperative robotic construction approach. Whereas multi-robotic material manipulation is well-established for repetitive pre-programmed tasks in assembly lines, its application in non-standard digital fabrication, derived from the inherent complexity of performing non-repetitive robotic movements in an altering building space, is still widely unexplored. Cooperating robots have been used in architectural research for filament winding (Parascho

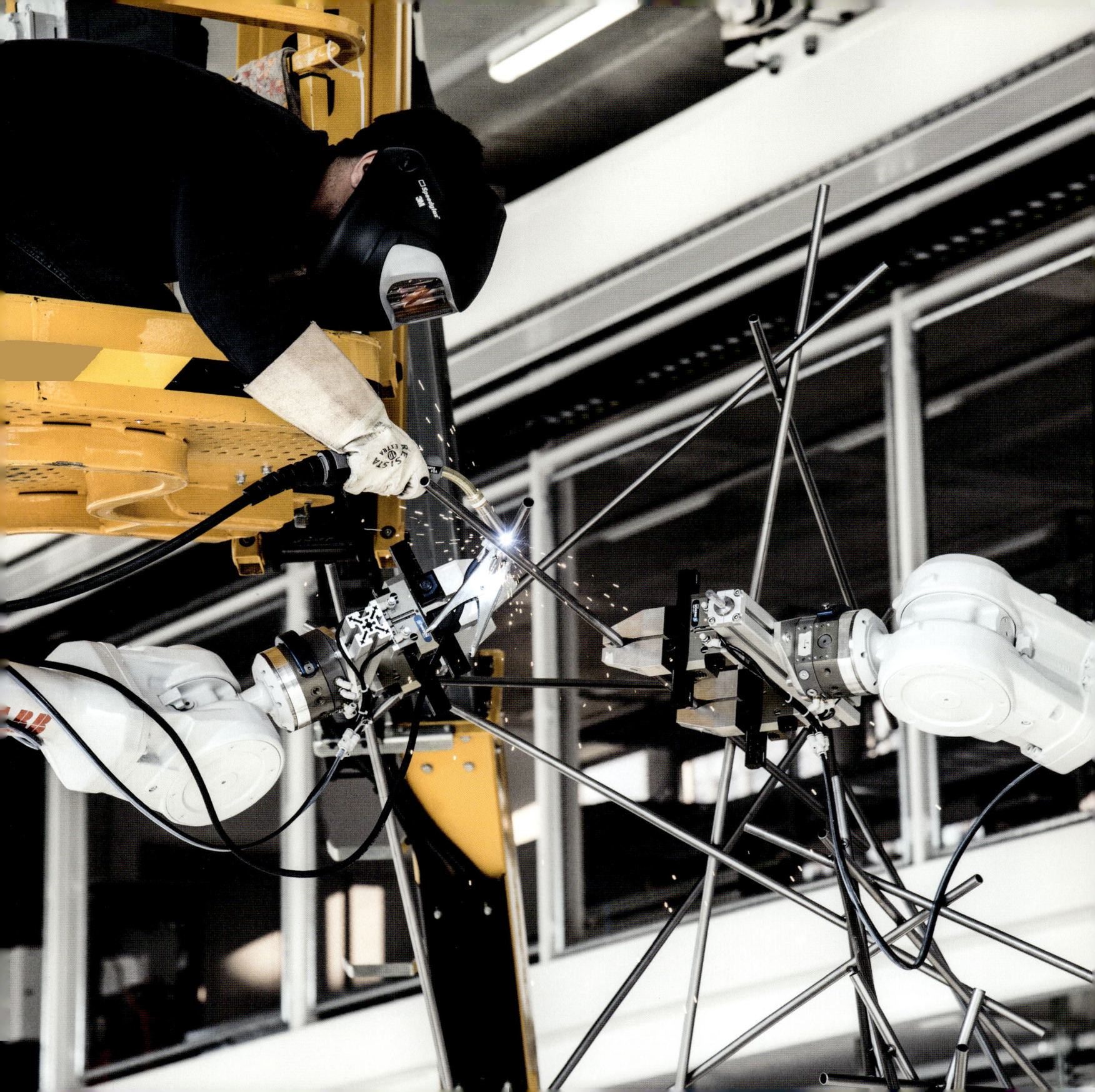

et al., 2015), hot wire cutting (Rust et al., 2016, Søndergaard et al., 2016) and metal folding (Saunders et al., 2016). While these applications hint towards the potential of using cooperating robots in architectural fabrication, they are usually manoeuvred at a safe distance from each other where the spatial configuration of the robotic arm is less of a concern than when operating at close proximity within the same fabrication space. Rather than merely focusing on the final pose of the robotic end effector when assembling an element, the research presented here considers the whole body of the robotic arm over time to guide building elements around material obstacles.

An investigation into possible assembly sequences for the realisation of space frame structures with multiple robots has shown that in principal only two cooperating manipulators are required to assemble stable, triangulated structures (Gramazio et al., 2014). This is based on the assumption that one robot can temporarily stabilise the structure while the other is picking and placing a new structural element. While one robot assembles a steel tube, the other briefly changes its function and acts as a structural support to balance the unstable assembly until it is triangulated and can be fixed. As a result, using digital design, robotic simulation and robotic fabrication in a negotiating manner, highly differentiated space frame structures can be erected without the need for additional support structures.

Computational design and fabrication simulation

Like other space frame structures, the system developed here is primarily characterised by the node. In order to allow for geometric flexibility in arranging the tubes at various angles, and to be able to later robotically fabricate them, the node is distinguished by the shifting of the tubes alongside each other around a shared centre point. As a result, two tubes connect at one point. While this shifted node offers a high degree of freedom in respect of possible spatial arrangement and robotic fabrication, it also presents structural challenges. In contrast to traditional space frame systems that join multiple structural members at a singular spherical point, the reciprocity of this expanded node induces flexural rigidity in the system, leading to a structure with a greater stiffness.

1. Tubes are welded manually at their connection points after each assembly step.

2. Cooperative robotic assembly of a spatial metal structure.

3. Assembly of the structure.

4. Build-up sequence of the structure.

Images: Gramazio Kohler Research, ETH Zurich, 2016.

3

4

Each newly added tube connects at each side to two neighbouring elements with the objective of assembling reciprocally closed nodes. These configurations are able to take bending forces, although every constructive joint between two tubes is hinged in a static sense. As a result, each tube, once assembled into a tetrahedral configuration, is comprised of at least four connections, making two reciprocal nodes with its neighbours. During the build, the number of connections to an individual tube increases over time, subsequent to the adding of neighbouring tubes requiring structural support points. This means that, in a final assembly, a tube can have up to eight connection points at each end, depending on the overall configuration. This novel construction system leads to geometric dependencies that require the use of computational design to explore possible spatial arrangements and to identify a fabrication sequence that considers the build-up of the structure into stable configurations accordingly.

The overall spatial organisation of the construction system is based on tetrahedra. A tetrahedron creates the minimum stable space frame structure. Combining a multitude of tetrahedra into larger, interconnected structures allows for the creation of complex load-bearing assemblies while assuring the structural integrity of the individual tetrahedron and, as such, the controlled assembly of tubular elements into spatial aggregations. When designing such an arrangement, the order of placing tubular elements has to be defined. This is directly related to the later construction of the structures. The fabrication space changes over time. Therefore it is crucial to define where and when to place the next building element and to which tubes it can connect, so that the computational design tool can find the appropriate geometrical solution for the tubular arrangements.

An important aspect of the design process, aside from the definition of the spatial arrangement, is the creation of the robotic movements that allow the integral verification of the fabrication feasibility. As described above, two cooperating robots are used to assemble the structures in a highly constrained three-dimensional space. A series of tests has shown that defining collision-free robotic movements is a challenging task that needs to be addressed at an early design stage. On one hand, this originates from the need to manoeuvre building elements into openings and gaps of already built parts to create the interlocking reciprocal joints, while avoiding collisions between the robot and the structure. On the other hand, the construction environment changes over time, as a result of the sequential and spatial build-up of the structure and of the continuously altering configurations of the robots, which limit each other's operational range.

Rather than only calculating the final pose of the tool centre point (TCP), the approach required designing the robotic configurations, translated into axis rotations, determining the entire spatial arrangement of the robot over time. For this reason, path planning strategies and robotic simulation tools that linked to the computational design were investigated. The proposed solution makes use of a robotic simulation platform (Coppelia Robotics, 2012) that uses the power of sampling-based path planning algorithms (Kavraki Lab, 2012) to generate collision-free trajectories. A software tool was created in a CAD environment (McNeel, 2013, McNeel, 2015) in order to integrate robotic simulation capabilities directly into the computational design process. The robotic trajectories can be generated by defining a start configuration of the robot and a desired end pose of the TCP, by outlining the robot's joint metrics (to set the joint constraints) and by setting a series of rapidly exploring random tree (RRT) algorithm-specific values, such as the sampling resolution. Following this method, a spatial configuration can be evaluated when designing it, which can be adapted if needed. For example, if no solution is found, the gripping pose can be shifted or the geometry of the structure can be altered.

Building a 4m-tall structure

To test the fabrication approach, a physical prototype was realised with two cooperating robot arms (Fig. 3). The structure is comprised of 72 steel tubes, creating 23 non-regular tetrahedra, concatenated into a spiral configuration growing from the ground to a height of 4.2m. The prototype is the first structure built in the Robotic Fabrication Laboratory (RFL), a test bed for large-scale robotic fabrication research at ETH Zurich. It consists of four 6-axis industrial robots mounted on a 3-axis gantry system that can cooperate on architectural fabrication tasks within a maximum building volume of 43 x 16 x 6m.

The structure was built from the ground up, pushing the vertical building envelope of the fabrication system (Fig. 4). Whereas from afar the construction process appeared as a surface-based assembly, the steel tubes were actually guided into the interlocking reciprocal node configurations in a truly spatial manner (Figs. 5 and 6). The non-intuitive robotic trajectories generated in the design environment were sent via a custom CAD-to-robotic-controller interface. In order to be able to create collision-free robotic movements, the model of the simulation had to match the actual physical set-up. While performing the first tests, collisions occurred – for example, because the kinematic model was not paired with the simulation or because the physical envelope of the IO box, mounted on the back of the robot, was ignored.

The 16mm diameter steel tubes were pre-cut at the required length and picked up by the robot from a pneumatically actuated pick-up station. The building elements were then guided through the building space, avoiding contact with physical obstacles, to their designated location within the structure. Once the element was in place, the robot changed its function from a tube manipulator to a tube holder. The element could then be joined to its neighbours by the manual welding of four spots around the connection (Fig. 1).

5

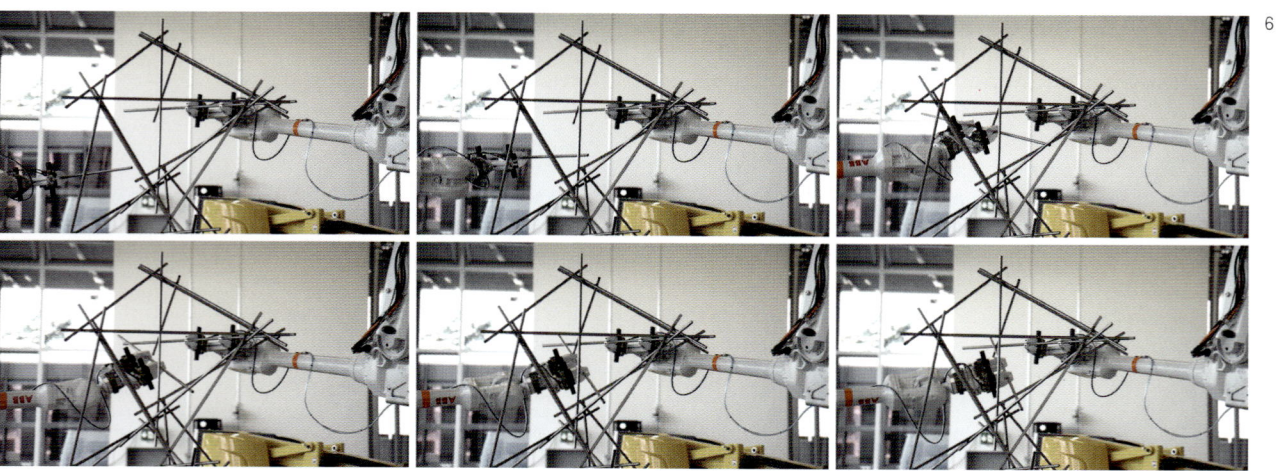

6

During the build, tolerances of up to approximately 5mm occurred at the joint between two building elements. The reason for this is that the robotic system, with its unprecedented large workspace, is still subject to initial adjustments and the large-scale metrology system of the RFL is not yet in operation; in addition, some of the tubes were slightly bent. However, since the welded joint can accommodate a few millimetres of tolerance and because each new structural element was placed according to the digital blueprint rather than based on what had already been built, the tolerances did not accumulate over time, which enabled a successful welding of the entire structure.

Successful cooperative fabrication

The work presented here successfully demonstrates the ability of cooperating robots to hold building elements in space, allowing for the building of non-regular spatial metal structures by integrating computational design, robotic simulation and digital fabrication. However, several aspects of the project require further development. Firstly, the settings of the simulation parameters still require several manual steps and knowledge from the designer about the functionality of the algorithm. Simplifying and further automating the integration of this process with the computational design environment would allow the user to interact more intuitively with the tool when designing robotic movements. Secondly, as described, the welding was manually performed, and further testing to transfer this joining to a robotic method is required. Finally, sensing the spatial arrangement of the structure while building on it would allow compensation for tolerances (for example, from bent tubes or errors that occur during the construction).

The realisation of the architectural scale physical prototype displays some of the potential of cooperatively building space frame structures with two robotic arms. Cooperative construction expands the capacity of digitally designed and robotically fabricated architecture. Here, multi-robotic cooperation is not merely used to distribute the workload between individual machines, but to perform building tasks a single robot (or a human) could not accomplish. The integration of path planning and robotic simulation capabilities within the computational design allowed for the generation and later physical realisation of intricate reciprocal space frame configurations. The project demonstrates the possibility of methods of computational design that take into account the full spatial movement of robots when materialising architecture and, as such, fosters a shift from layer- and surface-based assembly approaches towards truly spatial aggregations. As such, the construction system presented here is not limited to discrete metal tubes. Combining computational design with this increased three-dimensional autonomy of cooperative robotic construction potentially leads to novel architectural construction systems in general.

Acknowledgements

The research presented here is part of and supported by the research programme NCCR Digital Fabrication, funded by the Swiss National Science Foundation. A special thanks goes to Dr. Thomas Kohlhammer for assisting with the structural design and geometry, Marc Freese for collaborating on the topic of robotic simulation, Philippe Fleischmann and Michael Lyrenmann for their support in the RFL and Matthias Danzmayr and Marco Palma for contributing to the realisation of the prototype.

References

Chilton, J., 2000, *Space Grid Structures*, Oxford, Architectural Press.

Coppelia Robotics, 2012, *V-rep*, available at http://www.coppeliarobotics.com/ (accessed 25 September 2016).

Gramazio, F., Kohler, M. and Willmann, J., 2014, 'Robotic Metal Aggregations' in *The Robotic Touch*, Zurich, Park Books, p.430-439.

Gramazio, F., Kohler, M. and Willmann, J., 2014, 'Aerial Constructions' in *The Robotic Touch*, Zurich, Park Books, p.400-409.

Kavraki Lab, 2012, *Open Motion Planning Library*, available at http://ompl.kavrakilab.org/ (accessed 30 September 2016).

Parascho, S., Knippers, J., Dörstelmann, M., Prado, M. and Menges, A., 2015, 'Modular Fibrous Morphologies: Computational Design, Simulation and Fabrication of Differentiated Fibre Composite Building Components' in Block, P., Knippers, J., Mitra, N.J., Wang, W. (eds.), 2015, *Advances in Architectural Geometry 2014*, Switzerland, Springer, p.29-45.

McNeel, 2013, Grasshopper, available at http://www.grasshopper3d.com/ (accessed 30 September 2016).

McNeel, 2015, Rhinoceros, available at https://www.rhino3d.com/ (accessed 30 September 2016).

Rust, R., Jenny, D., Gramazio, F. and Kohler, M., 2016, 'Spatial Wire Cutting: Cooperative robotic cutting of non-ruled surface geometries for bespoke building components' in Chien, S., Choo, S., Schnabel, M.A., Nakapan W., Kim, M.J. and Roudavski, S. (eds.), 2016, *Living Systems and Micro-Utopias: Towards Continuous Designing, Proceedings of the 21st International Conference of the Association for Computer-Aided Architectural Design Research in Asia CAADRIA 2016*, 000-000, Hong Kong, The Association for Computer-Aided Architectural Design Research in Asia (CAADRIA), p.529-538.

Saunders, A. and Epps, G., 2016, 'Robotic Lattice Smock: A Method for Transposing Pliable Textile Smocking Techniques through Robotic Curved Folding and Bending of Sheet Metal' in Reinhardt, D., Saunders, R., and Burry, J. (eds.), 2016, *Robotic Fabrication in Architecture, Art and Design*, Switzerland, Springer, p.79-91.

Søndergaard, A., Feringa, J., Nørbjerg, T., Steenstrup, K., Brander, D., Graversen, J., Markvorsen, S., Bærentzen, A., Petkov, K., Hattel, J., Clausen, K., Jensen, K., Knudsen, L. and Kortbek, J., 2016, 'Robotic Hot-Blade Cutting: An Industrial Approach to Cost-Effective Production of Double Curved Concrete Structures' in Reinhardt, D., Saunders, R. and Burry, J. (eds.), 2016, *Robotic Fabrication in Architecture, Art and Design*, Switzerland, Springer, p.150-164.

5. Shifted node. The node consists of up to ten tubes and seven reciprocal connections.

6. Sequence of the placement of a tube. The geometrically complex connections require a specific spatial trajectory for the robots for each tube.

Images: Gramazio Kohler Research, ETH Zurich, 2016.

INFINITE VARIATIONS, RADICAL STRATEGIES

MARTIN SELF / EMMANUEL VERCRUYSSE
Architectural Association, London

The AA's satellite campus out in Hooke Park, Dorset, is the headquarters of its Design+Make programme and operates as a laboratory for architectural research through 1:1 fabrication. In an environment that combines forest, studio, workshop and building site, the large-scale fabrication facilities act as a testing ground where students devote time to advanced speculative research through a hands-on approach.

Designing and building architecture in the woods: within an idyllic forest ecosystem that is both material library and site, the programme explores how natural materials, craft knowledge and new technologies elicit exciting and unpredictable architectures while implying a deep connection between site, construction and tree species. It provokes a critical approach to designing and manufacturing – one which encourages a symbiotic relationship with the variability found in nature.

Design+Make's position, embedded within the forest, nurtures the students' attitude towards design, imbuing it with an expanded sense of material implications. They are exposed to the long-term investment of time and energy required for timber growth and the forestry processes required to manage it. This living material is formed by its spatial and environmental conditions, and the management of a forest is in many ways an act of design where it is possible to guide the structure of the trees it contains. In this way, design thinking begins under the canopy of the forest itself. The forest's delicate experiential qualities are due in no small part to its infinite variability and, rather than merely being a context for the work, the forest itself, with its material and structural diversity, becomes the inspiration for a way of working.

Digital design and fabrication tools are often used to develop non-standard series of components from standardised materials. Timber is usually considered as a rectilinear material, often reduced to sheets, planks or beams before having a complexity returned to it by milling procedures. And yet trees already present a naturally formed non-standard series – each is wholly unique. The Design+Make programme provokes an alternative conception of material form in which inherent irregular geometries are actively exploited by non-standard technologies.

Woodchip Barn

In a standing tree, the naturally occurring branching forks exhibit remarkable strength and material efficiency, being able to carry significant cantilevers with minimal material. Deriving non-standard timber components from wood's inherent forms, the truss of the Woodchip Barn is presented as a unique timber structure that makes full use of the capabilities of new technologies such as 3D scanning and evolutionary optimisation of the placement of each discrete component within a structurally determined arch, along with customised robotic fabrication. The rationale for this approach is that the diverse characteristics of onsite material can be exploited directly without wasteful industrial processing, while simultaneously providing fertile territory for an unconventional design attitude. The Woodchip Barn employs twenty beech forks within an arching Vierendeel-style truss. The building provides 400m³ of storage for biofuels and will enable the Hooke Park estate to use its own timber for renewable heat production.

While timber has seen a resurgence as an advanced architectural material, the complex and organic forms pursued are generally not attributable to the geometric and anisotropic structural properties of wood. Instead, fabrication processes generate complex components from standardised wood products to ensure consistency. An ambition for the project was to exploit the moment-resisting capacity of tree forks. In a standing tree, the naturally occurring forks exhibit remarkable strength and material efficiency[1], and before processing already present what digital tools are commonly employed in pursuit of: a non-standard series.

The Hooke Park woodland was first surveyed for trees with appropriately forked trunks, resurrecting the historic strategy of shipbuilders travelling into the woods equipped with a set of templates that described the specific forms they required to construct various components[2]. An initial photographic survey of 204 standing beech trees provided approximate two-dimensional fork representations with enough detail to make informed decisions about which trees to cut down. From an analysis of this database, a shortlist of 40 forks were selected for felling, from which 25 were successfully harvested. A detailed photogrammetric 3D scan was made of each of these in order to capture their complex forms. From the resulting surface mesh geometry, medial curves were extracted for each fork using a polygon-based method in which transverse sections were cut through each one at regular intervals to obtain the outer profile of their geometry. Following this, local best-fit diameters and centroids were calculated for each profile's section.

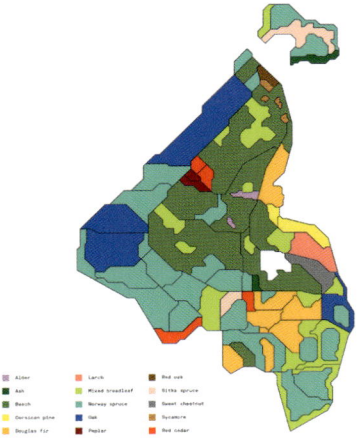

The structural form of the arching truss was determined, in discussion with the Arup team, to have the appropriate inverted-catenary form for a compression structure and a cross-sectional geometry which could accommodate the dimensions and angles of the sourced tree forks. The choice of an equilateral triangular section of typically 90cm side dimensions was found to work well by both providing stability to the arch and being a size on which the forks could be fitted. The structure is composed of two planar inclined arches in a distorted Vierendeel configuration that exploits the moment capacity of the forked junction. The structure lands at four points, the front slightly wider than the rear, with four inverted tripod legs supporting the robotically fabricated mid-section.

The positioning of each forked component within the truss was determined iteratively using an organisation script that sought an optimal arrangement of the components to best satisfy structural and fabrication criteria. This was achieved through evolutionary and simulated-annealing procedures carried out in the Galapagos solver within the Rhino-Grasshopper environment. Within the optimisation, there were two levels of position adjustment: the global swapping of components between possible locations in the structure, and the local shuffling of components in which each element was slid along the target arch curves to best find its location. The key criterion was to minimise deviation of the forks' medial curves from the target curves of the idealised arch centrelines. Further criteria were applied

1. Timber is usually considered as a rectilinear material – its irregular forms reduced to standard sections. The work undertaken proposes an alternative concept of material form in which inherently irregular geometries are directly exploited by non-standard technologies. Image: Valerie Bennett.

2. Design+Make projects attempt to exploit the inherent characteristics of the approximately 16 tree species found within Hooke Park. The Tree Fork Truss project was developed from the naturally occurring form of 25 distinct beech trees. Image: Zachary Mollica.

3

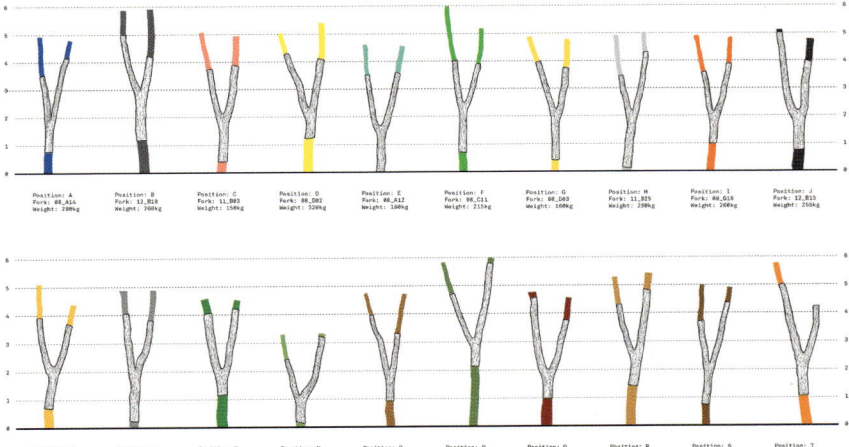

4

to place the larger diameter trees where axial forces were greatest and to deal with specific geometric constraints (for example, at the points where the truss bifurcated to form its legs). The optimisation was improved by indexing the component set according to the geometric strategy and by sequencing the placement so that the most critical positions were populated first[3].

The outcome of the optimisation process was a three-dimensional arrangement of the tree fork geometries in which the key setting-out nodes were coincident with the underlying target tree curves. The combination of this nodal data with the element medial curves and diameter data was used to derive the digital fabrication information for the machining of connecting features into the raw tree forks using a router spindle on Hooke Park's Kuka KR-150 6-axis robot arm. The connections were configured to achieve transfer for compression forces through timber-to-timber bearing and to reinforce these with steel bolts when additional tension or shear strength was required. The connection surface geometries varied in different parts of the structure and consisted of either planar face-to-face surfaces between elements along the chords or mortice-and-tenon joints in which a distorted elliptical cone geometry was found to best satisfy the structural and assembly constraints.

The robotic milling procedure consisted of first defining 3D volumes for router subtraction of connection shapes from the wood, then determining an appropriate robot toolpath to achieve that geometry. The key requirement was to produce precise relative positions of the machined surfaces such that dimensional accuracy during assembly could be achieved. Two strategies were developed to enable this. Firstly, a consistent referencing system was established which ensured that a tree fork component could always be correctly located in space in the virtual modelling environment, the machining cell and the

3. The planar geometry of a Japanese joint lends itself perfectly to the specific machining operations of robot and chainsaw.
Image: Valerie Bennett.

4. An idealised structural volume was established. A Grasshopper script was developed to allow it to be dynamically populated with real fork geometries. This image shows the 'trimmed' form of each fork contained within the final built structure.
Image: Zachary Mollica.

5. The robot arm machining one of each fork's two bearing surfaces. The digital model was translated into fabrication information with which a 6-axis robotic arm transformed each fork into a finished component.
Image: Pradeep Devadass.

5

ultimate assembly of the structure. This was achieved by physically drilling three reference holes on the truss that were tracked in the 3D models and subsequently used to support the fork during fabrication and assembly. The second strategy was to accept that it was difficult in practice to be sufficiently precise in modelling the exact surface geometry of the natural tree (an accuracy of +/-10mm was typical, rather than the +/-2mm required) and so to make the locating of the milled face independent of the outer tree surfaces. This was achieved by defining subtracting volumes larger than the tree as scanned and accepting some redundant milling of air rather than wood.

Following the fabrication of the fork components, the truss was pre-assembled in two halves in Hooke Park's assembly workshop. Again, drilled reference points were used to correctly locate the fork components within an erection jig whose support geometry had been extracted from the digital model. The precision of the robotic fabrication proved successful and only occasional manual woodworking was needed to achieve a well-fitting fully bolted assembly. This was further demonstrated when the two truss halves were crane-erected onsite and the full 25m arch was successfully de-propped. The building was completed with the addition of push-walls to contain the woodchip and a conventional timber-framed roof supported by the arching truss.

The building is presented as a demonstrator and validation of an approach proposed in various forms over recent years[4,5] in which new computation tools are applied to the configuration of material elements so that the inherent geometry of those elements is exploited. In this case, the underlying arch geometry was largely predetermined (i.e. anticipating typical geometries of the forks but not directly determined by them) and the optimisation was limited to locating components within that geometry. Thus a development of the method will be to enable the underlying structural form itself to self-organise through the varied components acting as agents towards a set of spatial and structural goals.

Advanced and bespoke system operations

Other strategies are now in place to enhance this approach, enabling more complex structural experiments. For instance, establishing the horizontal rotational seventh axis to operate in synchronisation with the 6-axis robot arm has been instrumental to advancing the manipulation of non-standardised timber. This configuration, capable of carrying large tree segments between two modified lathe end-stocks, means that the robot's end effector can access any point along the length of the tree log. The ability to carve a tree much more freely opens up new formal, structural and aesthetic potentials. The machining operations can be applied locally and the sculpted profile could be structurally optimised – analogous to the geometry of bone or open-grown trees – and gives timber as a material a new 'plasticity' (in the art history sense of the word) of form that is difficult to achieve with other materials.

The application of a variety of end effectors provides yet more possibilities for the manipulation of the material. The chainsaw – a tool not known for its exactitude – gains an augmented level of precision and control when wielded

6

6. The front half of the truss being assembled within the Big Shed. The finished elements of the truss were pre-assembled in two halves, each approximately 9 x 6m. A large jig allowed their accurate arrangement. Image: Zachary Mollica.

7. Precision augmentation of naturally formed geometries utilising robotically controlled chainsaw. Image: Zachary Mollica.

8. The natural forked geometry undergoes localised modification to facilitate a structural connection. Image: Valerie Bennett.

by the large Kuka KR150 robot. LiDAR scanning technologies form an essential component within these advanced system operations, not only providing a fully calibrated workspace but also crucially allowing operations on naturally formed geometries with surgical precision.

3D scanning allows us to treat something incredibly unique and complex in form in the same way that we might treat a standard plank of timber. The ability to scan the space of machining to align the worldview of the robot with the actual position of a non-linear object like a tree trunk allows for more flexible machining strategies, as the calibration becomes more organic. The digital model and the physicality of machining on this scale can converge with previously unimagined precision.

The innovative and radical nature of the approach employed at Hooke Park lies in the strategic precision with which Design+Make teams can augment the natural geometry grown there. The variability and complexity is natural – our machine strategies play to the beauty and strength of this complexity and follow its lead[6]. In this way, we are employing the tacit knowledge of a material on which craft relies, while exploring the possibilities afforded by the pinpoint precision of the technological eye and hand of scanner and robot.

The aim is to use robotic technology not forcefully, for power, repeatability or wilful formalism, but delicately, for the strategic augmentation of a natural and complex logic. It is with this attitude that we have established the campus as a 'continuous laboratory', where Design+Make operates as an agency of architectural innovation and presents a unique and alternative vision for architectural education.

Acknowledgements

The Woodchip Barn was designed and built by Design+Make students (Mohaimeen Islam, Zachary Mollica, Sahil Shah, Swetha Vegesana and Yung-Chen Yang) under the tutelage of the authors, with Charley Brentnall and Toby Burgess. It was fabricated and constructed with the support of SummerBuild volunteers, site project manager Jack Draper and the Hooke Park forestry and workshop team (Christopher Sadd, Charlie Corry Wright and Edward Coe). Pradeep Devadass oversaw the robotics development, and engineering support was provided by Arup (Francis Archer, Naotaka Minami and Coco van Egeraat).

Notes

1. Slater, D. and Ennos, R., 2014, 'Interlocking Wood Grain Patterns Provide Improved Wood Strength Properties in Forks of Hazel (Corylus avellana L.)' in *Arboricultural Journal, January 2015*, Berlin, Springer-Verlag.

2. Matthew, P., 1831, *On Naval Timber and Arboriculture*, Cambridge, Harvard University.

3. Mollica, Z. and Self, M., 2016, 'Tree Fork Truss: Geometric Strategies for Exploiting Inherent Material Form' in Adriaenssens, S., Gramazio, F., Kohler, M., Menges, A. and Pauly, M. (eds.), *Advances in Architectural Geometry 2016*, Zurich, p.138-153.

4. Jannasch, E., 2013, 'Embodied Information in Structural Timber' in Cruz, P. (ed.), *Structures and Architecture: New Concepts, Applications, and Challenges*, London, CRC Press, p.2176-83.

5. Schindler, C., Tamke, M., Tabatabai, A., Bereuter, M. and Yoshida, H., 2014, 'Processing Branches: Reactivating the Performativity of Natural Wooden Form with Contemporary Information Technology' in *International Journal of Architectural Computing*, Vol. 12, No. 2, p.101-116.

6. "I don't want to change the material, I want to follow its lead," Giuseppe Penone quoted in interview with Robert Enright, 'The Perfection of the Tree and Other Material Concerns' in *Border Crossings*, December 2013, available at http://bordercrossingsmag.com/article/the-perfection-of-the-tree-and-other-material-concerns (accessed 29 December 2015).

AUTOMATED DESIGN-TO-FABRICATION FOR ARCHITECTURAL ENVELOPES
A STADIUM SKIN CASE STUDY

JAMES WARTON / HEATH MAY
HKS LINE
RODOVAN KOVACEVIC
Southern Methodist University

Working with stadium architecture

The stadium roof structure surveyed in this paper is comprised of approximately 70,000 unique panels with over 500,000 square feet of surface area. These panels are uniquely articulated and cut to specification using a 3-axis CNC-coined die-punch machine and fabricated from titanium anodised aluminium. The panels are shop-fabricated and pre-assembled into mega-panels according to Zahner's proprietary ZEPPS process. Two key building components were isolated for development of a complete file-to-factory workflow. The panelised geometry and perforation patterns are fully automated and implemented within the examined project. The second component studied within this paper is a 3D printed fixation detail proposed as an alternative to the ZEPPS solution.

Both the exterior envelope's aluminium panels and the hypothetical node connections are discussed in terms of the challenges and constraints unique to their respective geometry, fabrication process and performance criteria. The design-to-fabrication workflow is described,

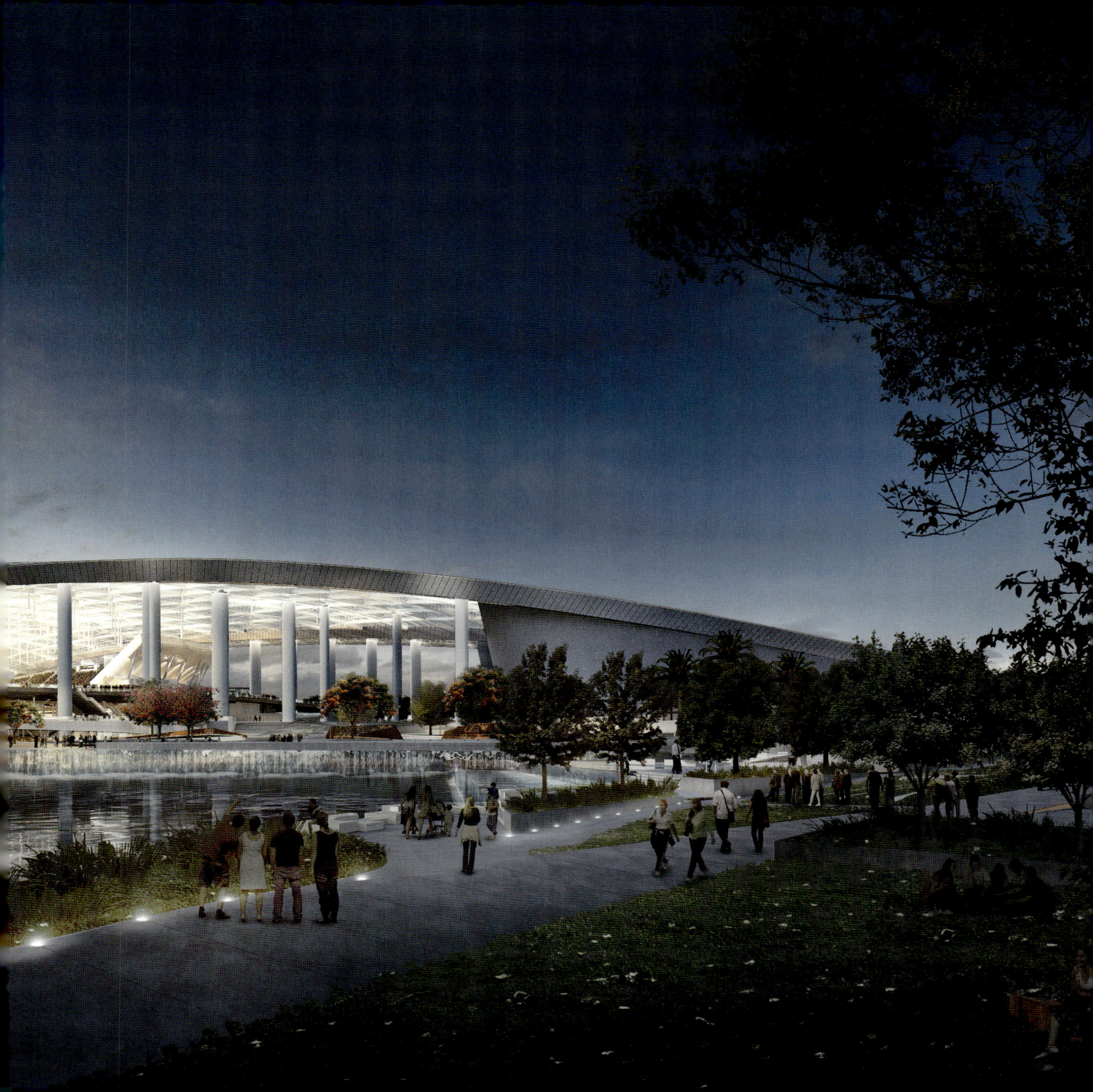

demonstrating improved efficiencies at various stages between design development and shop fabrication. In addition, an alternate node design is developed to accommodate the constraints of additively manufactured alloys and compared against the accepted proprietary solution. The resulting analysis demonstrates the advantages of the 3D printed alternatives, the potential of 3D printing as a fabrication method at this scale of execution and key developments that must be realised in order to achieve feasibility.

The workflow adopted leverages a customised C++ framework and implements open source libraries such as OpenGL for visualisation and Array Fire v3.2.2 for GPU-based image processing and matrix operations. The framework presented is not conceived as an autonomous design tool, but rather as a vehicle for the exploration and interpretation of computationally intensive procedures. This case study demonstrates the effectiveness and performance improvements of this workflow over visual programming approaches such as Grasshopper.

Project scope and mounting challenges

Advancement in computational design tools has led to an observable proliferation of architecture exhibiting greater degrees of geometric complexity, variability and differentiation. Although these tools increasingly enable the automation of design, coordination and documentation, the workflow connectivity between design and fabrication is typically severed, disrupting the linkages required to automate the manufacturing of these highly differentiated building systems. As the scale of a project and the scope of differentiated elements increase, computational overhead and connectivity challenges become increasingly evident.

One of the key obstacles examined within this paper relates to communication and connectivity between designer and fabricator. Common practice typically requires that both parties exchange dimensioned documents for all unique building components within a system. In cases where fabricators rely heavily on automation and leverage CNC manufacturing processes, the translation of diagrammatic or representational documents and even the processing of CAD drawings can be prone to errors and may lead to significant delays in production. This processing time often leads to a bottleneck impeding uninterrupted machine operation. The case study presented demonstrates an alternative approach for documentation in order to minimise time required for the shop-drawing review process, document processing and technical formatting for CNC instruction.

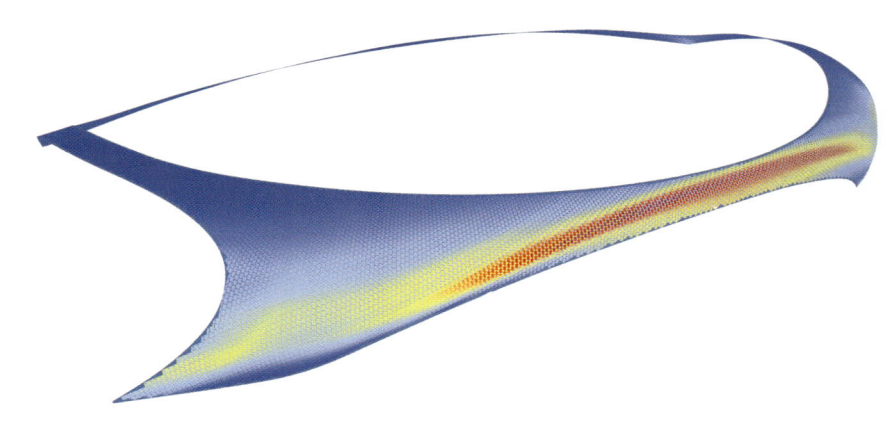

Another area of implementation directly affected by scope relates less to logistics within professional practice and can be exclusively attributed to increased computational overhead. The case study presented highlights deficiencies within widely adopted visual programming methodologies and offers an alternative workflow, targeting improved processing performance. The third area of investigation applies the experimental workflow to a speculative fixation detail based on additively manufactured building components. The objective for this area of research is to assess the viability of the proposed methodology within a more complex geometric scenario, while establishing the foundation for further case study development, targeting additively manufactured building components and assemblies. Within this examination, limitations to the status quo are discussed, with emphasis placed on trajectories for future research in additive manufacturing and its potential as a direct design-to-fabrication process.

Addressing design complexities and increasing demand for computational performance

These objectives are examined through the lens of a single architectural project. Firstly, through a tessellated double-curved cladding system, and secondly through a speculative structural node which addresses additional design complexities and an increased demand for computational performance.

Because large stadiums commonly exhibit a relatively high degree of geometric complexity and scope, this typology represents an ideal candidate for the case study's development and implementation of direct design-to-fabrication methodologies. The envelopes of these large-scale sports and entertainment venues often challenge traditional means of documentation and necessitate alternative approaches to design development and project deliverables. Computational overhead places an excessive demand on commonly used open source and proprietary software platforms, often yielding sluggish responses during the design process and limiting opportunities to pursue a broader range of iterations. Furthermore, labour-intensive document processing and set-up times attributed to the reformatting of CAD data into machine instruction often lead to operational bottlenecks and prolonged production time. Even in cases where the shop-drawing documentation process is fully automated, as described by Front Inc. (Levelle et al., 2017), the production bottleneck is pushed further downstream, placing an excessive burden on fabricators.

The case study presented is based on the current stage of design and development for an NFL football stadium and future home to the Los Angeles Rams, designed by HKS Architects (Fig. 1). The project will be located in Inglewood, California, and was recently awarded to Turner Construction Company, with construction completion scheduled for November 2019. There are a broad number of situations and systems within this project that demonstrate the variability and geometric complexity applicable to this research objective; however, for the purposes of this paper, portions of the stadium's envelope have been isolated for examination. The outer layer of stadium skin surveyed is comprised of approximately 35,000 unique panels covering nearly 275,000 square feet of surface area (Fig. 2). Zahner was retained for pre-design and design assist services and worked directly with the authors during the development of the project's cladding system.

The panels comprising the tessellated surface geometry have been isolated as the primary design element for investigation and development of a complete file-to-factory workflow. The vast majority of these panels are flat triangles and, based on the currently adopted fixation strategy, require four fasteners along each edge. According to this method, the examined areas would require over 376,000 fixation points. Each panel is uniquely articulated and cut to specification using a 3-axis CNC-coined die-punch machine and fabricated from 0.125in-thick titanium anodised aluminium sheets (Fig. 3). The skin is perforated with up to eight circular hole sizes, ranging from 0.375in to 1.25in at 0.125in increments. The perforation sizes correspond to a global grayscale image mapped to the domain of the stadium skin's parent design surface. Once the perforated panels are fabricated, they will be shop-assembled into unitised mega-panels according to Zahner's proprietary ZEPPS process.

According to the current scheme, panels will be attached to curved aluminium ribs and straight linear cross-members (Fig. 4). The second component to be examined is based on an alternative fixation strategy proposed in lieu of the continuous edge frames previously described. The alternative strategy relies on a 3D printed node using powder-bed-supported additive manufacturing technology, and leverages a 0.375in-thick stress-skinned concept proposed during the early stages of design development. This concept requires only three fasteners per panel (Fig. 5) and would employ a quad-mesh-based structural system. The torsion-free purlins would follow only two of the three triangular edge grids defined by the panelisation geometry. Although the two systems have not been fully engineered for comparison, it has been documented that triangular framing systems require more members than quadrilateral systems[1] and are often heavier (Pottman et al., 2014).

This concept and the corresponding six-point fixation method were abandoned due to the complexity required to fabricate an effective node connector that could accommodate the variability and deviation present within the panels[2]. The schemes considered were based on commonly accepted manufacturing processes such as plate assembly, casting or milling and were not suitable for non-standardised conditions at the scale of production required – and if standardised would feature a high degree of assembly and local adjustment. It was concluded that an operable or flexible node connector comprised of numerous elements would require too

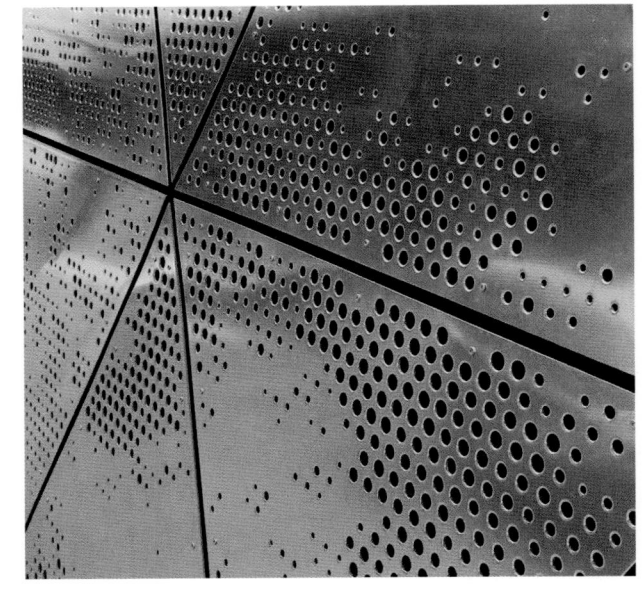

3

1. NFL stadium featuring double-curved cladding system.

2. The 274,962 square foot outer region of the stadium envelope comprised of approximately 35,000 panels. Heat map indicates deviation between panel normal and surface normal of design surface at node.

3. Close up of aluminium panels showing coined die-punch perforation.

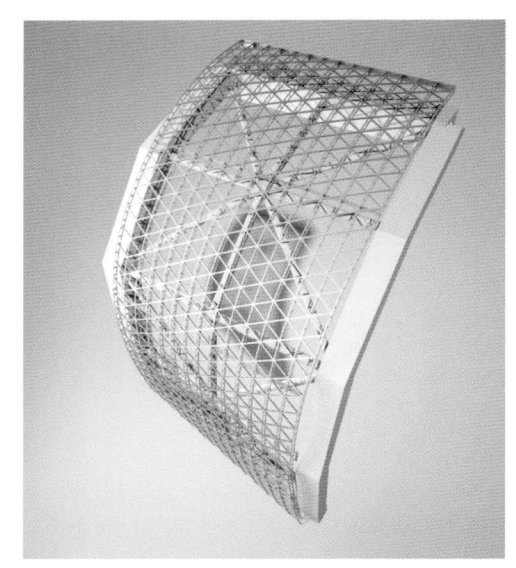

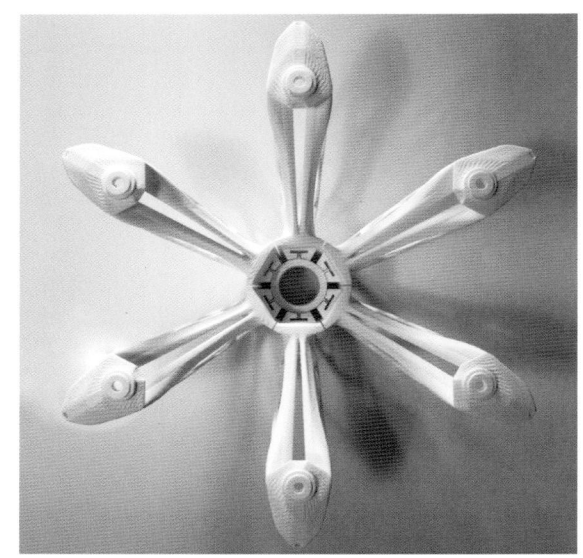

many degrees of freedom and site adjustment, whereas a mass-differentiated element comprised of primarily fixed branches could provide a more viable conceptual solution. Provided additive manufacturing was feasible in terms of construction cost and schedule, this method would satisfy the level of geometric variability needed while enabling the initial purlin strategy comprised of fewer structural members. In addition, this manufacturing method is more aligned with the development of direct design-to-fabrication processes, due to the ability to fully automate each part's production as a single process. On this basis, the authors propose the development of a node definition that satisfies the complexities of the valence 6 vertices typical of all triangular freeform meshes, while leveraging the advantages of a quadrilateral substructure.

In order to minimise file conversion and processing time associated with design documentation and CNC-based fabrication processes, physical drawings and CAD files have been bypassed as a form of communication. During early coordination discussions between the design team, façade consultants and structural engineers, the metal panel fabricators asserted that one of the primary bottlenecks for fabrication and coordination is associated with the set-up and translation of CAD files, such as .dxf or .dwg, into machine instruction. Rather than rely on over 75,000 individual 2D drawings to dimensionally describe each panel, a text-based file format containing all dimensional criteria was adopted. Through adequate file nomenclature, tokenisation and formatting, the fabricator could automate the translation of these files directly into machine instruction.

To ensure proper coordination, file formatting conventions were established. Each panel was described in both world/project and local/machine space coordinate systems. Node centrepoint positions, corner positions and panel orientation vectors were provided in project space for coordination. For fabrication, local node positions, trimmed corner positions, fastener positions and perforation centrepoint positions with their specified diameter were provided within local machine coordinates. Local machine coordinates were established based on an origin point and x-axis coincident with the first node position and first panel edge respectively.

Since the constraints for individual panels were driven by Zahner's proprietary assembly process, rationalisation of the design surface was performed by Studio NYL, a façade design and engineering consultant contracted by Zahner. Ultimately, the proposed proprietary system would yield an increase of 12.8% more panels over the in-house panelisation routine, totalling 3,571 additional panels within the surveyed region alone. The governing criteria for these panels was to minimise deviation from an ideal equilateral triangle cut from a 48in-wide sheet and oriented consistently within each sheet to maintain uniform material grain.

The host or parent design surface modelled in Rhinoceros and shared among consultants serves as the primary design input for the stadium's skin. Once the surface is tessellated into panels, the node centrepoints are extracted using Grasshopper and formatted into text files corresponding to eight regions or zones delineated by the cladding fabricator. These node positions are then loaded

4. Isolated structural bay showing substructure and framing strategy with continuous rolled extrusions along primary grid line and segmented straight framing members between.

5. Panel layout showing fastener locations for both substructures. Fixation points for ZEPPS vs. additively manufactured node with six branches.

into a custom application developed by the authors in C++. The initial data extracted from the model is limited to eight lines of instruction per panel. For improved computational performance, the C++ framework leverages multi-threaded functions and implements open source libraries, such as Armadillo v5.600.2 (Sanderson et al., 2016) and ArrayFire v3.2.2 (Yalamanchili et al., 2015), for GPU-based matrix operations and image processing. OpenGL is also implemented for visualisation purposes.

As the panel identification and its node positions are read from a source file for each of the regions, this data is stored and a new panel object is defined. The application calculates the transformation from 3D world coordinates into 2D machine space and stores instances of each becoming part of the panel object's properties. After being described according to a local coordinate system, the panel corner positions and edges are defined according to a predetermined edge offset parameter. Then the panel is subdivided with a perforation grid unique to each panel's geometry, and the fastener positions are located so that they coordinate with this grid. A separate routine is then performed to map pixel values from the global design image into the panel's perforation grid (Fig. 6). Grayscale values from the design image, ranging between 0-255, are then remapped to correlate with hole sizes corresponding to one of eight die-punch tools available during the fabrication process. Once a panel's fabrication data are fully defined, a text file is generated containing a comprehensive geometric description of the panel. A graphic interface provides a searchable visual display of all the panels within the currently processed batch, alongside a global view of the stadium indicating the active panel's location. Various display states are also available that describe overall system mappings of area deviation, angle deviation and panel opacity. The average file size is 47KB and file sizes range between 4 and 145KB, dependent upon the quantity of perforations specified for each panel.

A similar design methodology is applied to the additively manufactured node definition and utilises the same code libraries developed in C++. An additional layer of data management is incorporated to ensure that neighbouring panels are properly associated with their common node. Vector trees for each node are calculated based on their respective vertex normal, with branching elements connecting adjacent panel fixation points to the intersecting sub-framing below (Fig. 7). Once the branching topology and configuration is defined, the node object is instantiated and a bounding volume for each vector is constructed. Ultimately, these branching volumes will be sized iteratively using FEA of the linear elements of the vector tree and procedurally defined load cases. The FEA solvers have been developed and validated, yet at present have not been fully integrated. Subsequent operations are performed to provide the required mesh density and smoothing necessary for fabrication. Ultimately, the constructed volume can be utilised as an initial design space for topology optimisation. To test this approach, a generic node was generated and topology optimisation was performed using SolidThinking Inspire.

Since the implementation of this building component is speculative and requires assumptions based on an earlier schematic design scenario, performance concerns

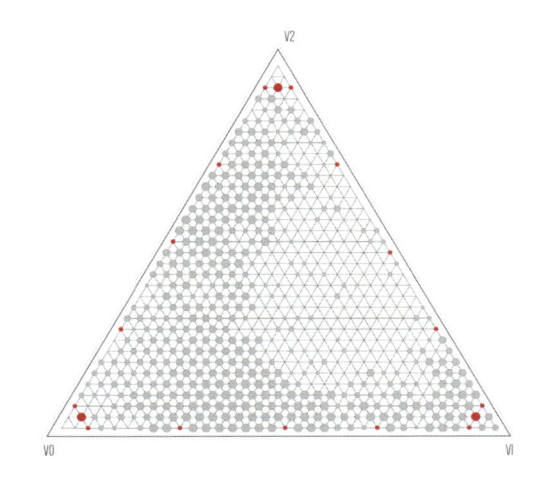

6. Panel layout showing subdivision grid and image mapped perforations. Fixation points for ZEPPS framing vs. AM node with six branches are delineated in small and large red dots respectively.

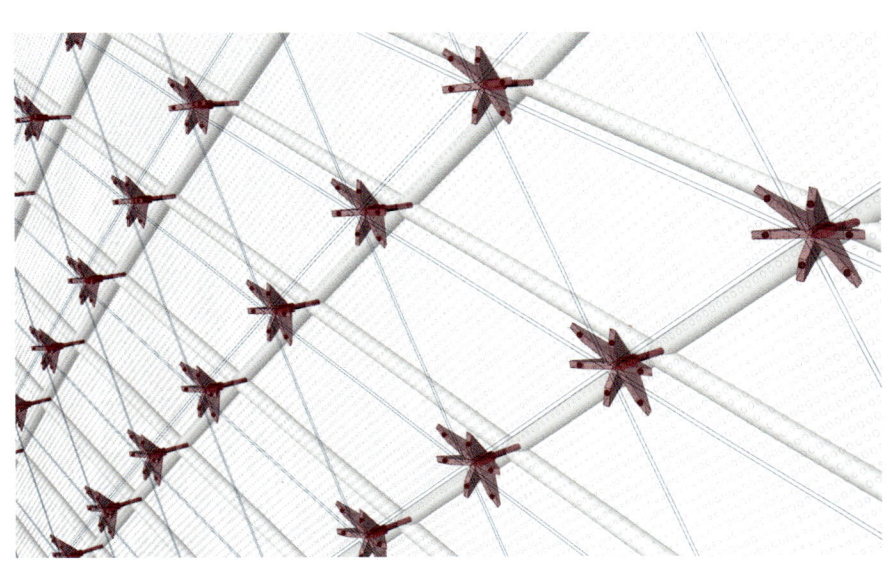

7. Panelised surface geometry with nodes constructed specific to each juncture.

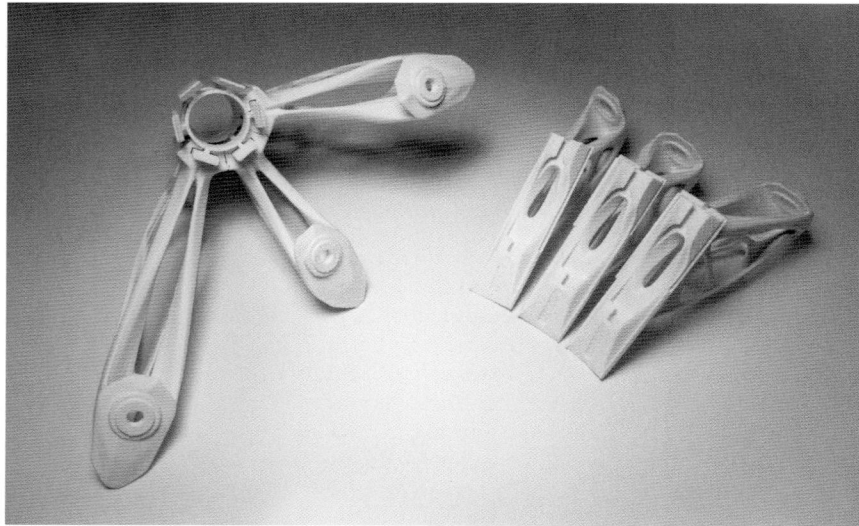

8. AM node prototype subdivided into separate branching elements. Each branch is indexed according to its position relative to the building envelope. A unique locking key prevents each component from being assembled out of sequence.

have focused on internal workflow development and prototyping constraints. Cost and production feasibility assessment is underway with the assistance of Concept Laser. It should be noted here that Concept Laser produces additive manufacturing systems and is not a for-service parts fabricator. Fabrication constraints for the proposed nodes are based on the use of their X line 2000R. Concept Laser claims that this is the largest build volume currently on the market for a powder-bed-supported additive manufacturing system which utilises a laser heat source for production of metal components. They are currently promoting this machine as one of the key components within their model operation for lean additive manufacturing, which they refer to as the 'AM Factory of Tomorrow'. The build volume available for the X line 2000R is 800 x 400 x 500mm, allowing an average of two nodes per build and an approximate production time of 18 hours. Due to the required build time, a more compact arrangement would be ideal, but to achieve this, the node must be subdivided into its constituent parts, similar to the example shown in Fig. 8. Further subdivision may prove advantageous for production, but this decision would have inevitable impacts on production time and coordination due to the additional assembly required. It is assumed, however, that the assembly could be managed in tandem with subsequent print processes.

Looking at performance gains

While it is generally accepted that lower level programming languages such as C++ provide superior performance over higher level programming languages, a benchmarking trial was established to test the authors' assumptions[3]. Since Grasshopper offers limited profiler stats, a precise measure of computational performance is not immediately available. Due to this limitation, a trial was conducted recording overall calculation times rather than conducting a piece-wise process comparison. Only the two most computationally intensive functions – calculation of the subdivision-based perforation grid and image-mapped perforation size – were implemented within Grasshopper. Transformations from world to local coordinate systems and evaluating fastener positions were omitted. These outcomes were then compared to implementation written in C++ using both single-threaded and multi-threaded programmes, executing the entire procedure required to define and document the perforated panel system. To further simplify the comparison, the input data describing each panel's world coordinates was internalised within the .gh definition, rather than read from an external source file. Each trial was conducted five times, with the resulting averages recorded as graphs (Figs. 9 and 10). The performance gains achieved using the proposed methodology are evident, although it was noted that the overall computation time does not increase at a linear rate. This may indicate an area for future work, and further study is required to optimise the programme for increased scope.

Once all the relevant data for each panel are calculated, the panel object can then be documented in several ways, depending on the file format needed. The primary means of fabricator communication is managed though discrete data files containing all the relevant information needed for coordination and fabrication. These files are then compiled into a database and prepared for translation into CNC instruction or G-Code, which will be automated by the fabricator using their in-house post-processor. This communication process has been coordinated directly between the authors and the fabricators expected to complete this project. Preliminary mock-ups have been produced to test the hypothesis and work will begin in 2017 to test viability of implementation at scale. As the project moves into construction, and the building envelope is finalised, a comparison between the as-built structural framing and a system which incorporates the proposed 3D printed node can be evaluated.

The future experimental workflow development

Conventional methods of communication between the designer and fabricator present logistical challenges to a complete direct-to-fabrication workflow. Despite the advantages present with increased scope and scale of economy, conventional means of exchange impede or diminish the ability to fully realise these advantages. The authors propose an experimental workflow that mitigates some of the concerns regarding design development, documentation, fabricator and construction coordination and production feasibility. The alternative means of documentation rely on a workflow where graphical diagrams and representational drawings, either physical or digital, are omitted as the primary way to convey information. Computational performance gains are demonstrated using this proposed workflow, and applicability to both 2D and 3D building components has been demonstrated.

Furthermore, integration of this workflow to design for additively manufactured components increases opportunities for design optimisation. The integration of additively manufactured structural components shows great promise beyond localised structural optimisation and simplified assembly. The proposed approach to node design would enable further system-wide structural optimisation within freeform architectural envelopes, which could yield an overall reduction of parts and framing members. Continued research and development as this project nears completion will demonstrate the viability and quantifiable measure of this hypothesis. There is, however, a great deal of advancement required to realise these potentials in practice. The availability of for-service fabricators with the resources to produce the parts described is very limited and presently not sufficient for production at the scope required for the envelope presented.

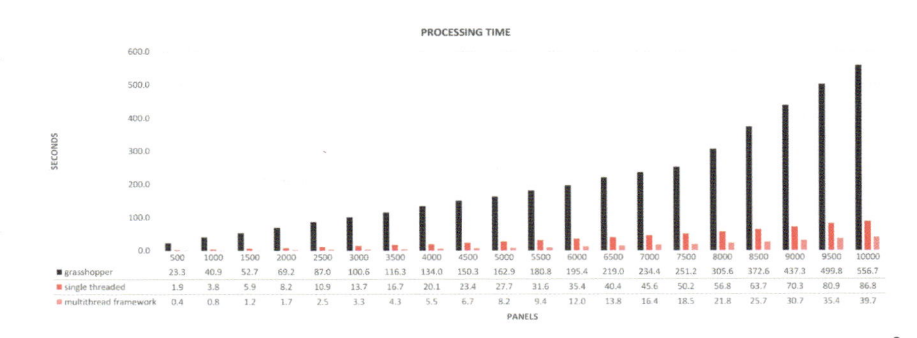

Notes

1. A more substantial disadvantage is that in a triangle mesh a typical vertex is incident with six edges and thus is significantly more complex than the valence four vertices typical for quad meshes (see Fig. 3). Generally triangle meshes require more parts and are heavier than quad meshes (Pottman et al., 2014).

2. For structural elements like nodes, beams and frames, however, the tolerances are often tighter and the geometry of these structures is often more complex than that of the outer skin. Therefore optimising freeform structures for repetitive elements is highly challenging and sometimes impossible. This complicates logistics and increases production cost, and is a typical feature of freeform shapes in architecture (Pottman et al., 2014).

3. System specifications:
Workstation name: PWDA1378
CPU specification: Intel(R) Core(TM) i7-5930K CPU @ 3.50GHz
Memory: 63.902698 GB

9. Performance comparison showing improved computation speeds.

10. Comparison showing single threaded vs. multi-threaded performance.

ROBOTIC WOOD TECTONICS

PHILIP F. YUAN / HUA CHAI
Tongji University

As the only naturally reproducible green building material, wood has become the first choice when addressing environmental concerns. With the rapid development of laminated wood technologies and other production techniques, modern wood has become a high performance material with a large scale and low weight-to-strength ratio which demonstrates great potential in the future development of the construction industry (Menges, 2011). Digital design has marvellously expanded the scope of wood structure application. While the growing trend for research in robotic fabrication has accelerated the development of mass customisation concepts in architecture, the mass customisation of geometrically complex wooden elements has become one of the major concerns in terms of robotic wood fabrication research and wood-producing industry (Buri & Weinand, 2011). The capacity of current CNC-milling-based non-linear wood component fabrication methods, which not only consume a lot of time but also produce a lot of material waste, is falling out of line with the rapid development of digital design technology (Brell-Cokcan et al., 2009). The 'Robotic Wood Tectonics' project of 2016 DigitalFUTURE Shanghai explored the combination of

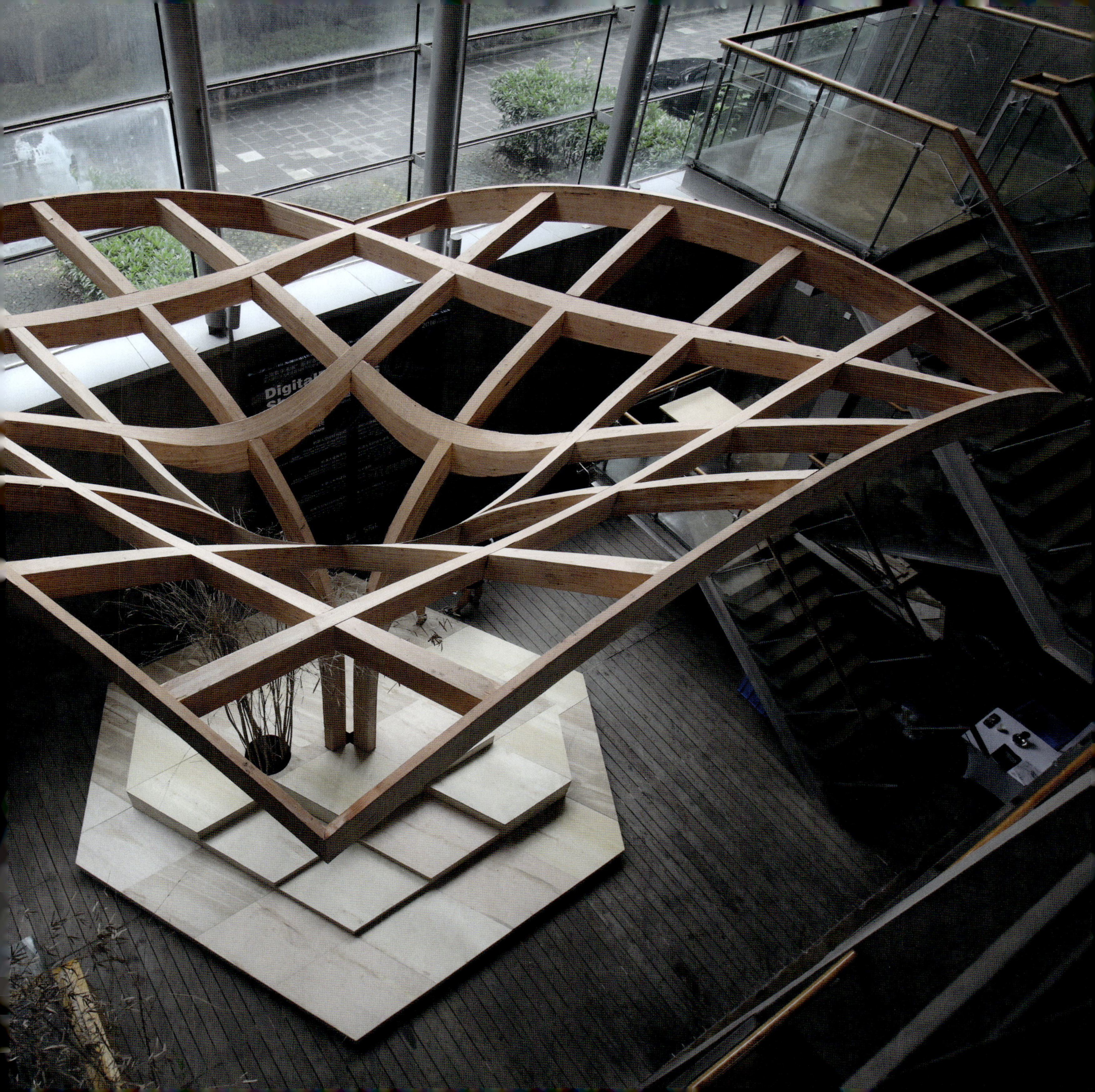
Digital

2

1. Final full-scale pavilion. Image: Lin Bian.

2. Robotic fabrication process.

3. The production of glued laminated beams. Image: © SUZHOU CROWNHOMES CO. LTD.

4. The robotic effectors. Image: © College of Architecture and Urban Planning, Tongji University.

robot wire-cutting technology and traditional woodcraft to produce geometrically complex wooden elements – without the immense material consumption of a CNC milling production process – in a full-scale wood pavilion. Furthermore, this project explored the extent to which this approach has the capacity to mass customise large-scale architectural wood elements, which would be critical to the robust processes demanded by the manufacturing industry. This project aims to demonstrate innovative robotic wood tectonics – an integrated working process from design to fabrication.

Research context

In the wood manufacturing field, milling currently seems to be the only way to deal with geometrically complex wood components. Built projects such as Centre Pompidou Metz and the Nine Bridges Golf Club by Shigeru Ban were constructed using a milling approach with indispensable technical support from Designtoproduction. In addition to defects, waste and processing time, data transformation between the design and manufacture stages in CNC milling remains a major constraint, and in fact these issues constituted a large part of Designtoproduction's work (Scheurer, 2010). Indeed, these defects are more obvious when they relate to factors like design changes.

With the rising trend of research in robotic fabrication, some research institutions are trying to explore new possibilities of wood manufacture by employing industrial robots in the fabrication process, which has proved to have a great impact on the design thinking. With robots, the design information can be transformed into a fabrication toolpath directly in real time without the complex data transformation of CNC. The combination of a conventional mechanical bandsaw and robotic wire-cutting technology, where the bandsaw plays the role of wire, is one of the feasible solutions that has to some extent been researched and demonstrated. In the paper 'Bandsawn Bands: Feature-Based Design and Fabrication of Nested Freeform Surfaces in Wood', researchers from Greyshed and Princeton University (Johns & Foley, 2014) for the first time utilised a robotically operated bandsaw to cut a series of curved strips, which, rotated and laminated, can approximate doubly-curved and digitally defined geometry. Using a robotic bandsaw was demonstrated as a materially efficient technique for designing and fabricating freeform surfaces within the constraints of irregular wood flitches. On the other hand, RMIT University (Williams & Cherrey, 2016) has further studied the robustness of this new craft with regard to speed, accuracy and material finish in the mass customisation of ruled surface production, shown in the paper 'Crafting Robustness: Rapidly Fabricating Ruled Surface Acoustic Panels'. This has demonstrated the feasibility of this approach in robotic fabrication of double-curved non-standardised wood elements in furniture and decoration.

Research questions

As demonstrated above, what is not considered in previous studies is the 'crisis of scale' of digital mass customisation, which has been proven to work effectively at the small scale of industrial design and fabrication but has not performed well at the full scale of construction. When it comes to full-scale architectural wood components, the speed, accuracy and effectiveness of this robotic bandsaw cutting method remain unclear. This project is trying to figure out whether this new robotic craft is capable of and appropriate for the mass customisation of full-scale architectural wood components. The research question is studied in detail through some sub-questions:

1. How to negotiate between technical issues like speed, accuracy and stability to ensure the optimum fabrication results;
2. How traditional mechanical tools and the knowledge of materials can be used in guiding robot fabrication; and
3. How the full application of existing wood manufacturing technology might improve the practical significance of the state-of-the-art robot technique.

3

Furthermore, this research tries to figure out how this new robotic wood technology might affect the design process to achieve an integrated design process from design to fabrication as a new form of robotic wood tectonics.

Research methods

This research is carried out through the design and fabrication of a full-scale wooden pavilion. The material properties, structure performance and fabrication constraints are integrated into the design process, while both industrial prefabrication and digital robotic fabrication are employed in the fabrication stage.

Fabrication-oriented form-finding

Based on a structural performance form-finding method, this project takes the Rhinoceros plug-in Rhinovault (Rippmann et al., 2012) and the Grasshopper plug-in Millipede (Michalatos & Kaijima, 2007) as form-finding tools, where the former is used to find a reasonable form of compression-only structure while the latter is applied to optimise the size of the structural elements. The initial geometry of the timber structure is first generated through the form-finding process in Rhinovault (Rippmann & Block, 2013), and is then translated into a grid-beam system in which the beams are all full size, with lengths varying from 5.8m to 7.5m. According to structural simulation, the beam sections are optimised to a constant thickness of 100mm, and varied height ranges are from 120 to 200mm. The top and bottom surfaces of each structural element are all designed as ruled surfaces in order to be fabricated with wire-cutting technology. Finally, the geometric system is divided into four layers while beams of different layers are connected with the most traditional mortise-tenon joints.

Digital fabrication

In order to make the process material-efficient, the three-dimensional curved beams are expected to be cut from two-dimensional curved beams with minimum volume that are able to accommodate the desired beam.

Taking glued laminated wood as the structural material, the raw beams are produced in a factory with the existing glued technology under the guidance of a CNC template (Fig. 3).

The bandsaw end effector is a modified 14in bandsaw reinforced with a welded steel frame and installed on a hanging KR120 KUKA robot to conduct the ruled surfaces fabrication (Fig. 4). In contrast to the wires in wire-cutting, the bandsaw blades have a certain width which gives more complicated constraints to both the desired surface curvature in the design stage and the blade's forward speed and direction in the fabrication process. The saw blade must always be strictly perpendicular to the forward direction. Small surface curvature and high speed may block the saw, and even broke saw blades. During the fabrication test, a traditional carpenter was employed to provide guidance on the mechanism of the bandsaw – an undoubtedly important part of the transmitted knowledge of traditional craft and material performance being added to the robotic fabrication process. After several tests, the 13mm-wide blades were employed to meet the requirement of desired surface curvature and ensure fabrication efficiency.

Following the fabrication tests, the robot movements were simulated within Rhino. Then generated toolpaths were converted to KRL for the KUKA robot with the Grasshopper plug-in KUKA PRC (Braumann & Brell-Cokcan, 2011) (Fig. 5). During the fabrication process, the raw beams are fixed to two adaptable tables, which

4

5

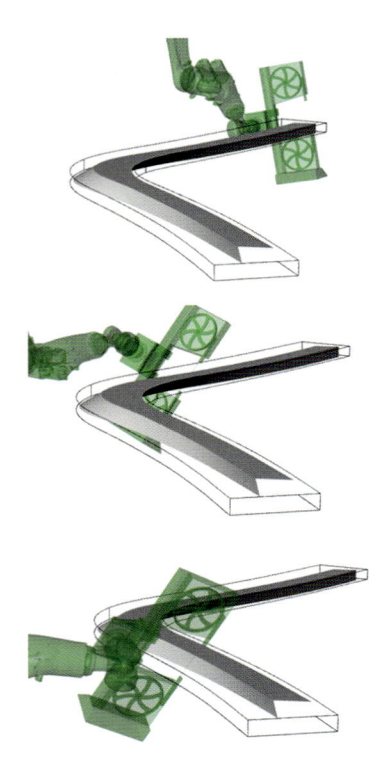

can easily meet the need for beams with different curvatures. The desired beams are cut out by the hanging robot using four cuts (respectively, the top and bottom surfaces and the two ends). With a reasonable speed of 5-8m per hour, the time taken for each beam can be restricted to within three hours, i.e. significantly shorter than the milling method. By equipping the same robot with a 24,000rpm spindle, the slots on the beams were milled after the bandsaw cutting process. The fabrication process of all 16 beams was completed in three weeks with great accuracy and efficiency.

Site assembly
Due to the employing of a mortise-tenon joint system, the site assembly process was simplified to putting wood beams in place in accordance with the design order (Fig. 6). The entire assembly process was completed by five workers within two days.

The wood pavilion appears as a mushroom structure with a height of 7m and a maximum cantilevered span of 4.5m. The combination of mechanical bandsaw technology and robotic wire-cutting technology effectively guarantees fabrication accuracy and form smoothness (Fig. 1).

Research evaluation
This project explores the entire process chain, from form-finding and optimisation to fabrication, which results in a technologically and aesthetically successful prototype. The resulting pavilion is efficient in terms of structural performance and rich in aesthetics, indicating the novel design possibilities of technology.

As this project demonstrates, the robotic bandsaw performs with a high material efficiency in both the design and fabrication stages. This is because the bandsaw has the smallest possible kerf of any mechanical wood-cutting method, which also ensures that the process is swifter than the CNC milling process. The 6-axis industrial robot allows the fabrication not only of two-dimensional curved surfaces, but also of high quality three-dimensional ruled geometries through the continuous rotation of the blades, which apparently have a higher resolution than the traditional milling geometries created from CNC. The robotic bandsaw applied in the project has demonstrated its capacity for the mass customisation of full-scale geometrically complex wooden components, and the ability to further adapt to the requirements of industrial mass production. Although this technique has great advantages in material efficiency, there are still some deficiencies to be improved. It is undeniable that there is still a waste of material due to the volume difference between two-dimensional raw beams and the desired three-dimensional beams. The waste may be minimised through the optimisation of gluing technology or by employing a more precise CNC template to guide the material distribution to minimise the volume difference between the raw and desired beams. In addition, there is also room for optimisation in terms of speed control. Due to the continuous change in beam thickness during the fabrication process, an automatic speed control system (to adjust the speed according to the resistance that the blade is facing in real time) will contribute to both the fabrication results and the life of the blade itself.

Conclusions
This project presents robust robotic wood tectonics capable of full-scale wood component fabrication. This technology – with its high efficiency in material and time, as well as the capability for the mass customisation of geometrically complex wood – has thrown the traditional 'subtractive' mode of CNC milling into question. Oriented by the fabrication technology, this project demonstrates an entire integrated process for digital wood architecture, from form-finding

5. Robotic fabrication simulation.
Image: © College of Architecture and Urban Planning, Tongji University.

6. Site assembly.
Image: Xie Zhenyu.

6

and form optimisation to digital fabrication. The design therefore is not only determined by the physical mechanism of form-finding, but is also defined by the fabrication constraints. Meanwhile, the fabrication process is not merely state-of the-art research, but also tries to make full integration with the existing wood production method much more valuable in practice. The final outcome is the result of constant negotiation between design expression and fabrication constraints. In addition, while the project is an attempt to provide innovative technical support for modern wooden architecture, it also aims to make this fabrication method the driving factor in the design process. Given the great differences from traditional wood tectonics, this innovative method can be considered as representative of the new robotic wood tectonics.

In future research, this novel technology is going to be improved in terms of efficiency, stability and integration with existing design methods and industrial production approaches. On the other hand, as the tectonics applied in this project are only applicable to specific geometry, new tools will be required for the continuous expansion of the capacity and scope of robotic wood tectonics.

References

Braumann, J. and Brell-Cokcan, S., 2011, 'Parametric Robot Control: Integrated CAD/CAM for Architectural Design' in *Proceedings of the 31st Annual Conference of the ACADIA*, p.242-251.

Brell-Cokcan, S., Reis, M., Schmiedhofer, H. and Braumann, J., 2009, 'Digital Design to Digital Production: Flank Milling with a 7-Axis CNC-Milling Robot and Parametric Computation: The New Realm of Architectural Design' in *27th eCAADe Conference Proceedings*, p.323-329.

Buri, H.U. and Weinand, Y., 2011, 'The Tectonics of Timber Architecture in the Digital Age', *Mississippi Valley Historical Review*, 39 (2).

Johns, R.L. and Foley, N., 2014, 'Bandsawn Bands: Feature-Based Design and Fabrication of Nested Freeform Surfaces in Wood' in *Robotic Fabrication in Architecture, Art and Design 2014*, Springer.

Menges, A., 2011, 'Integrative Design Computation: Integrating Material Behaviour and Robotic Manufacturing Processes in Computational Design for Performative Wood Constructions' in *Proceedings of the 31st Annual Conference of the ACADIA*, p.72-81.

Michalatos, P. and Kaijima, S., 2007, 'Structural Information as Material for Design' in *Expanding Bodies: Art, Cities, Environment, Proceedings of the 27th Annual Conference of the ACADIA*, p.84-95.

Rippmann, M. and Block, P., 2013, 'Funicular Shell Design Exploration', Conference of the Acadia, 2013.

Rippmann, M., Lachauer, L. and Block, P., 2012, 'Interactive Vault Design' in *International Journal of Space Structures*, 27(4), p.219-230.

Scheurer, F., 2010, 'Materialising Complexity' in *Architectural Design*, 80, p.86-93.

Williams, N. and Cherrey, J., 2016, 'Crafting Robustness: Rapidly Fabricating Ruled Surface Acoustic Panels' in *Robotic Fabrication in Architecture, Art and Design*, Sydney, Springer International Publishing, p.294-303.

RAPID ASSEMBLY WITH BENDING-STABILISED STRUCTURES

JOSEPH M. GATTAS / YOUSEF AL-QARYOUTI / TING-UEI LEE
School of Civil Engineering, University of Queensland, Australia
KIM BABER
School of Architecture, University of Queensland, Australia

This project seeks to enhance press-fit fabrication techniques through the use of hybrid material construction technology and bending-stabilised forms. It overcomes certain press-fit limitations and undertakes a systematic improvement to connection design, which in combination with material and form enhancements allows for an increase in spanning capacities and robustness of press-fit structures, an increase in the reliability and precision of assembled geometry and retention of the critical press-fit benefits of lightweight, high-speed and uncomplicated construction.

Press-fit connection techniques streamline digital construction methods through elimination of mechanical fixing components and thus enable rapid construction of complex three-dimensional geometries. However, the reliance on dimensional tolerance and oversizing, in lieu of mechanical fixing, causes an inherent instability in press-fit connections in the direction of component insertion. This can be partially abated with increased tightness between parts and/or a 3D interlock, but such measures can also offset the ease of assembly and structural performance.

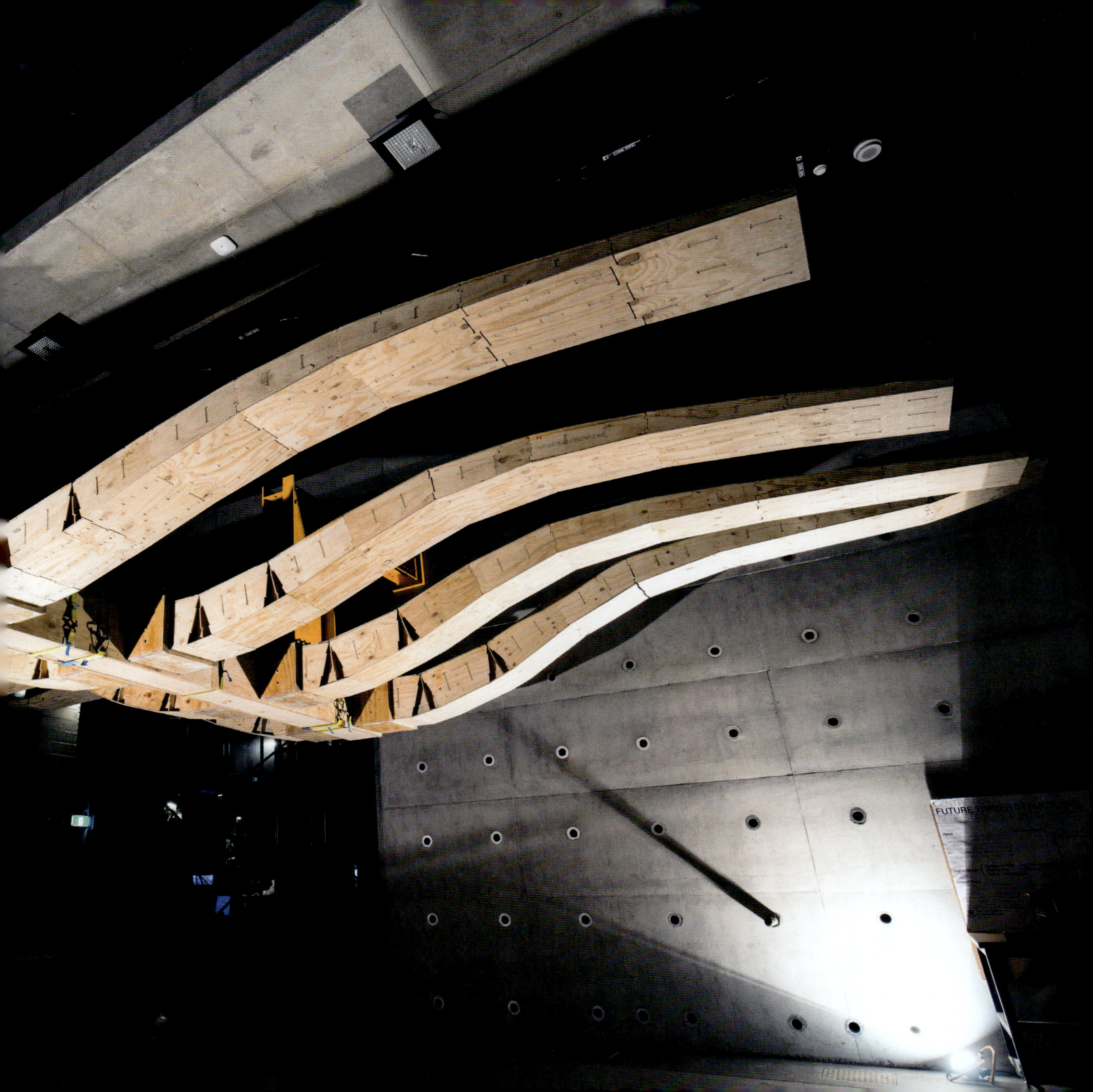
FUTURE

The project aims to address existing press-fit limitations via three key advancements in fabrication: (a) the introduction of material hybridity with the combination of glass fibre-reinforced plastic (GFRP) skin and plywood sandwich segments; (b) the introduction of bending-stabilised geometry to the overall assembled configuration; and (c) the utilisation of rotational press-fit joints between structural components. The project is of particular significance due to the combined benefits of these advancements working to create a solution in which any curved profile can be manufactured without the need for moulding or propping.

While the technology may be applied to a range of geometric configurations, the project investigates two specific applications: a tied arch and a cantilever structure, shown in Figs. 2 and 3. Both applications are used to demonstrate the benefits of the three key fabrication advancements, but additional post-fabrication analysis was undertaken for each specific structural type. The arch was tested to failure to demonstrate the suppression of press-fit pop-off instability and the corresponding strength and robustness of the assembly method; and the cantilever was 3D-scanned to demonstrate the extreme versatility, speed and accuracy of the assembly method.

Press-fit construction

Sophisticated digital design processes can reduce a complex structure to a complete set of individual elements suitable for fabrication with the use of automated workshop machines (Gramazio & Kohler, 2008). A key capability in digitising the complexity of traditional construction is the introduction of integral mechanical attachments in place of conventional mechanical fastening systems such as screws and nails. Such integral attachments are particularly suited to timber construction, as their design can draw on a rich history of traditional wood-working joints (Robeller et al., 2015). A correspondingly wide range of integral attachment types is thus seen across recent timber works (Menges, Schwinn & Krieg, 2016). Beyond the streamlining of digital construction methods, the inclusion of integral mechanical attachments can produce structures that possess extreme fabrication and assembly speeds. For example, the 'Instant House' clad frame structure was assembled in four days from 984 plywood components (Sass & Botha, 2006) and the 'Plate House' modular sandwich structure was manufactured in five hours and assembled in seven hours from 150 cardboard components (Gattas & You, 2016).

A fundamental type of integral timber connection is the press-fit (or friction-fit) joint. It consists of a male tab and female slot and enables precise alignment and assembly of components, but contains an inherent instability in the direction of component insertion. This can be partially abated through a fine control of part tolerance to achieve a friction-only fit (Robeller, 2015), or through interlocking geometry which prevents the movement of two parts in all but one direction (Robeller & Weinand, 2016), but such measures can also offset the ease of assembly. In terms of structural capacity, press-fit structures can possess compressive capacity approaching that of the glued sections, but can also be subject to a catastrophic 'pop-off' failure mechanism where sudden loss of friction cohesion causes an explosive bifurcation and complete disassembly (Al-Qaryouti et al, 2016).

Hybrid construction

Fibre-reinforced polymer (FRP) composites have obtained wide acceptance in civil engineering and digital fabrication communities in recent years, due to their high strength-to-weight ratio (Teng, Yu & Fernando, 2012) and versatile construction options (Parascho et al, 2015). Timber materials, and more particularly engineering wood products (EWPs), have similarly seen increased recent uptake for broadly similar reasons to FRP; their high machinability and lightness make EWPs well-suited for modern prefabricated structures and robotic construction methods.

The use of hybrid FRP-timber structures has been rather limited compared to hybrid FRP-concrete and FRP-steel structures, due to a range of factors including economics, durability and fire performance. However, recent work has hinted at the potential benefits of such material hybridisation. FRP can reinforce weak sections of EWP beams (Raftery & Hart, 2011) and is thus able to upgrade low-quality timber resources for high-performance structural use, minimising the overall system cost (Fernando et al., 2015).

The project seeks to explore the combined value of press-fit and hybrid FRP-timber construction technologies. It will be seen that, with such a combination, a novel fabrication system can be developed that possesses a number of advantageous geometric, structural and constructability innovations that are not available in existing systems which utilise these construction techniques in isolation.

1. Suspended cantilever structure.

2 & 3. FRP-reinforced plywood cantilever structure, utilising a continuous tensile skin and bending-stabilised rotational joints, increases the spanning capacity and robustness of lightweight press-fit fabrications.

Rationalisation

A principal aim of the project is to increase structural strength, stability and robustness of press-fit fabrication methods, so it is useful to consider their inherent limitations. Consider the press-fit plywood beam constructed from three segments which are themselves constructed from core and face plates. The need for discrete segments and plates arises from the use of a 2D sheet material with finite size.

If the beam is loaded with end moments as shown, a range of structural behaviours manifest, both favourable and unfavourable. When internal stress acts in a direction that is perpendicular to or along the press-fit direction of insertion, the joint acts favourably (for example, shear stress through joints 1, 2 and 4). Notionally, compressive stress through joint 5 should also act favourably. However, the lack of joint rotational resistance could cause the entire compressive face to act in a manner analogous to a compressive beam with lateral spring support. This is a highly unstable configuration that can lead to panel fragmentation or pop-off failures. Furthermore, when stress acts opposite to the direction of insertion (for instance, tensile stress through joint 3), there is no structural capacity at all.

4

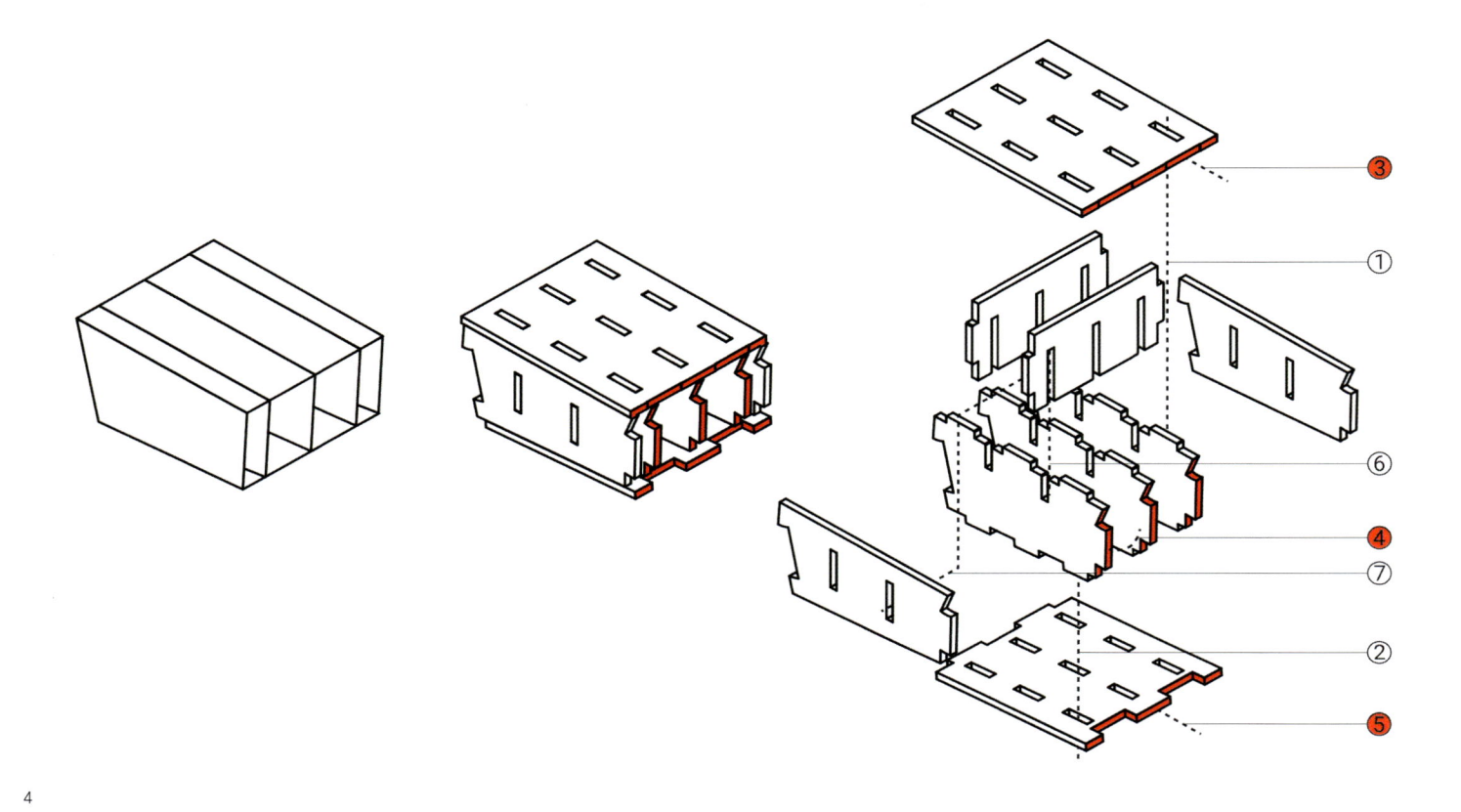

Consider now its modified press-fit beam. Three key innovations have been introduced which together act to eliminate many of the weaknesses seen above.

- Material hybridity: a GFRP is introduced as a continuous tensile skin, providing a stress transfer mechanism at joint 3 and negating stress concentrations in joint 1, thus enabling the use of thinner, lighter and more economical plywood grades.
- Bending-stabilised geometry: a positive curvature is introduced into the beam, which acts to reduce the effective buckling length of the compressive face and introduces an inclined compressive stress component. These act to suppress pop-off and fragmentation failures respectively.
- Rotational press-fit connection: the GFRP skin creates a hinge mechanism which can be used for a novel rotational press-fit connection. This retains a shear stress transfer capability but with a rotational direction of insertion that matches displacements from the applied moment loading, i.e. segments can be rotationally 'folded' together, rather than axially 'pushed' together. The beam can therefore self-assemble if subjected to a bending load.

As will be demonstrated, these key innovations serve to resolve many of the structural weaknesses, while retaining the speed and accuracy of typical press-fit construction. There is also one further benefit that arises from the above three innovations acting in concert: any curved profile can be manufactured without the need for moulding or propping, as the structure can fold from a flat state. The fabrication process in each of the structural applications that were investigated demonstrates this final capability.

4. Plywood sandwich segment, isometric view. Joints 1, 2, 6 and 7 are conventional press-fit connections. Joints 3, 4 and 5 (red) form the rotational press-fit connection and control surface curvature, which required specific adaptation and are unique to this project.

5. CNC router cutting the seven 9mm-thick plywood segments.

5

Fabrication

A key fabrication aim for the project is the ability for the segments to 'self-assemble' from a flat state, which will now be described in detail. A target beam profile is specified with a control spline and depth offset. This is discretised into segments by subdivision of the control spline. Segments in regions of positive curvature ($\alpha < 180°$) can unfold without issue onto a flat surface when inverted, but segments in regions of negative curvature ($\alpha > 180°$) would be unable to readily unfold, necessitating the introduction of additional 'wedge' segments. Each segment profile is then translated into a complete plywood sandwich structure encoded with all necessary press-fit joints. For example, its dark grey segment is shown in isometric view in Fig. 4. It is composed of core plates and face plates with the same joints 1 to 5, as described previously, and with additional joints 6 and 7 for cross plates and facing plates respectively.

The rotational press-fit connection determines the overall surface curvature by controlling the relative inclination between adjacent segments. The connection is composed of joints 3, 4 and 5, with design considerations required for each. Joint 3 shifts the rotation point from plate centrelines to the outer skin and is composed of an inclined press-fit joint with slight front and back offsets suitable for a 2.5-axis cutter. The staggered tab locations allow the joint to fold without collision. Similarly, joint 4 is an inclined press-fit and acts to enforce the transverse alignment of inside face skins, and by extension precise centreline alignment of core plates. This alignment is important for thin core plates to avoid eccentric loading. Finally, joint 5 has press-fit tabs formed along arcs centred about the rotation point, and so can travel through the complete folding motion without collision.

Tied arch and structural performance

A 3m-wide symmetrical arch structure was constructed using the above fabrication process. It consisted of seven segments, each of which was assembled from 9mm-thick plywood plates cut on a CNC router (Fig. 5). Assembled arch segments were placed end-to-end on a flat surface (in the arch's inverted orientation) and bonded to a continuous GFRP skin. The need for chemical adhesion in the fabrication process does slow down the overall construction time due to placement and curing, although this is offset by the GFRP providing a simple three-dimensional coordination of the segments, virtually eliminating the need for further consideration of set-out or construction sequencing; the only required alignment is readily achieved on a flat surface through the transverse alignment of segment edges.

Once cured, the structure can be folded into its final shape with extreme rapidity, as all six segment joints are actuated with a single bending load. This was induced with a single post-tensioned tie, providing a line of force between end segments. The arch was then tested to failure, with an actuator again applying a force between end segments. The arch was designed so that maximum moment occurred at the arch peak and GFRP material tensile failure occurred prior to the onset of the suppressed compression face instabilities. Fibre rupture occurred first, as predicted in the theoretical analysis.

Cantilever structure and construction performance

A cantilevering branch structure was designed to explore the versatility, speed and accuracy of the assembly method. The design of the curvilinear branches was via a digital model with multi-scale parametric control over the geometry of the individual 'branches' as well as the generation of all component parts with integral mechanical attachments. Eight counter-balancing cantilevers were constructed from 710 9mm-thick plywood plates, which were assembled, bonded to GFRP, folded into branches (Fig. 7) and attached to a central suspended spine (Fig. 1). The final and longest branch was 5m long (10.5m tip-to-tip across the branch pair), tapered from a depth of 370mm to 30mm and weighed 70kg. The fabrication phase was just two weeks, with the structure exhibited at the official opening of the University of Queensland Centre for Future Timber Structures.

The structure was measured using a 3D laser scanner, and collected point cloud data were processed using a surface error minimisation routine with the Galapagos

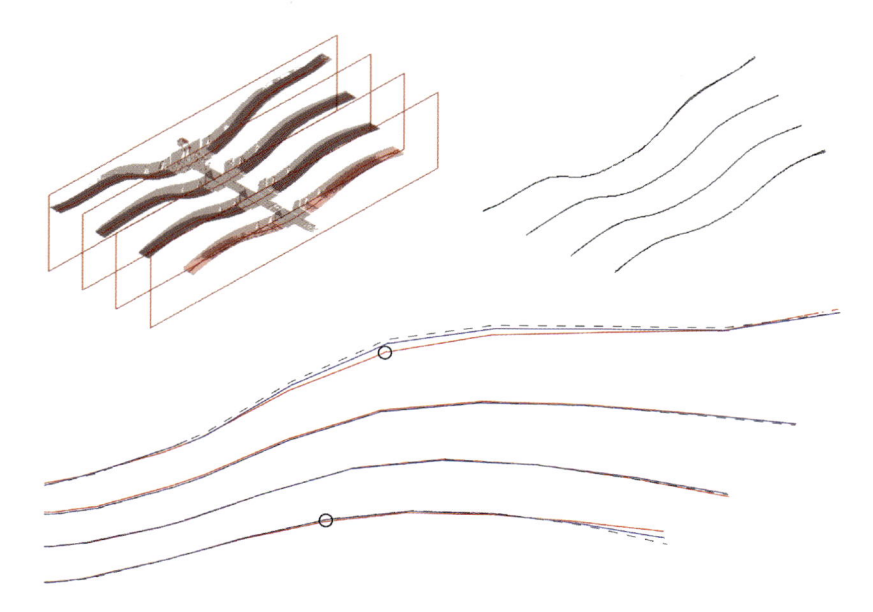

6

plug-in for Rhino/Grasshopper. Surface error was measured as absolute distance between the bottom scanned surface and the underformed design geometry (Fig. 6), with deformations due to self-weight neglected due to the extremely light weight of the structure. Six out of eight branches had an average absolute surface error of less than 4mm. The remaining two branches had larger errors, but in both cases this could be traced to a single rotational press-fit connection that failed to completely interlock due to a slight misalignment of the hinge point. As 70 out of 72 rotational press-fits engaged without issue, and as the majority of branches were constructed almost exactly as designed, it can be concluded that the new fabrication method has achieved a significant level of reliability with regard to the geometric precision between the digital model and the built artefact. It is hypothesised that such precision is the result of the self-assembling capacity of the structure, i.e. the cantilever self-weight induces a bending load that acts in conjunction with the FRP hinge mechanism to compress the press-fit joints to the maximum tolerance tightness and thus minimise insertion errors.

Capacity and construction benefits

The project has demonstrated a hybrid material and press-fit fabrication technique that can produce key benefits in both structural capacity and ease of construction. The increase in tensile stress transfer from the hybrid material and the improved compressive stability of the modified global geometry act together to significantly enhance the spanning capacity of members subjected to bending loads, while maintaining a very lightweight structure. The additional use of FRP as a flexible hinge and planar alignment mechanism, combined with the use of rotational press-fit connections for precise curvature control between adjacent segments, was seen to create a robust and versatile construction method.

Two sample applications of the tied arch and an array of cantilevers were explored. Both were fabricated in a condensed timeframe and served to demonstrate the structural and construction benefits respectively of the new fabrication technique. Testing of the tied arch structure confirmed that the bending-stabilised geometry resisted the press-fit 'pop-off' failure mechanism, thus preventing catastrophic failure of large-scale press-fit structures. Surface measurement of the cantilever structure showed that the FRP hinge successfully acted as a precise guide for the rotational assembly of adjacent segments to produce a stable and highly accurate overall form.

While the two applications illustrated at this time may be considered 'components' of larger structures, it is envisaged that, through the successful demonstration of the combined innovations, the technologies developed in this project can enable a significant range of possible formal configurations. As the developed improvements to spanning capacity and robustness of press-fit fabrication systems occur alongside the new method for rapid assembly of long-span bending structures, more ambitious applications for larger press-fit structures used in permanent building applications could become a reality.

6. Comparison of design and manufacture geometry. Clockwise from top left: point cloud data and sectional planes; bottom surface sectional profiles; comparison of design (dotted), left cantilever (red) and right cantilever (blue) geometries for each branch pair. Circled locations are rotational press-fits that failed to completely engage.

7. Proposed cantilever and rotational press-fit assembly sequence shown starting from a flat surface, using the self-weight of the structure and the FRP hinge to enable self-assembly.

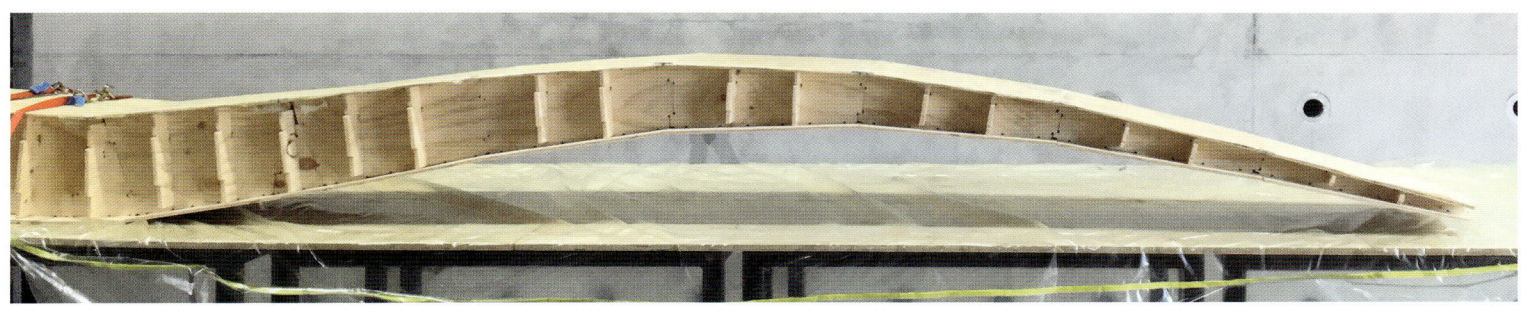

Acknowledgements

The authors would like to thank project team members Sam Butler, Jordan Hunter, Stephen Joseph, Sophie Sachs, John Stafford, Jonathan Tan, Crystal Wang and Shuwei Zhang. This work was supported by the University of Queensland Centre for Future Timber Structures and the Australian Research Council Discovery Early Career Researcher Award DE160100289. The software used for the fabrication process described in this project is freely available for download (Gattas, 2014).

References

Al-Qaryouti, Y., Gattas, J., Shi, R. and McCann, L., 2016, 'Digital Fabrication Strategies for Timber Thin-Walled Sections' in *Proceedings of the 5th International conference on Mobile, Adaptable and Rapidly Assembled Structures*, Siena, Italy.

Fernando, D., Gattas, J.M., Teng, J.G. and Heitzmann, M., 2015, 'Hybrid thin-walled members made of FRP and timber', presented at the *12th International Symposium on Fiber Reinforced Polymers for Reinforced Concrete Structures (FRPRCS-12) and The 5th Asia-Pacific Conference on Fiber Reinforced Polymers in Structures (APFIS-2015) Joint Conference*, China.

Gattas, J.M., 2014, Digital fabrication toolbox, available online at http://joegattas.com/digital-fabrication/.

Gattas, J.M. and You, Z., 2016, 'Design and digital fabrication of folded sandwich structures', in *Automation in Construction*, p.63, 79-87.

Gramazio, F. and Kohler, M., 2008, *Digital Materiality in Architecture*, Baden, Lars Müller Publishers.

Menges, A., Schwinn, T. and Krieg, O.D. (eds.), 2016, *Advancing Wood Architecture: A Computational Approach*, Routledge.

Parascho, S., Knippers, J., Dörstelmann, M., Prado, M. and Menges, A., 2015, 'Modular Fibrous Morphologies: Computational Design, Simulation and Fabrication of Differentiated Fibre Composite Building Components' in *Advances in Architectural Geometry 2014*, Springer International Publishing, p.29-45.

Raftery, G.M. and Harte, A.M., 2011, 'Low-grade glued laminated timber reinforced with FRP plate' in *Composites Part B: Engineering*, 42(4), p.724-735.

Robeller, C., 2015, 'Integral mechanical attachment for timber folded plate structures', PhD thesis, EPFL ENAC, Lausanne, Switzerland.

Robeller, C., Stitic, A., Mayencourt, P. and Weinland, Y., 2015, 'Interlocking Folded Plate – Integrated Mechanical Attachment for Structural Wood Panels' in *Advances in Architectural Geometry 2014*, Springer International Publishing, p.281-294.

Robeller, C. and Weinand, Y., 2016, 'A 3D cutting method for integral 1DOF multiple-tab-and-slot joints for timber plates, using 5-axis CNC cutting technology' in *Proceedings of the World Conference on Timber Engineering WCTE 2016*, Vienna, Austria.

Sass, L. and Botha, M., 2006, 'The Instant House: A Model of Design Production with Digital Fabrication' in *International Journal of Architectural Computing*, Volume 4, p.109-123.

Teng, J.G., Yu, T. and Fernando, D., 2012, 'Strengthening of steel structures with fiber-reinforced polymer composites' in *Journal of Constructional Steel Research*, 78, p.131-143.

A PREFABRICATED DINING PAVILION
USING STRUCTURAL SKELETONS, DEVELOPABLE OFFSET MESHES AND KERF-CUT BENT SHEET MATERIAL

HENRY LOUTH / DAVID REEVES / SHAJAY BHOOSHAN / PATRIK SCHUMACHER
Zaha Hadid Architects
BENJAMIN KOREN
One to One

92 Nodes
5mm Steel Plate

169 Beams
Aluminum and Steel 140x40x5

184 Node Cover Plates
CNC Stainless Steel 2mm

80 Loops
Thermo Ash Planks 130x26
Lasercut Steel Sheet 1.9mm
Lasercut Tricoya Sheet 12mm

3943 Individual Elements
12.5km Length Laser Cut Elements

31 Floor Fill Panels
Thermo Ash Planks 130x26
Lasercut Steel Sheet 1.9mm
Lasercut Tricoya Sheet 12mm

Edge Band

This project focuses on the role of computational geometry within computer-aided architectural design and construction workflows, i.e. computational geometry as a mediating device between architectural, engineering and construction logics. While the scale of a dining pavilion is relatively modest, the intention is to utilise this research for wider application in larger construction-scale projects. In this regard, the project operates within a tight time-bound, multiple-stakeholder, collaborative and bespoke production pipeline, as typically necessitated by architectural projects.

Digital workflows

Workflows in architectural design can be characterised by two paradigms – one drawing-based, the other model-based. The drawing paradigm is popularly known as Computer Aided Design (CAD) and the model paradigm as Building Information Modelling (BIM). While both drawings and models encode 2D and 3D geometry, a model also contains meta-information about the encoded geometry – its material specification, role in and processes of assembly, etc. Also, the drawing paradigm, especially Computer Aided Geometric Design (CAGD), can support the creation of a wider range of (arbitrarily) complex geometries and their processing for Computer Aided Manufacturing (CAM). An essential aspect of CAGD, as used in disciplines such as automotive or product design, is the abstraction of the complex physical phenomena and machine parameters associated with manufacturing methods into geometric properties and constraints. Famous examples include the automobile, aircraft and shipbuilding industries motivating the development and use of Bézier curves and surfaces, physical splines and developable surfaces (Bézier, 1971, Sabin, 1971, De Casteljau, 1986, Pérez & Suárez, 2007, Pottmann & Wallner, 1999), etc.

This project aims to apply these operative principles from the automated fabrication industry in architectural design and assembly. Thus the project primarily focuses on developing structural and construction-related meta-information for complex geometries – in other words, augmenting complex CAGD objects with construction-specific information, thus enabling the research to be incorporated within larger, more complex

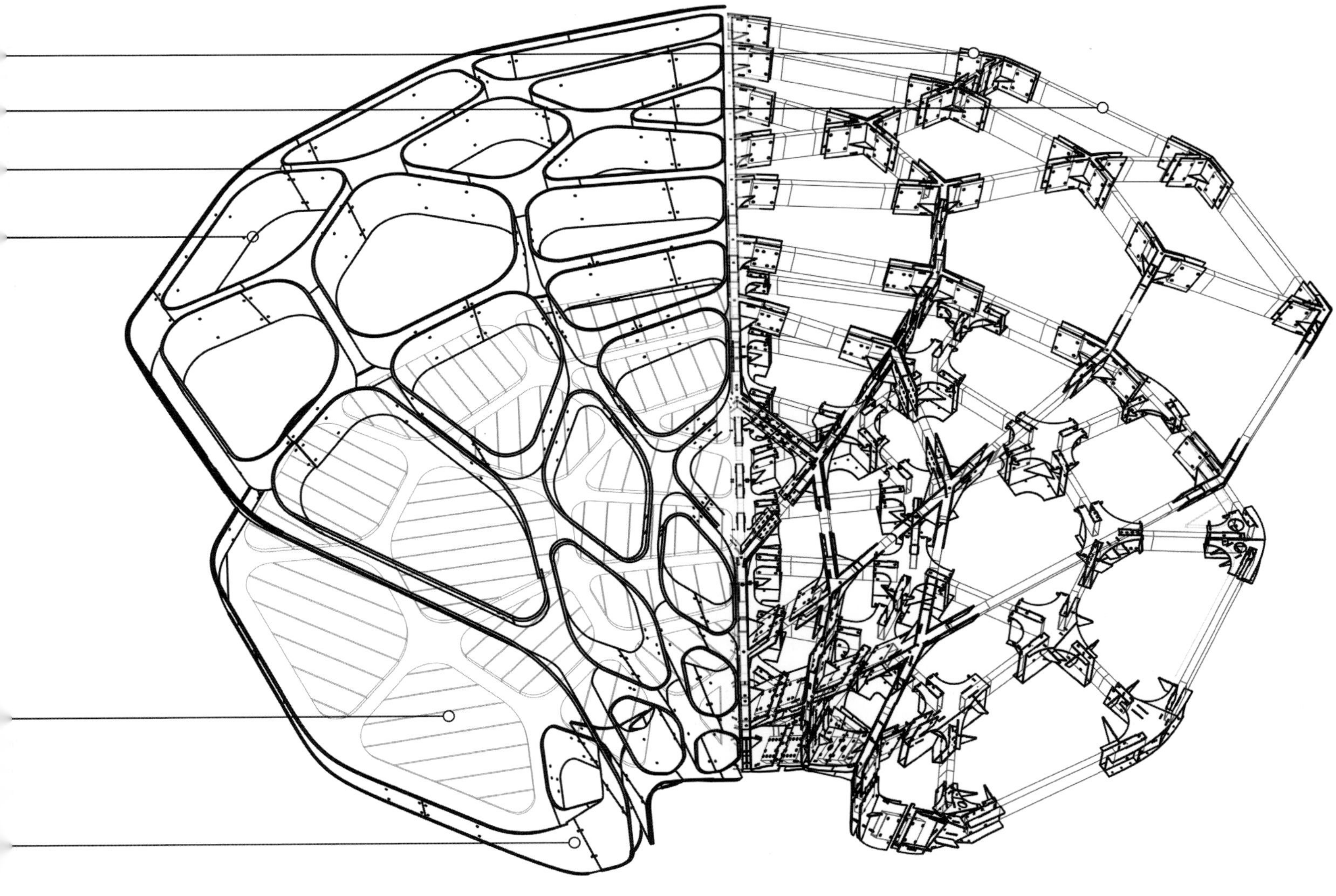

projects. Recent developments in the application of discrete differential geometry to architectural design – so-called architectural geometry (Pottmann, 2007) – share some of these aims. These developments have contributed to the popularisation of the CAGD paradigm, at least in architectural projects with high geometric complexity, such as the Heydar Aliyev Centre by Zaha Hadid Architects (Veltkamp, 2010, Janssen, 2011).

A review of applicable methods of architectural geometry and/or CAGD to address these contextual aspects of the research is described below. The development of bespoke implementations thereof and assimilation of the various state-of-the-art methods into a cohesive and flexible design workflow is described subsequently.

The design brief of the project proposes manufacturing an economical, prefabricated pavilion using off-the-shelf parts and/or laser-cut components. The structural skeleton is to be realised using standard hollow sections. Furthermore, the skeleton is to be adequately covered along the top and bottom, and the walls of the cells described by the edges of the skeleton (Fig. 2). In view of such a brief, the dominant design concerns relate to the development of geometries that are lightweight and can be made from flat-sheet materials.

Lightweight construction

The earliest practice of a deliberate focus on the economical use of material via a geometric understanding of structure and effective channelling of (axial) structural forces can be attributed to the Gothic period (Tessmann, 2008, Heyman, 1966). The earliest mathematical treatment of economic (timber-framed) structures is widely credited to engineer A.G. Michell (Michell, 1904). Michell used geometric principles involved in the static equilibrium of funicular frames (Maxwell, 1870, Rankine, 1864) to establish his solution for the layout of materially economic timber trusses. Recently, William Baker and his colleagues at SOM Architects (Beghini et al., 2014, Baker et al., 2013) have shown the compatibility of these geometric methods of structural design with numerically based methods of material reduction – so-called topology optimisation (Rozvany, 2001, Bendsoe & Sigmund, 2013).

Torsion-free beam network and developable surfaces

In view of the brief above, the critical fabrication constraints (expressed geometrically) are to ensure that the joint geometries are torsion-free, or extrudable, and the surfaces – top and bottom covers, and walls of the cells – are developable. Extrudability of the vertices ensures that the edges of the mesh can be uniformly offset, and thus the derived beam network can be of uniform thickness. This makes the edge-layout amenable for realisation using standard hollow sections of aluminium. Developable surfaces retain a variety of applications in sheet- and plate-based industries including architecture because they can be isometrically mapped onto a plane (Lawrence, 2011). The chosen method of forming sheet material is kerf-cutting and bending for the node covers and cell walls.

Interactive design

The early design method adopted for the project aims to build upon interactive benefits of the subdivision mesh modelling approach (Shepherd & Richens, 2010, Bhooshan & El Sayed, 2011). There have also been prior attempts to combine this user-friendly representation of geometry with numerical modelling techniques to physically realise them with fabric (Bhooshan & El Sayed, 2012), curved-crease folded metal (Bhooshan, 2016b, Louth et al., 2015) and 3D printing (Bhooshan, 2016a). This is in line with our intention to augment easy-to-use CAGD objects with construction-related information. Thus the extension of this approach to address skeletal geometries forms the last significant context of the research.

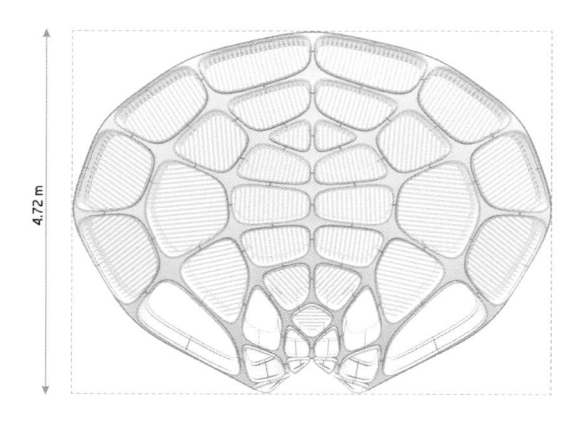

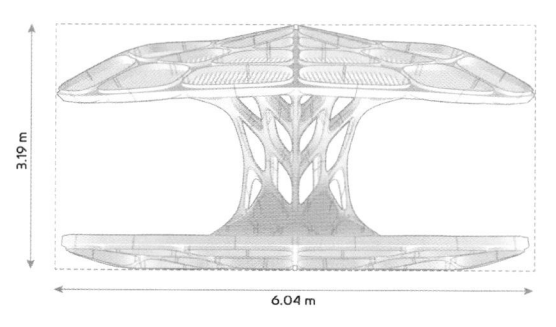

1. Overview of element quantities. Image: Ben Koren, One to One and Zaha Hadid Architects.

2. Volu overview. Overall bounding dimensions. Image: Courtesy Zaha Hadid Architects.

3. Coarse mesh development. Top: Topology optimisations under varying load cases. Middle: Coarse mesh, corresponding subdivision surface, topology optimised (TO) result, iso-surface of TO result. Bottom: Manual quad re-meshing of TO result. Image: Courtesy Zaha Hadid Architects.

Material economy

In view of the explicit desire for lightweight construction, the so-called equilibrium modelling methods (Lachauer, 2015) are of particular relevance. These methods attempt to synthesise surfaces that explicitly avoid bending and thus are well-aligned with the fundamental tenets of lightweight structures (Schlaich & Schlaich, 2000, Bletzinger & Ramm, 2001). The spatial constraints and the client-related history of the project do not allow for the application of these principles to shape the initial geometries. As such, the global shape of the pavilion can be considered predominantly invariant or given.

Operating within these constraints, the development of the layout of the structural skeleton is informed by topology optimisation (TO). The gradated material densities associated with a TO solution are interpreted as discrete beam/node elements that serve as a general arrangement suitable for further optimisation under spatial and fabrication constraints (Beghini et al., 2014).

Data-structures for design production

The other significant research question is the development of an appropriate data structure that assimilates the various design, fabrication and downstream production requirements. Furthermore, the intention is to be able to drive production geometries and information to the fullest extent possible, as opposed to the common practice of handing over design geometry to production specialists for shop drawings and post-rationalisation (Romero & Ramos, 2013, Sanchez-Alvarez, 2010, Peña de Leon, 2012).

Thus the development of the torsion-free beam network, developable surfaces and cell walls is a dominant research question. The geometric properties of the so-called edge-offset mesh (EO mesh) (Pottmann & Wallner, 2008, Liu & Wang, 2008) and the requisite properties of the control net of Bézier surfaces that make the surfaces themselves developable (Lang & Röschel, 1992) are the most relevant prior works in this regard. The work of Bouaziz et al. (2012) and Attar et al. (2010) are also relevant with regard to iterative solutions of multiple constraints.

Design pipeline

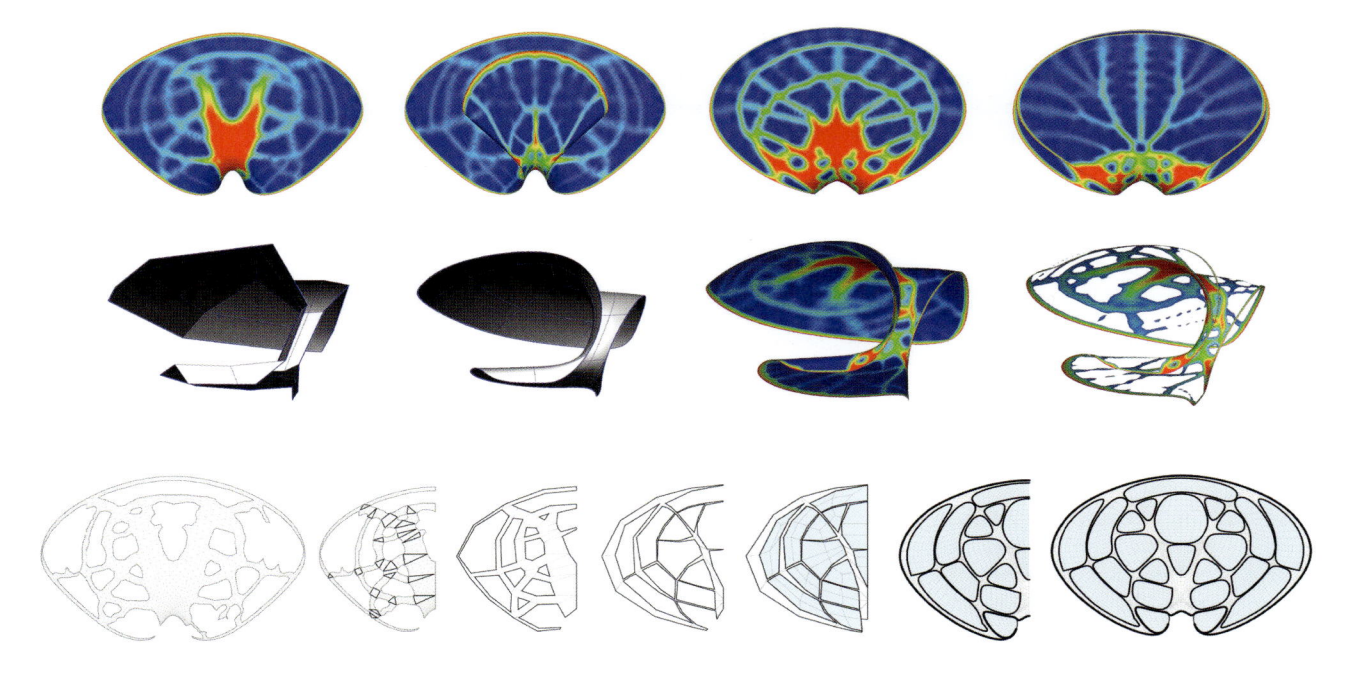

The design pipeline builds upon the subdivision mesh modelling approach (Shepherd & Richens, 2010, Bhooshan & El Sayed, 2011), capturing shape features (Leyton, 1988) of the TO encoded using a set of primitives into a coarse mesh representation for downstream processing (Fig. 1). This manual reinterpretation extracts node location, node connectivity and relative node offset weighting (Blum & Nagel, 1978) from the TO results by isolation of probable singularities and constructing a mesh comprised of predominantly quadrilateral faces (Fig. 3). While this intuitively suggests that the geometry retains an economy of homogeneous material from the TO, the topology is further evaluated and design options developed with respect to a range of potential composited construction methods of fabrication. Structural build-ups considering material properties and assembly techniques are correlated to topological features to embed fabrication assumptions into a construction-relevant expression of the TO, arriving at a segmental linear beam type that retains developability (Fig. 4).

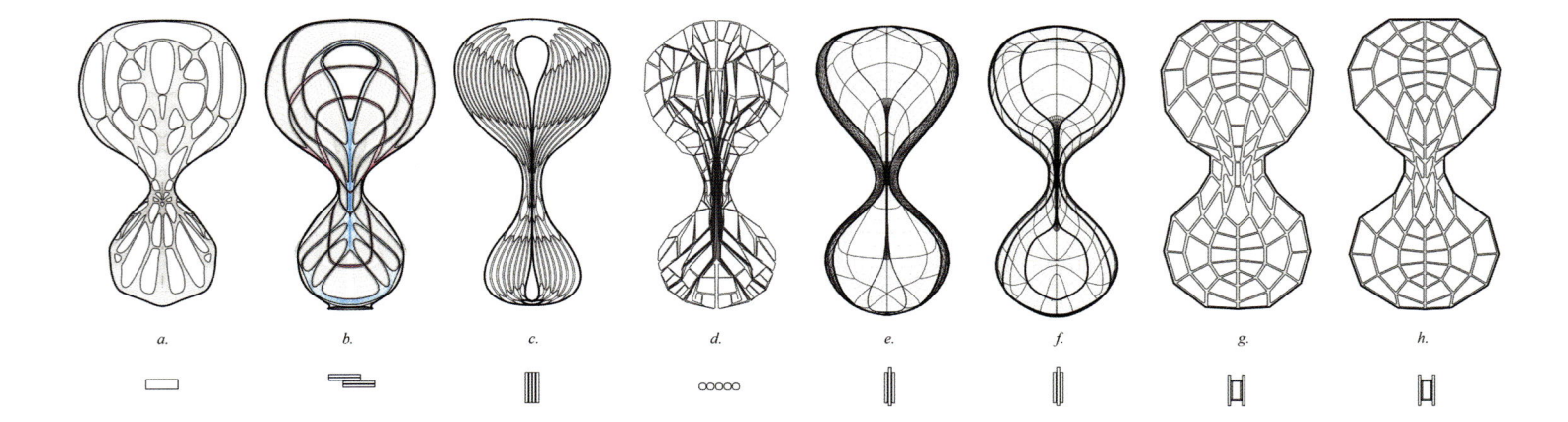

Network extraction

The density distribution given by TO serves as a guide for the reconstruction of the implied structural network as a coarse polygon mesh, M (Fig. 3). This representation is well-suited to subsequent design development, as its discretisation corresponds directly with components of the physical assembly (edges to beams and vertices to nodes). Furthermore, it allows for the application of established numerical methods to solve for geometric constraints related to fabrication.

Offset mesh

To define the depth of the structure, an offset mesh M' is numerically derived from M such that corresponding edge pairs are co-planar. Together, M and M' implicitly represent the 'beam mesh' M_B – a non-manifold planar quad mesh (PQ mesh) whose faces define the symmetry planes of beams in the resulting structure. By asserting planar quadrilateral faces within M_B, developability of derived components is guaranteed (as per Killian et al., 2008, Lang & Roschel, 1992). This also implies torsion-free nodes, since each non-manifold edge in M_B is the common axis at which all of its adjacent planar faces intersect (Wallner & Pottmann, 2008). From an implementation standpoint, the implicit representation of M_B via M and M' is favourable, as it avoids the increased complexity of topological navigation introduced by these non-manifold edges.

Perturbation

Offsetting the vertices of M along their normals provides an initial approximation of M'. For most cases of M, however, this produces non-planar faces in M_B, requiring that the PQ criteria be solved for numerically. A projection-based approach (Bouaziz et al.) is used to minimally perturb the vertices of M' such that the faces of M_B are planar. The vertices of M are excluded from perturbation, as they define the inner surface of the pavilion and are constrained by additional design considerations such as furniture placement and walkability.

For a given quadrilateral face in M_B, the constraint boundary for planarity is defined by the nearest point of intersection between the two diagonals. The projection vector can therefore be calculated as half of the shortest vector between them. This formulation is analogous to the planarity energy gradient defined by Poranne et al. (2013).

While projecting to the nearest constraint boundary ensures that vertices of M' are minimally perturbed, the solver does not necessarily converge for all cases of M. In cases where M exhibits strong discontinuities in local curvature, non-manifold edges of M_B tend towards zero length, which is unsuitable for the intended application as they represent the axes of structural nodes. To mitigate the collapse of node axes in M_B, an additional

4. Topology morphology by fabrication constraint. (a) Initial TO topology. (b) Lapped timber bands. (c) Striated bundles. (d) Bundled pipes. (e) Laminated plate custom edge. (f) Laminated plate outriggers. (g) Beam net open spine. (h) Beam net closed spine.
Image: Courtesy Zaha Hadid Architects.

constraint is introduced which tries to maintain a constant distance between each vertex in M' and all edges incident to the corresponding vertex in M. For most cases of M, the vertex-offset constraint partially opposes planarity and therefore must be assigned a smaller relative weight (roughly one order of magnitude) to ensure convergence of the PQ criteria within acceptable tolerance while preventing degeneracy of node axes in M_B.

In cases where the majority of vertices in M exceed valence 4, vertices in M' often become over-constrained and planarity cannot be achieved for all faces in M_B. This motivates the iterative revision of input mesh M to find an acceptable balance between preserving structural features generated via TO and satisfying fabrication constraints related to developability and the use of standardised elements.

Fabrication and assembly context

Given that the project delivery period from concept to prototype is approximately four months, time was a key consideration regarding approach, design logic and assembly. The concurrent timeline for design and fabrication suggests the development of a method to facilitate team interoperability, whereby data are preserved during exchange, enabling parallel design exploration during the set-up of parametric assembly associations. The distinct advantage of this is the extraction of relevant machinable parts during the early design process, which promotes a more constructive prototyping and feedback period.

The design logic of the structure, comprising uniform cross-sections of segmented lengths, is indicative of expediting engineering load calculations and member sizing for a 'worst case' scenario rather than of individual beam performance. Additionally, use of 'off-the-shelf' sheet and hollow sections compatible with ubiquitous 2-axis cutting technologies eliminated time-intensive milling techniques from design consideration (Scheurer, 2013), constraining the domain of geometric possibility to developable surfaces (Lawrence, 2011). Subsequently, all parts in the prototype were cut using 2-axis cutting methods, predominantly by laser.

Similarly, the assembly methods are consistent with an accelerated manufacture via prefabricated, mechanically fastened elements, and are further considered as a temporary travelling structure constrained by a limited installation period and the potential for numerous installation and de-installation cycles.

Structuring fabrication information

The half-edge data structure was used to represent M and M' throughout design development. While the advantages of this data structure are well-documented within the context of discrete differential geometry (Botsch et al., 2002), this project extends its use as a means of structuring fabrication information.

In developing detailed production geometry from M_B, components of the assembly were bound to the elements of the input mesh M. Beams were associated with edges, nodes and cover plates with vertices and loop panels with half-edges. The individual components of each node (steel plates and fasteners) were further distributed to half-edges around the corresponding vertex.

In this sense, fabrication information (be it geometric or otherwise) is treated as a collection of mesh attributes – analogous to colours, normals or texture coordinates typically found in mesh representations used within computer graphics. This greatly aids the procedural development of detailed production geometry, as fabrication information can be efficiently queried and cross-referenced locally. It also maintains a direct link between design geometry and production information, enabling a higher frequency feedback loop between designer and fabricator.

Relevant assembly details

The pavilion consists of linear segments of hollow section beams mechanically fastened to built-up steel plate nodes to form a raised dining platform oversailed by a 4m cantilevered shading canopy. Loops of kerf-cut sheet are bent and suspended from beam face centres. Flooring panels comprised of wood planks scribed to profile are suspended from beam centres in the platform. Node covers are patterned and face-fastened to the structure, obscuring the structural beams (Fig. 1).

Details address issues of prefabrication including installation sequence, material workability, geometric tolerance and lifetime performance. For example, exposed face-fastening loops and node covers in lieu of concealed hangar elements enable the localised changeability of parts and minimise the composite area of the cladded structure cross-section, tending toward the perception of a lighter, more slender pavilion. Similarly, mechanical fastener joining, in lieu of friction-fitting via slotting, tabbing or clipping, facilitates ease of workability, increases allowable in situ adjustment and promotes the independence of parts from neighbouring element dependencies, further assisting in situ fitting of parts.

The 92 self-similar, individually unique nodes further categorise and are parametrically modelled in response to neighbouring geometric conditions. The typical node is a pre-assembled, welded composite of plate steel corresponding to the mesh intersection planes of incoming half-edges. Each of the floor nodes and foundation nodes introduces a planar top and bottom plate respectively. The boundary nodes are clad with a continuous boundary edge band, inheriting the same fastening procedure as typical nodes.

Auxetic material

Material flexibility and hand-bending in the pavilion is accomplished primarily through kerfing patterns corresponding to the scale of the bend radius in each loop (Fenner, 2012). During prototyping, torsional deformation and subsequent 'oil canning' (Kalpakjian, 2008) developed in 'worst-case' node covers with extreme angles located at the top and bottom of the trunk. An exploration into auxetic materials (Konaković, 2016) proceeds to introduce local discontinuity and bi-directional flexibility in the 2mm plate (Fig. 5).

Assembly process

The design of the pavilion assemblies anticipates a number of factors, including: a remote installation, a limited four-day install period, the use of traditional tools, a commencement from a completed list of parts, a confined exhibitor space and an install in conjunction with local labourers unfamiliar with the design logic. The initial prototyping and test fittings undergo a contrasting set of circumstances which include: a span of six weeks, the use of specialised tools and hoists, a sequence of assembly that corresponds to parts manufacture, a project-dedicated workshop and an assembly team knowledgeable in each aspect of the design.

Exacerbating the disparity of assembly processes, elements were fabricated in order of increasing complexity to allow for extended design and prototyping considerations. Effectively, the pavilion platform and canopy nodes and beams were produced while details of the trunk transitions were resolved, designed and manufactured. The workshop was able to undertake continuous manufacture and compress the delivery timeline by constructing the pavilion discontinuously, not from the ground up as is done onsite.

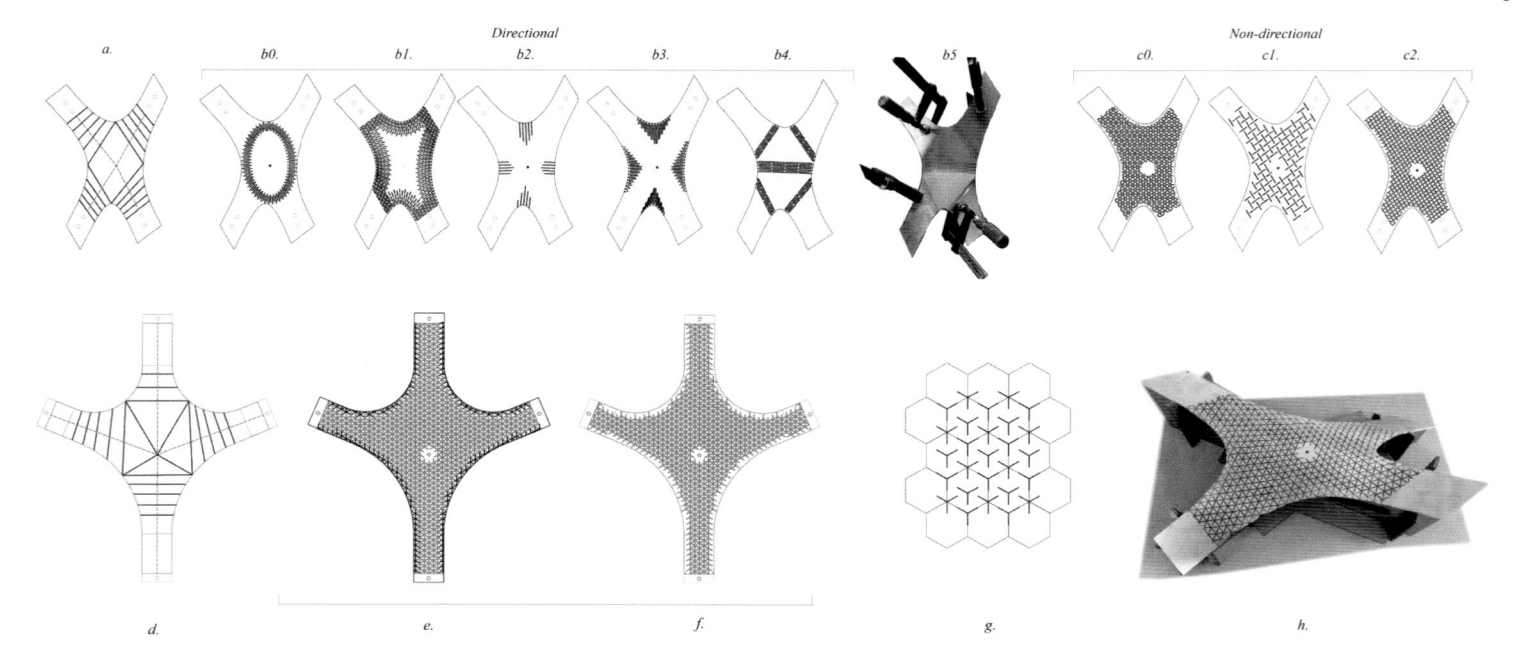

The factory sequence assumes that a partially assembled canopy is positioned to minimise the total number of connection points subject to live-loading at any given time during assembly. The onsite sequence proceeds without hoists, from node to next neighbouring node, from the ground up. Consequently, each canopy node connection withstands live-loading and rotation due, in part, to compounding bolthole tolerances, resulting in a deflection effect in aggregate upon the canopy (Fig. 6).

Compatibility of method

The prototype beam configuration presented here suggests the potential incompatibility of a discretised node-beam type structure proceeding from a conceptual TO analysis. Specifically, beam elements are not aligned to principal curvature directions of the surface using TO analyses in the same way that stress accumulation and fall-off are not gradated in beam element assemblies. Such a geometric constraint is not represented within the TO process, as its benefits are not directly structural, but rather are constrained to a chosen fabrication method for a prescribed loading condition. Converging upon apparent lightness is therefore not a cumulative result of material reduction techniques.

Workarounds

While the use of auxetic material in the node covers provided a workaround for delivering doubly-curved surfaces in partially torsioned materials using 2-axis cutting, it neither supports the geometric principles of developability nor is particularly suited to exterior environments.

5. Auxetic studies in node covers. (a) Prototypical node showing rulings, (b0) radial, (b1) radial edge constrained, (b2) re-entrant corner, (b3) dense re-entrant corners, (b4) ruling aligned, (b5) mock-up, (c0) diagrid, (c1) coarse grating, (c2) dense grating. (d) Actual node element showing rulings, (e) "Y" ragged boundary, (f) "Y" smooth boundary, (g)"Y" pattern grid set-out, (h) Actual node mock-up.

6. Assembly onsite.

Images: Ben Koren, One to One and Zaha Hadid Architects.

Design assumptions

The relative newness of working with the data structure and the speed of delivery assumed to execute the project result in a loss of expression in some design elements such as the uniformity of structural cross-sections, resulting in undifferentiated expression of load, the inherited typical detail at the boundaries resulting in perceived boundary edge thickness, as well as the bounding box approach to preliminary costing resulting in the perceived flatness of the platform and rear of trunk.

While the benefits of design geometry processed as mesh attributes is apparent in a self-referential setting, it underscores the schism between design and fabrication design as incompatible workflows with regard to anticipated input geometry at each stage. The structuring of data in this regard seeks to reorder delivery workflow to assume fabrication-relevant information at the outset of design rather than part-way through, and highlights a requirement to merge early-stage design with fabrication intelligence.

This paper presents a scalable pipeline for the design and production of freeform multi-layer support structures that exhibit a high degree of material economy. While this is demonstrated at the scale of a dining pavilion, the process is governed by the consideration of material and fabrication constraints which are even more critical when designing large-scale support structures. As such, long-term objectives will focus on extending the proposed methods so that they can be leveraged within full-scale architectural applications. To this end, the pavilion serves as a relevant example, as it operated within a tight time-bound, multiple stakeholder, collaborative and bespoke production pipeline, as is typically necessitated by architectural projects.

The most critical limiting factor when scaling up is the translation of the TO density distribution to an appropriate discrete representation of M. Currently, this remains a manual step in an otherwise procedural design process, creating a feedback bottleneck between early- and late-stage design – the severity of which will only increase with the complexity of the project. As such, efforts are being made to automate this step by leveraging techniques from machine learning, image processing and character animation to procedurally extract relevant features from the TO in a format suitable for subsequent processing and early design exploration consistent with current workflows.

More immediate future work will focus on the delivery of a second iteration of the pavilion intended for exterior use. This calls for the revision of materials and assembly methods/details with respect to durability, which presents a new set of fabrication constraints to be geometrically represented within the design model. The use of variable cross-sections among edge members will be of primary importance, allowing for further expression of structural performance via material economy. While this would suggest additional complexity during both design and production, it is anticipated that the impact will be significantly mitigated through the use of the half-edge representation. By defining dimensional attributes per edge, unique elements of the assembly can be resolved with respect to one another through efficient topological queries – an operation supported by the chosen data structure.

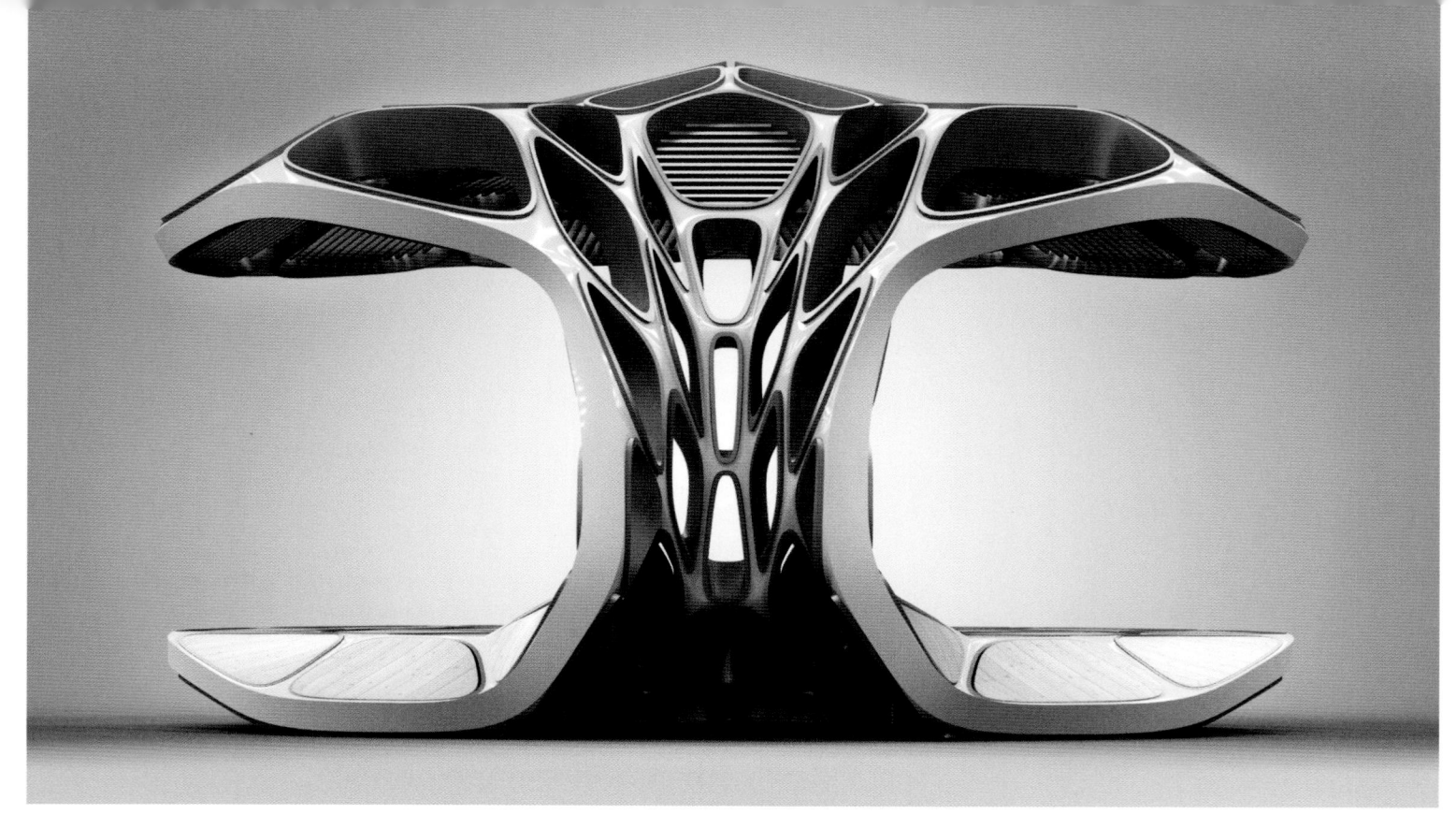

Overall dimensional constraints imposed by the context of the original prototype are significantly relaxed for the second iteration, allowing various formal aspects to be revisited as well.

Specifically, curvature in the transitions from the trunk to the floor and ceiling can be more evenly distributed, reducing problems related to numerical convergence during subsequent rationalisation. Further effort will be made to better understand and formalise this relationship as a means of informing design exploration.

References

Attar, R., Aish, R., Stam, J., Brinsmead, D., Tessier, A., Glueck, M. & Khan, A., 2010, 'Embedded Rationality: A Unified Simulation Framework for Interactive Form Finding' in *International Journal of Architectural Computing*, 8 (4), p.399-418.

Baker, W.F., Beghini, L.L., Mazurek, A., Carrion, J. & Beghini, A., 2013, 'Maxwell's Reciprocal Diagrams and Discrete Michell Frames' in *Structural and Multidisciplinary Optimisation*, 48 (2), p.267-277.

Beghini, L.L., Carrion, J., Beghini, A., Mazurek, A. & Baker, W.F., 2014, 'Structural Optimisation using Graphic Statics' in *Structural and Multidisciplinary Optimisation*, 49 (3), p.351-366.

Bendsoe, M.P. & Sigmund, O., 2013, *Topology Optimization: Theory, Methods, and Applications*, Springer Science & Business Media.

Bezier, P.E., 1971, 'Example of an Existing System in the Motor Industry: the Unisurf System' in *Proceedings of the Royal Society of London A: Mathematical, Physical and Engineering Sciences*, Vol. 321, No. 1545, The Royal Society, p.207-218.

Bhooshan, S., 2016a, 'Collaborative Design – A Case for Combining CA(G)D and BIM', *Architectural Design*.

Bhooshan, S., 2016b, *Interactive Design of Curved Crease Folding*, University of Bath.

Bhooshan, S. & El Sayed, M., 2012, 'Sub-division Surfaces in Architectural Form-finding and Fabric Forming' in Orr, J.J., Evernden, M., Darby, A.P. & Ibell, T. (eds.), *Second International Conference on Flexible Formwork*, p.64-74.

7. Volu overview. Artistic impression. Image: Courtesy Zaha Hadid Architects.

Bhooshan, S. & El Sayed, M., 2011, 'Use of Sub-division Surfaces in Architectural Form-finding and Procedural Modelling' in *Proceedings of the 2011 Symposium on Simulation for Architecture and Urban Design*, Society for Computer Simulation International, p.60-67.

Bletzinger, K.U. & Ramm, E., 2001, 'Structural Optimization and Form-finding of Lightweight Structures' in *Computers & Structures*, 79 (22), p.2053-2062.

Blum, H. & Nagel, R., 1978, 'R. Shape Descriptors Using Weighted Symmetric Axis Features' in *Pattern Recognition* 10, p.167-180.

Bouaziz, S.. Deuss, M., Schwartzburg, Y., Weise, T. & Pauly, M., 2012, 'Shape Up: Shaping Discrete Geometry with Projections' in *Computer Graphics Forum*, Vol. 31, No. 5, p.1657-1667.

Botsch, M., Steinberg, S., Bischoff, S. & Kobbelt, L., 2002, 'Openmesh – a Generic and Efficient Polygon Mesh Data Structure' in *Proceedings of OpenSG Symposium*, 2002.

De Casteljau, P. de F., 1986, *Shape Mathematics and CAD*, Kogan Page.

Fenner, P., 2012, 'Laser-cut Lattice Living Hinges', available at http://def-proc.co.uk/b/pivkg/ (accessed October 2016).

Heyman, J., 1966, 'The Stone Skeleton' in *International Journal of Solids and Structures*, 2 (2), p.249–279.

Janssen, B , 2011, 'Double-Curved Precast Load-bearing Concrete Elements', Masters Thesis, TU Delft.

Kalpakjian, S. & Schmid, S., 2008, *Manufacturing Processes for Engineering Materials*, 5th Ed, Pearson Education, (7).

Kilian, M., Flöry, S., Chen, Z., Mitra, N.J., Sheffer, A. & Pottmann, H., 2008, 'Curved folding' in *ACM Transactions on Graphics (TOG)*, Vol. 27, No. 3, ACM, p.75.

Konaković, M., Crane, K., Deng, B., Bouaziz, S., Piker, D. & Pauly, M., 2016, 'Beyond Developable: Computational Design and Fabrication with Auxetic Materials' in *ACM Transactions on Graphics (TOG)*, 35 (4), p.89.

Lachauer, L., 2015, *Interactive Equilibrium Modelling – a New Approach to the Computer-aided Exploration of Structures in Architecture*, Zurich, ETH Zurich, Department of Architecture.

Lang, J. & Röschel, O., 1992, 'Developable (1, n) – Bézier Surfaces' in *Computer Aided Geometric Design*, 9 (4), p.291-298.

Lawrence, S., 2011, 'Developable Surfaces: their History and Application' in *Nexus Network Journal*, 13 (3), p.701-714.

Leyton, M.. 1988, 'A Process Grammar for Shape' in *Journal of Artificial Intelligence*, 34 (2), p.213-247.

Liu, Y. & Wang, W., 2008, 'On Vertex Offsets of Polyhedral Surfaces' in *Proceedings of Advances in Architectural Geometry*, p.61-64.

Louth, H. et al., 2015, 'Curve-folded Form-work for Cast, Compressive Skeletons', in *Proceedings of the SIMAUD 2015 Conference, Alexandria, USA*, available at http://simaud.com/proceedings/download.php?f=SimAUD2015_Proceedings_HiRes.pdf.

Maxwell, J.C., 1870, 'On Reciprocal Figures, Frames, and Diagrams of Forces' in *Transactions of the Royal Society of Edinburgh*, 26 (01), p.1-40.

Michell, A.G.M., 1904, 'The Limits of Economy of Material in Frame-structures' in *The London, Edinburgh and Dublin Philosophical Magazine and Journal of Science*, 8 (47), p.589-597.

Peña de Leon, A., 2012, 'Rationalisation of Freeform Façades: a Technique for Uniform Hexagonal Panelling' in *Proceedings of the 17th International Conference on Computer Aided Architectural Design Research in Asia/Chennai*, p.243-252.

Pérez, F. & Suárez, J.A., 2007, 'Quasi-developable B-spline Surfaces in Ship Hull Design' in *CAD Computer Aided Design*, 39 (10), p.853-862.

Pottmann, H., 2007, *Architectural Geometry*, Bentley Institute Press.

Pottmann, H. & Wallner, J., 1999, 'Approximation Algorithms for Developable Surfaces' in *Computer Aided Geometric Design*, 16 (6), p.539-556.

Pottmann, H. & Wallner, J., 2008, 'The Focal Geometry of Circular and Conical Meshes' in *Advances in Computational Mathematics*, 29 (3), p.249-268.

Pottmann, H., Liu, Y., Wallner, J., Bobenko, A. & Wang, W., 2007, 'Geometry of Multi-layer Freeform Structures for Architecture' in *ACM Transactions on Graphics (TOG)*, 26 (3), p.65.

Poranne, R., Ovreiu, E. & Gotsman, C., 2013, 'Interactive Planarization and Optimization of 3D Meshes' in *Computer Graphics Forum*, Vol. 32, No. 1, Blackwell Publishing Ltd, p.152-163.

Rankine, W.J.M., 1864, 'Principle of the Equilibrium of Polyhedral Frames' in *The London, Edinburgh and Dublin Philosophical Magazine and Journal of Science*, 27 (180), p.92.

Romero, F. & Ramos, A., 2013, 'Bridging a Culture: the Design of Museo Soumaya' in *Architectural Design*, 83 (2), p.66-69.

Rozvany, G.I.N., 2001, 'Aims, Scope, Methods, History and Unified Terminology of Computer-aided Topology Optimization in Structural Mechanics' in *Structural and Multidisciplinary Optimization*, 21 (2), p.90-108.

Sabin, M.A., 1971, 'An Existing System in the Aircraft Industry. The British Aircraft Corporation Numerical Master Geometry System' in *Proceedings of the Royal Society of London, Series A, Mathematical and Physical Sciences*, 321 (1545), p.197-205.

Sanchez-Alvarez, J., 2010, 'Practical Aspects Determining the Modelling of the Space Structure for the Free-form Envelope Enclosing Baku's Heydar Aliyev Cultural Centre' in *Symposium of the International Association for Shell and Spatial Structures (50th, 2009, Valencia), Evolution and Trends in Design, Analysis and Construction of Shell and Spatial Structures: Proceedings*, Editorial Universitat Politècnica de València.

Schlaich, J. & Schlaich, M., 2000, 'LIGHTWEIGHT STRUCTURES' in *Widespan Roof Structures*, Thomas Telford Publishing, p.177-188.

Scheurer, F., 2014, 'Digital Craftsmanship: From Thinking to Modelling to Building', available at https://gsappworkflow2014.files.wordpress.com/2014/09/scheurer-fabian_digital-craftsmanship22.pdf (accessed October 2016).

Shepherd, P. & Richens, P., 2009, 'Subdivision Surface Modelling for Architecture' in *Proceedings of the International Association for Shell and Spatial Structures (IASS) Symposium 2009*, Valencia.

Tessmann, O., 2008, *Collaborative Design Procedures for Architects and Engineers*, University of Kassel.

Veltkamp, M., 2010, 'Structural Optimization of Free-form Framed Structures in Early Stages of Design' in *Symposium of the International Association for Shell and Spatial Structures (50th, 2009, Valencia), Evolution and Trends in Design, Analysis and Construction of Shell and Spatial Structures: Proceedings*, Editorial Universitat Politècnica de València.

OPEN CAGE-SHELL DESIGN AND FABRICATION (HEALING PAVILION)

BENJAMIN BALL / GASTON NOGUES
Ball-Nogues Design Studio

Breaking boundaries in CNC steel tube rolling

Healing Pavilion, completed in December 2016, explores the boundaries and possibilities of CNC steel tube rolling. Inspired by the prowess of thin structural shells, this project translates the robust double curvature inherent in such forms into a dynamic cage-like array. By delving into the nuances and challenges of bending and rolling tube steel, the design adopts the surface form of a shell while introducing a level of transparency and controlled irregularity only possible through working with a network of individual components. Each tube has a unique three-dimensional curvature and is located at a fixed distance relative to its neighbour.

The pavilion balances structural load paths and assembly considerations with a rigorous exploration of patterning and layering. In addition to creating a space for shade and respite, the porous, shifting grid of steel tubing allows the reading of the complex form to fluidly adjust in relation to its background. The double curvature of the form demonstrates the physical limits of the CNC steel bending and rolling technology. That double curvature allows for structural shape efficiency, which creates natural rigidity through non-planar arcs. With just five construction details for the entire project, this final incarnation isolates and streamlines the design and construction process to tackle structure and the interstices between structural components simultaneously. The five structural details consist of:

1. Where the curved tubes meet the anchor plate at the base;
2. Where the tubes are mitred;
3. Where the tubes are spliced;
4. Where the tubes are capped; and
5. Where the tubes are spaced.

The successful translation of the digital design into a physical fabrication workflow without substantial variation from a digital ideal stands as the key driver defining the success of the project. *Healing Pavilion* combines a commitment to meaningful place-making with a deeply experimental fabrication goal.

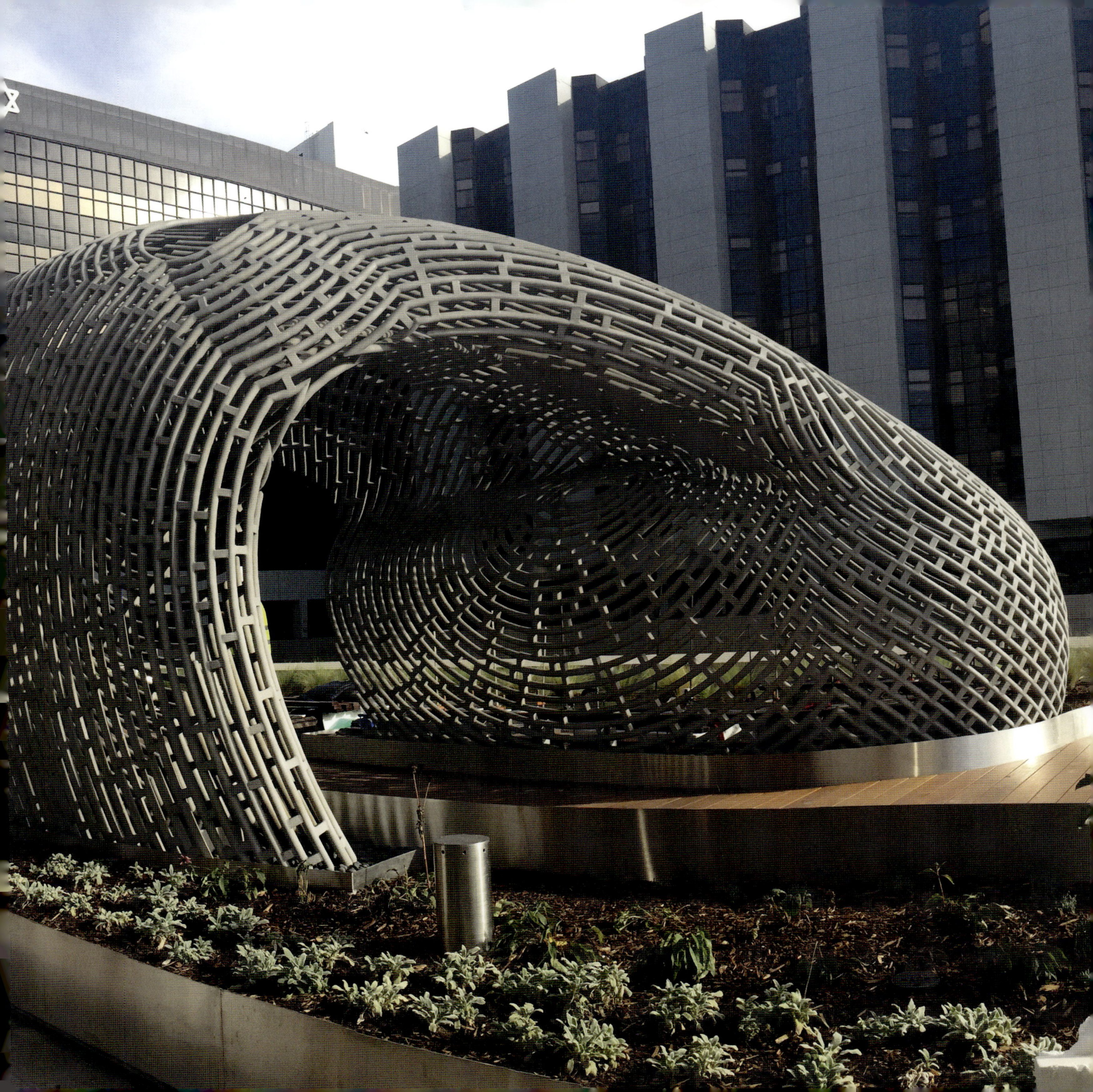

1. Photograph of the pavilion installed onsite.

2. The pavilion being craned onto structural table at site.

Images: Ball-Nogues Studio, 2016.

Research context

Several context considerations define the parameters of the design for *Healing Pavilion*. The site represents the first contextual influence. Commissioned by Cedars-Sinai Hospital in Los Angeles, California, the project forms a key element in a larger garden renovation of the complex's plaza level. The structure performs as both a shading device and a transporting beacon for hope and contemplation – a place to take one's mind away from illness. The design fits within a larger proposal executed by AHBE Landscape Architects and responds to existing circulation paths, water features and vegetation.

The context also acts as a source of design constraint. Flanked by hospital towers on three sides, the sensitive location demanded that no field welding or finishing could occur onsite. The project had to be completed within seven months, installed in a single day, require no routine maintenance and meet seismic, wind load and ADA requirements. In order to address ventilation and noise concerns, the pavilion was fabricated and finished in its entirety offsite. Overall dimensions of the form were kept within the size specifications needed to qualify as an oversize load for transport. The piece was driven as a singular object via a flatbed truck over city streets to the site and then craned into place (Figs. 2 and 9).

The size of the foundation system and the intricacy of the pavilion itself had to be optimised according to the budget. These budgetary factors and the nature of the material itself serve as another contextual factor informing the design. Choosing to work with stainless or corten steel would have made the project too expensive. The decision to explore the process of computerised bending techniques also revealed that malleable mild steel would be easier to weld and bend than corten, stainless or aluminium. Additionally, the context introduced a weight limitation to the pavilion. Placed upon an extant concrete structure, the project distributes its 2,722kg weight over a 'steel table'. The steel table, designed as a customised platform, receives concentrated seismic and gravity loads and transfers them to specific locations on the existing concrete facility structure. More atmospheric aspects of the context guide several defining formal moments in the project. Several openings, including an oculus framing the open sky, follow the path of the sun and orientation of the semi-enclosed space towards the street.

The Healing Pavilion design process

Several questions guided the research and design process of *Healing Pavilion*. The first concept question – how to make a shell using the logic of a cage – drove the number of studies that followed. Once this key research idea concretised, this line of formal inquiry raised the next issues: what kind of machine could bend and roll steel tubing? How to identify its limitations and keep these limitations within a smooth flow of seismic and gravity-induced stresses? This period of investigation led to more targeted research into the minimum diameter of tubing, its wall thickness and its maximum length, whether or not square or round spacers were preferable and the tolerances introduced by the machine.

After developing a better understanding of the CNC rolling machine, research shifted into the successful interfacing of the digital and physical realms. How to choreograph the optimal workflow became a crucial phase of the research agenda. This working process encompassed the translation of the pavilion out of the design softwares Rhino and Grasshopper and into the structural engineering program SAP for finite element analysis (Fig. 3) and finally into Solid Works for production. From there, discerning whether or not the work of precision steel tube rolling, where each tube is unique, could be distilled into a repeatable fabrication process that yielded predictable results within acceptable tolerances structured the next set of inquiries. With the fabrication phase regularised and the results at last predictable, the final component for evaluating the project focused on whether the production process outlined above matched the final workflow used to execute it.

Working through design iterations

Healing Pavilion relied on numerous digital iterations and physical mock-ups to reach its final form. Scripted in Grasshopper, the parametric model facilitated rapid revisions between Ball-Nogues Studio and Buro Happold Engineering. For the sake of expediting the structural analysis of the form, no manual modelling occurred during this phase. The engineers would identify initial undesired stress concentrations by running preliminary analysis models and respond with sketches and three-dimensional model iterations. This feedback identified where additional structural members might be needed and where certain areas would require reorientation of tubing and modifications of curvatures. Ball-Nogues Studio then adapted the digital model accordingly and the structural analysis began again. The feasibility of fabrication played a principal role in the design process. In each design iteration, digital adjustments to the curvature of the tubes factored in the rolling machine's minimum radius.

After digital analysis, the project's research methodology shifted into physical mock-ups to test the plausibility and difficulty of producing such expressive geometries. Working closely with the fabricator Plas-Tal Manufacturing (based in Santa Fe Springs, California) and their CNC tube-bending subcontractor, Caroll Racing Development (Orange County, California), the Studio began testing one-to-one sections of the pavilion. This fluid interfacing and feedback loop between the physical and digital was not unique to the philosophy or working practice of the Studio, but still proved unusual in the number of iterations needed to test different aspects of the design. Each mock-up isolated a different question in the fabrication process, from curvature and assembly to finishes and beyond.

The fabricator chose the first area of the computer model for mock-up. This full-scale swatch proved successful for the least complex section of the form (Fig. 5), but what about the most complex? This first mock-up provoked more questions than it answered. Ball-Nogues Studio

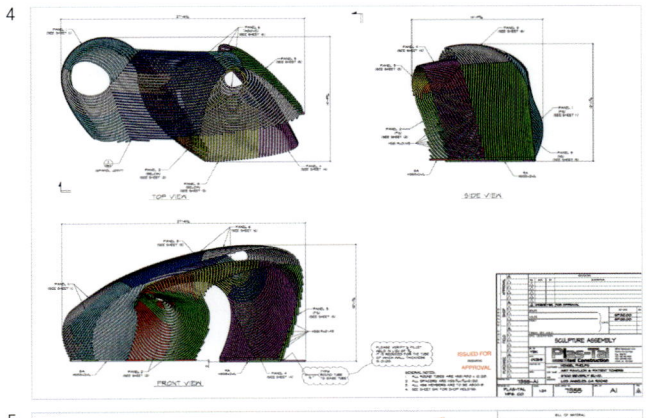

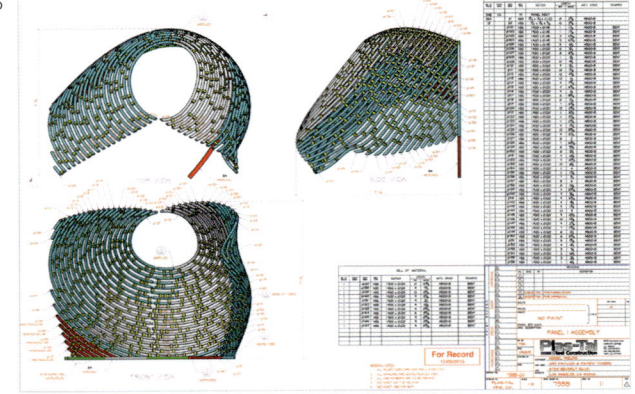

3. Shop drawing of area of pavilion with complex mitre joints and tight three-dimensional curvature chosen for mock-up.

4. Shop drawing of panel and splice layout.

5. Shop drawing of individual panel assembly, spacer layout and bill of materials.

Images: Plas-Tal Steel Construction, 2015.

picked an area (Fig. 3) with complex mitre joints and tight three-dimensional curvature to explore next. The result of this mock-up highlighted the CNC bending machine's capabilities and shortcomings, especially at tight-radiused areas. Traces of the machine's handling manifested as minor kinks in the tube steel. These visible pinches appeared every 10cm, exposing the incremental bending process and breaking the fluid reading of each curve, but it was especially evident on tight curves.

Before the process of bending tube steel was modernised into a computerised system, each hollow section was traditionally filled with sand to achieve maximum precision during hand-shaping. The sand acted to resist compressive forces and to keep the tube steel from collapsing. This same logic had to be introduced into the contemporary version of the bending process. The Studio worked with the fabricator to develop a custom mandrel that could fit inside the tubing and behave as a buffer during the bending, thereby softening the kinks. Accounting for this custom rod called for a special type of tube with a reduced interior weld. The typical way of forming tubing involves rolling a sheet of steel into shape and then welding the seams from the outside. This welding technique results in a considerable amount of internal slag, which acts as a barrier for fitting anything inside and makes the process of working with tube steel imprecise. By developing a more precise tube with minimal welding imperfections, a custom 3m-long rod with the mandrel attached could then be inserted into the 3m steel tube to mimic the analogue process of sand-based bending.

For the welding team to access their work from the ground during fabrication, the pavilion was made as distinct panels (Figs. 4 and 5) that could be positioned within reach of welders in the shop (Fig. 6) and connected seamlessly into one cohesive object. Certain panels proved more problematic than others. These areas of concern included places of tight three-dimensional curvature and where tubing needed to be massaged into place by hand to weld.

Because raw tube is made in standard dimensions that are shorter than the length of most of the curves in the project, most tubes required splicing. The splices were therefore located at the edges of panels.

Whether or not the physical results of the pavilion matched the digital model served as the main criterion for evaluating the outcome of the research. *Healing Pavilion* was fabricated from 851 linear metres of 5cm-diameter mild steel tube with 3mm wall thickness.

6

7

While the form and curvature of the pavilion oscillates and shifts, each tube and the space between each tube are always the same. In order to maintain a high level of precision across the shell, the tubes were assembled over large metal fixtures to ensure proper alignment and consistent spacing during welding. The fixtures were waterjet-cut from plate steel and welded to a single base plate (Fig. 7). These fixtures controlled tolerances by correcting the unavoidable discrepancies between the ideal curvatures in the digital model and the rolled tubes. The question of how to control for deviation in the steel also influenced the decision to finish the pavilion with a coating of ceramic alumina applied by thermal flame spraying (Fig. 8). This finishing technique was applied by melting the constituent materials and then atomising them with air. Developed for the non-skid tarmac surface of aircraft carriers, the finish has little precedent in architecture. Galvanising was ruled out because the size of the pavilion made it impossible to fit it in a typical bath; furthermore, the heat of galvanising could have distorted the tubes and therefore the shape of the shell. The granularity of the coating also masks minor welding imperfections and kinks that result from the bending process, further obscuring the footprints of fabrication.

In addition to finish considerations, the fixtures minimised the deformations in the tubes that typically occur from the heat of welding. Mild steel has a level of springback when rolled, as it tries to revert to its original condition. To compensate for this movement, each tube had to be clamped and adjusted by hand to fit within the rigid fixture. Once the tubes were tack-welded into place, 2,544 square spacers were inserted to keep the cage in a state of tension (and sometimes compression) to avoid deflection over time. With more than 3,000 360°

6. Welding spacers onto a panel #3 while resting on fixture.

7. Rolled tubes being assembled on CNC waterjet-cut fixture.

8. Ceramic alumina finish application being applied by thermal spraying.

9. Photograph of the pavilion arriving at site via oversized truck, ready to be craned into place.

Images: Ball-Nogues Studio, 2016.

8

structural fillet welds in total, each spacer standardises the distance from one tube to another and standardises the process of welding. While every spacer is the same, the irregularity of their placement relates to key stress points in the form and introduces a compelling patterning to the final design. The commercial project kept within its budget. This issue of budgeting limited the rounds and iterations possible during the engineering phase. In spite of this back and forth between digital and analogue processes, the pavilion reflects the digital model within less than 1cm of deviation.

Hands-on problem solving and optimising digital design

Healing Pavilion reconciles the output tolerance of digital machines with the allowable tolerances of the physical world. The project celebrates the concept of fidelity and hands-on problem solving. To translate ideal geometry from a software environment into the tolerances inherent in the output of a numerically controlled tube bender and then into a high fidelity final product meant that the machine's capabilities could not be taken for granted. Reading the available product literature would not answer the questions the project needed addressed. Instead, a specific machine had to be engaged with directly. By building an intimate relationship with the tool, one could identify its capabilities. These understandings helped to craft and optimise translations between one software system and another, as well as to predict the physical ramifications of such digitally based design decisions. Some insights throughout the research influenced the digital aspects of the project, while others directly impacted its physical construction. In a few instances, the machine was pushed too far and certain moments had to be resolved by hand, such as very small radii. Even so, the final incarnation of *Healing Pavilion* demonstrates the optimal interface between handcraft and the computer, and offers a contribution to the fields of design and fabrication that use CNC tube rolling.

The seamless physical execution of the digitally driven design created a calibrated interface for accurately automating the process of bending steel tubing. Designed as a surface and then adapted to the logic of a cage, the final shape retains no superfluous elements. The pavilion defends its structural integrity without conforming to a structural hierarchy. Each steel spacer and tube reflects an element through which loads and stresses flow, so the cage, once again, adapts to the performance of a shell. A sinuous bench of solid Ipe wood nestles into the organic shading structure. The ambitious structural endeavour never lost sight of the project's greater goal as a transporting space far removed from the stresses and stigmas associated with sickness. A space for sharing a moment with a loved one or simply sitting in contemplation, *Healing Pavilion* combines rigorous fabrication techniques with an inherent sensitivity to client and context (Fig. 1).

9

MAGGIE'S AT THE ROBERT PARFETT BUILDING, MANCHESTER

RICHARD MADDOCK / XAVIER DE KESTELIER / ROGER RIDSDILL SMITH / DARRON HAYLOCK
Foster + Partners

A home away from home

Located across Britain and abroad, Maggie's Centres were conceived as a place of refuge where people affected by cancer could find emotional and practical support. Inspired by the blueprint set out by Maggie Keswick Jencks, they place great value upon the power of architecture to lift the spirits and help in the process of therapy. The design of the Manchester centre aims to establish a domestic atmosphere in a garden setting.

The building is arranged over a single storey, the roof rising in the centre to create a mezzanine level, naturally illuminated by triangular roof lights and supported by lightweight timber lattice beams. The beams act as natural partitions between different internal areas, visually dissolving the architecture into the surrounding gardens.

It was vital to create an atmosphere that would make visitors feel at ease, as if they were at home. The use of exposed timber for the structural elements enabled the creation of a homely, domestic ambience throughout the centre, exploiting the warmth and softness of the material.

Using the practice's expertise in digital modelling and analysis, the structure is the protagonist – a cantilevered timber wing 'tiptoeing' lightly over the site. To that end, much work was undertaken to assess how the design intent could be realised with contemporary materials and digital fabrication methods. Investigations were carried out to explore the structural optimisation potential in minimising the material used. For the construction, an Airfix™ (Airfix, 2016) analogy was deemed desirable – a kit of parts fabricated offsite and assembled onsite, facilitating quick erection.

The result is an innovative use of a traditional material, taking advantage of a complete file-to-factory process to provide the driver of the building aesthetic.

Making design match function

Functionally, the building is laid out to provide accessible open spaces along either side of a central zone: public spaces to the west, with the more private cellular spaces on the east. The centralised horizontal core houses the building's services and an administrative zone on the

1

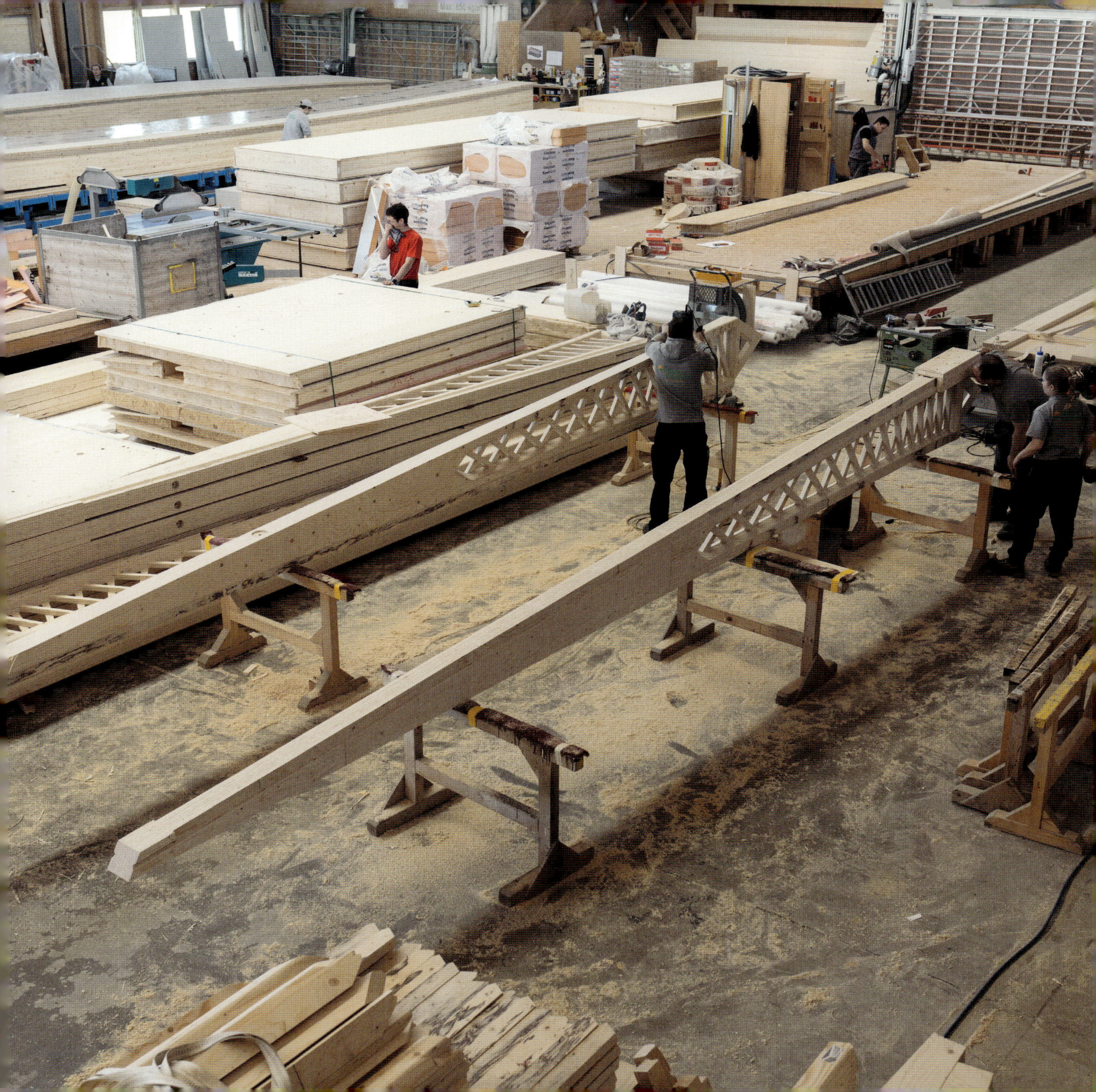

2

mezzanine deck. The southern end of the building extends to embrace a greenhouse – a celebration of light and nature – which provides a garden retreat, a space for people to gather, to work with their hands and enjoy the therapeutic qualities of nature and the outdoors. It is a space to grow flowers and other produce that can be used at the centre, giving the patients a sense of purpose at a time when they may feel at their most vulnerable.

Throughout the centre there is a focus on natural light, greenery and garden views, with a warm material palette. This spatial arrangement naturally led to a structural system where the primary support, a series of 17 identical frames repeated on a 3m grid, springs from a central spine, with a propped cantilevered roof on either side. Slender steel columns just beyond the façade make the entire structural system more efficient. These elements significantly reduce the bending moment in the overhead span, and remove the need for a deflection head at the top of the glass in the roof lights (Fig. 5).

Timber is the natural choice for this type of structure not only for its aesthetic value, cost and carbon efficiency, but also because it has high strength but low stiffness in comparison with steel. A propped cantilever benefits from exactly these properties – high strength for the large central bending moment, with low relative stiffness accounted for by the prop.

A more conventional approach might have used a glulam beam, although high self-weight is a drawback of this type of construction, resulting in large and heavy sections. In contrast, digital fabrication has allowed the timber to be provided exactly where required – at the top and bottom flanges for tension and compression – and the minimum material in the web to provide adequate shear transfer. Any portion that is superfluous to structural requirements has been removed.

Challenges and questions

Wood-based I-beams have many advantages, displaying high stiffness and strength for their low weight (Hermelin, 2006), and sustainably sourced timber has the added benefit of being more environmentally friendly than steel. The design intent and structural analysis inferred that the beam webbing could have a number of openings such that the structural behaviour is reflected in its form and materials. It is relatively easy to cut holes in timber webbing, further reducing the weight of the beam. However, the effect of this is to reduce the shear capacity of the member. A central issue was the study of the webbing shear capacity and how this was factored into the manufacturing of the Maggie's timber beams.

The choice between CNC-machined timber beams or handcrafted ones was made early in the design process. While handcrafted beams would permit individual web members to have their grain aligned to the forces they would experience, thus providing a clearer load path, the longer manufacturing time and the need for multiple connections between each diagonal proved prohibitive. Although digitally fabricating beams from an engineered timber such as laminated veneer lumber (LVL) meant that the grain orientation was fixed for each web member, requiring a denser web configuration, the faster manufacturing time, increased timber grade and ability to easily and accurately produce complex geometry was deemed far more beneficial to the project. This also helped to achieve the objective of an offsite fabrication, onsite assembly project.

The greenhouse 'cockpit' at the southern end of the building presented another structural challenge. In strong winds, the building would rack up to 15mm longitudinally. However, the triangulated geometry of the cockpit unintentionally acted to prevent this deflection, placing more load on the greenhouse timber members than they could handle, inducing buckling and thus shattering the glass. Thicker members would render the cockpit structure visually distinct and heavier in comparison to the rest of the building, and the option of making it an entirely separate structure was also deemed incompatible with aesthetic aims. Resolving this structural conundrum satisfactorily was critical to the success of the project and is outlined later in this paper.

Physical prototyping and seeking solutions

An integral aspect of the practice's working methods since its inception, physical prototyping was a key part of the design process. Full-scale elevations of the 8m timber beams were printed on paper and hung in the studio. The in-house 3D printing facilities produced many options of node, truss and beam details at multiple scales. Model makers created versions of the entire structure as well as focusing on details, again operating at many scales. Three 1:1 prototypes of the key triangular node were produced for evaluation purposes: one by the Foster + Partners' Modelshop team, and two by contractors bidding for the job: Blumer-Lehmann AG and Merk Timber. Upon appointment of Blumer-Lehmann AG, an entire full-size mock-up of the final truss was produced. Testing even extended to 3D printing and placing onsite 1:1 models of the ceramic tiles at the foot of each column. These prototyping methods were invaluable, as the process of fabricating full-scale mock-ups greatly influenced the final design.

The main timber structure is formed by a series of portal frames pinned at the base, with Y-shaped branches forming the apex. The frames carry both gravity and lateral loading in the transverse direction. Connections between members are achieved by means of hidden pre-embedded steel flitch plates (Fig. 3) with bolts and screws as fasteners (Bangash, 2009). Linear elastic static analysis in Oasys GSA (Oasys, 2012) was carried out for the basic load cases and superposition used to assess the load combinations.

An analysis of the stresses caused by wind load (sideways) and snow and dead loads (vertically) indicated where the timber could be optimised. The beams thus have a top and bottom flange, and diagonals through the web, which vary in density as the shear force varies along the section (Munch-Andersen, J. & Larsen, H. (eds.), 2011). The trusses taper in elevation as the bending forces reduce towards the cantilever tip, through the column to the pin connection at the ground and at the central node above the spine. This taper provides the slope of the roof. The bottom flange of the beam varies in width, reflecting the structural demands upon it. This can be seen in the contouring of the LVL layers on the bottom flange.

1. Fabricating the trusses at the Blumer-Lehmann workshop.

2. The greenhouse and cockpit, with moveable work table on rails.

3. Inserting the steel flitch plates.

Images: Nigel Young/ Foster + Partners.

In addition to the tapered form of the timber beams, with the shallowest ends corresponding to the points of minimum bending moment, the web also incorporates openings such that where shear demand is low, a higher percentage of material is removed, and vice versa (Williams, 2008). For a given web thickness, the shear demand was transformed into a net area required at each section so that the resulting stress did not exceed the material's capacity (American Foster & Paper Association, 2006). The analysis undertaken demonstrated that a trellis-like geometric arrangement would be suitable, and a script was created in Rhinoceros and Grasshopper that generated the webbing geometry. In the final design, the webbing is solid as the beam crosses the building envelope. This also provides greater support for the hogging moment above the steel prop.

There was much experimentation with the form of the webbing in the trusses. One option was explored that aligned curved timber members to follow the tension and compression stress lines within the beam. This would allow the members to work mostly axially. Despite producing an intriguing outcome, the fabrication constraints were judged too great, although this work has informed a separate research project currently being undertaken by Foster + Partners' Specialist Modelling Group.

A simpler solution was settled on whereby the truss webbing is made from a pattern of straight elements whose frequency varies to match the material required to resist shear forces. As the shear force increases, the area of material required to counter it increases. The angle of the roof means that the available cross-sectional area of the web decreases along its length, which creates a varying percentage of webbing that must be solid. Integrating this curve gives another curve whose slope is the required density. Distributing points evenly along this second curve and projecting them straight down defines the nodes of the struts. As the spacing varies, the angles change accordingly, ensuring the requisite amount of cross-sectional material is provided.

The node that links the beam and column trusses is a key connection in the entire structural system. It is at this node that the vertical loads from the roof – its self-weight and the snow loads – are transferred to the columns and subsequently down to the ground. Simultaneously, the node acts as a fixed portal frame haunch to provide the rigidity required to resist the horizontal wind forces acting across the structure and to bring these forces down to the ground as well. The forces at this critical connection resolve themselves into a set of pure axial stresses around the triangle, which provides the required rigidity and strength through the efficiency of its form.

Each timber lattice truss is comprised of four CNC machine-cut pieces that are glued together offsite to form one of the elements assembled onsite as the complete portal frame. Understanding the abilities of the 5-axis milling machine at Blumer-Lehmann's disposal was paramount. The limitations of drill bit size, cutting speed and cutting angle all informed final design decisions.

Offsite construction was essential to produce structural elements that were highly finished, precisely fabricated and could be assembled without need for tolerance adjustment onsite (Fig. 4). The process was also cost-efficient and enabled rapid and predictable construction to fit within the tight programme.

The greenhouse cockpit problem was resolved using Oasys GSA, Rhinoceros and Grasshopper. A viable solution was devised whereby the two cockpit supports were placed on springs, allowing vertical movement to cater for the racking of the building. The final solution utilises a cantilevered sprung RHS beam to support the cockpit. When the building racks in strong wind, the cockpit is free to move vertically so as not to absorb any load from the building.

Benefits of 3D modelling and CNC manufacturing

The project required close collaboration between multiple teams at Foster + Partners and the contractors involved. The firm's Specialist Modelling Group produced geometry with Rhinoceros and Grasshopper, which was analysed by the in-house structural engineering team using Oasys GSA, all the while liaising with Blumer-Lehmann and glass contractors Bennett Architectural Aluminium to ensure that architectural aims were met and manufacturing constraints were incorporated. The interaction and dialogue between designers and contractors was key – learning and understanding the limitations of the cutting equipment so that the design intent responded creatively to the manufacturing process. The back-and-forth of 3D information helped the design and construction process, with CAD models shared from architects to contractors and vice versa for review.

The diagonal arrangement of the trusses in plan across the central spine enables the primary timber structure to provide stability to the roof without the need for any additional bracing elements or stiffeners. The roof can

4. The final truss of the first portal frame is installed onsite.

5. The lattice trusses and skylights allow plenty of light into the building.

Images: Nigel Young/Foster + Partners.

4

5

act as a single diaphragm, transferring the wind loads into the trusses, which provide rigidity as a portal frame across the building. Along the length of the building, the diagonal trusses deliver load into the spine. In this way, the building's structure directly reflects the forces it resists.

The timber structure is sustainable and tactile, and was built quickly to a tight budget. The CNC-crafted LVL lattice beams are constructed from Kerto, a MetsäWood product. It is made from 3mm-thick rotary-peeled softwood veneers that are glued together. The spruce is sustainably sourced, using whole logs in the manufacturing process, with consequently minimal waste. The waste material generated by the milling of the trusses is used as fuel to heat Blumer-Lehmann's factory space (Fig. 1).

Removing material from the beam's webbing resulted in a truss that was a third the weight of a similar solid glulam beam. The behaviour of the web as affected by the removal of material was further investigated by a number of finite element analysis models in Oasys GSA in order to assess the maximum and minimum principal stress and the shear stresses at various locations in the web. These stresses compared favourably to the material strengths (ETA, 2010).

The use of 3D modelling and CNC manufacturing has unlocked new methods of working a traditional material. Crafting timber with these modern tools has resulted in an expressive structure that celebrates connections and details while evoking horticultural references such as the garden trellis (Gould, 2001).

A successful exploration of material qualities

The product of the twin desires of design intent and structural requirements, the Maggie's Centre in Manchester continues the long history of actively integrating the two within Foster + Partners' work.

With a focus on the process of design evaluation through full-size mock-ups and prototyping, using the full range of capabilities at the firm's disposal, the nature and fabrication of the final structure was evaluated at every step of the journey. Timber was chosen as the primary building material for its warmth and sculptural qualities, giving the building unique scale, depth and texture. There is no attempt at cladding or concealing the distinctive structure; the building is an open, honest exhibition of the material and its biophilic qualities.

The use of advanced manufacturing technologies allowed new ways of exploring the expressiveness of the material to be investigated. The exchange of 3D CAD models between teams within the office and external contractors for architectural, structural and fabrication review was also vital to the project's success and contributes to a timber structure that is entirely digitally fabricated using a file-to-factory process.

The project combines fundamental design philosophies from the earliest days of the practice – prefabrication, dry construction and the benefits of speed and quality that this process offers – with modern digital simulation and manufacturing technologies. The result is an innovative lightweight structure and therapeutic space that is a celebration of light and nature (Fig. 6).

Project Credits

Architects: Foster + Partners; Norman Foster, David Nelson, Spencer de Grey, Stefan Behling, Darron Haylock, Diego Alejandro Teixeira Seisedos, Xavier De Kestelier, Mike Holland, Richard Maddock, Daniel Piker, Harri Lewis, Elisa Honkanen
Client: Maggie's
Structural engineering: Foster + Partners; Roger Ridsdill Smith, Andrea Soligon, Karl Micallef, Mateusz Bloch
Environmental engineering: Piers Heath, Evangelos Giouvanos, Nathan Millar
Fire engineering: Thouria Istephan, Michael Woodrow
Landscape: Dan Pearson Studio
Timber fabrication: Blumer-Lehmann AG
Site area: 1,922m^2
Built area: 500m^2

References

Airfix, 2016, *Airfix*, available at http://www.arfix.com (accessed 1 October 2016).

American Forest & Paper Association, 2006, *National Design Specification: Design Values for Wood Construction*, Washington, American Wood Council.

Bangash, M.Y.H., 2009, *Structural Detailing in Timber*, Dunbeath, Whittles Publishing.

European Technical Approval No. ETA-03/0056, 2010, *Wood-based I-shaped Composite Beams and Columns for Structural Purposes*, Forestia AS.

Gould, M.H., 2001, 'A Historical Perspective on the Belfast Truss Roof' in *Construction History Volume 17*, p.75-87.

Hermelin, R., 2006, *Strength Analyses of Wooden I-beams with a Hole in the Web*, Masters, Lund University.

Munch-Andersen, J. and Larsen, H. (eds.), 2011, *Timber Structures – a Review of Meeting 1-43*, CIB W18.

Oasys Sofwtare, 2012, *GSA Version 8.6* (computer program), available from http://www.oasys-software.com.

Williams, P.A., Kim, H.A., Butler, R., 2008, 'Bimodal Buckling of Optimised Truss-Lattice Shear Panels' in *American Institute of Aeronautics and Astronautics*, Vol 46, No. 8, p.1937-1943.

6. The Manchester Maggie's Centre at dusk.
Image: Nigel Young/Foster + Partners.

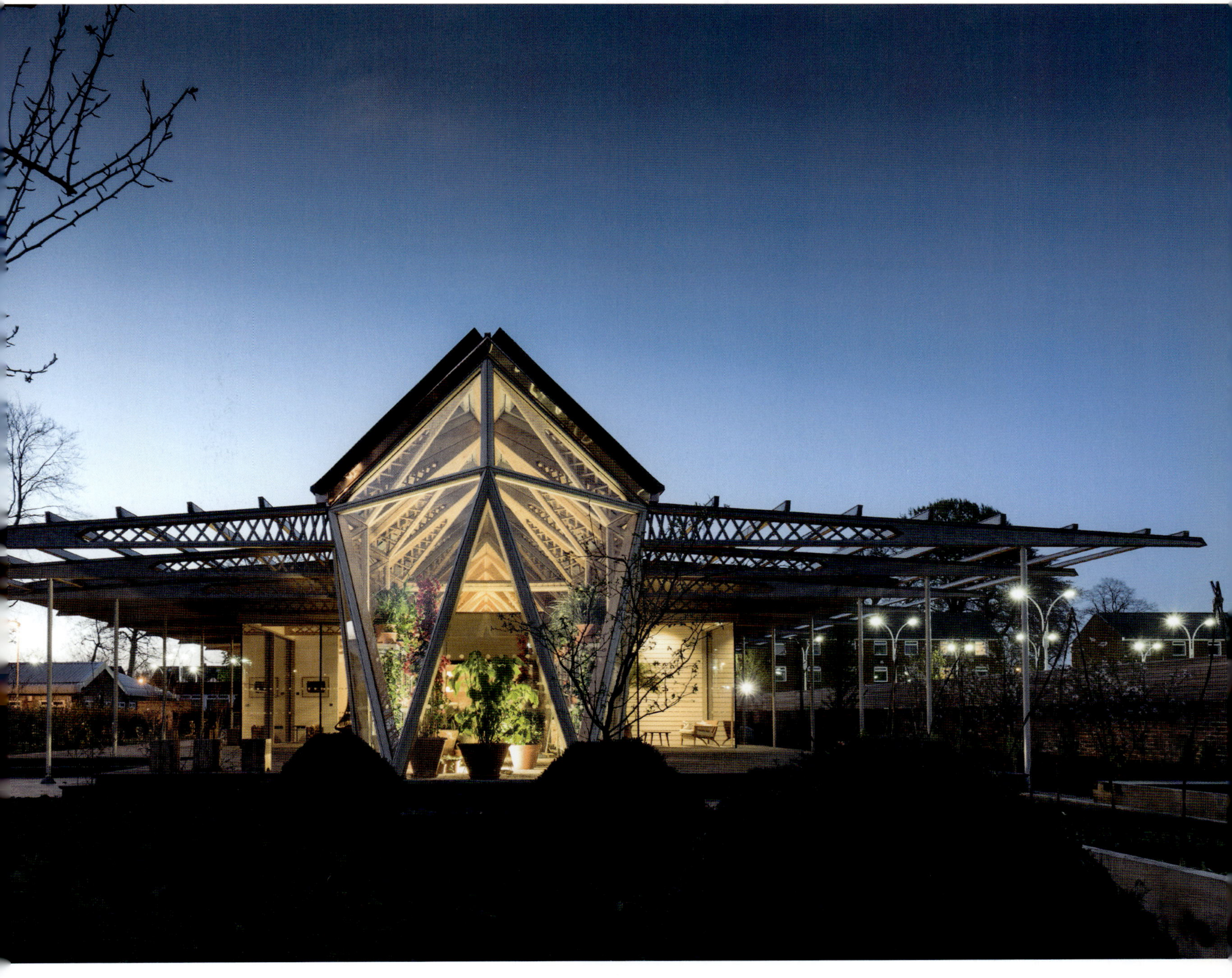

2

RETHINKING MATERIALISATION

INFUNDIBULIFORMS
KINETIC SYSTEMS, ADDITIVE MANUFACTURING FOR CABLE NETS AND TENSILE SURFACE CONTROL

WES MCGEE
University of Michigan, Taubman College of Architecture and Urban Planning | Matter Design
KATHY VELIKOV / GEOFFREY THÜN / DAN TISH
University of Michigan, Taubman College of Architecture and Urban Planning | RVTR

The work of the *Infundibuliforms* project aims to advance research in lightweight kinetic surfaces as systems that have the ability to create spatial enclosures with minimal amounts of material and that are capable of dynamic reconfiguration. This paper describes the iterative research and full-scale prototype evaluation of a cable-robot-actuated, geometrically deformable elastic net that has been developed through close coupling between geometric explorations in computational spring-based physics solvers and experimental additive manufacturing techniques for net- or mesh-based structures. The title of the project refers to the catenoid forms that define the geometry of a surface; the term 'infundibula' is most commonly used to refer to funnel-shaped structures in biotic systems and plant morphology.

The work advances three parallel trajectories in computational, fabrication and geometric research:

- The development of dynamic models to both simulate and control the operation of a physical tensile system in real time.

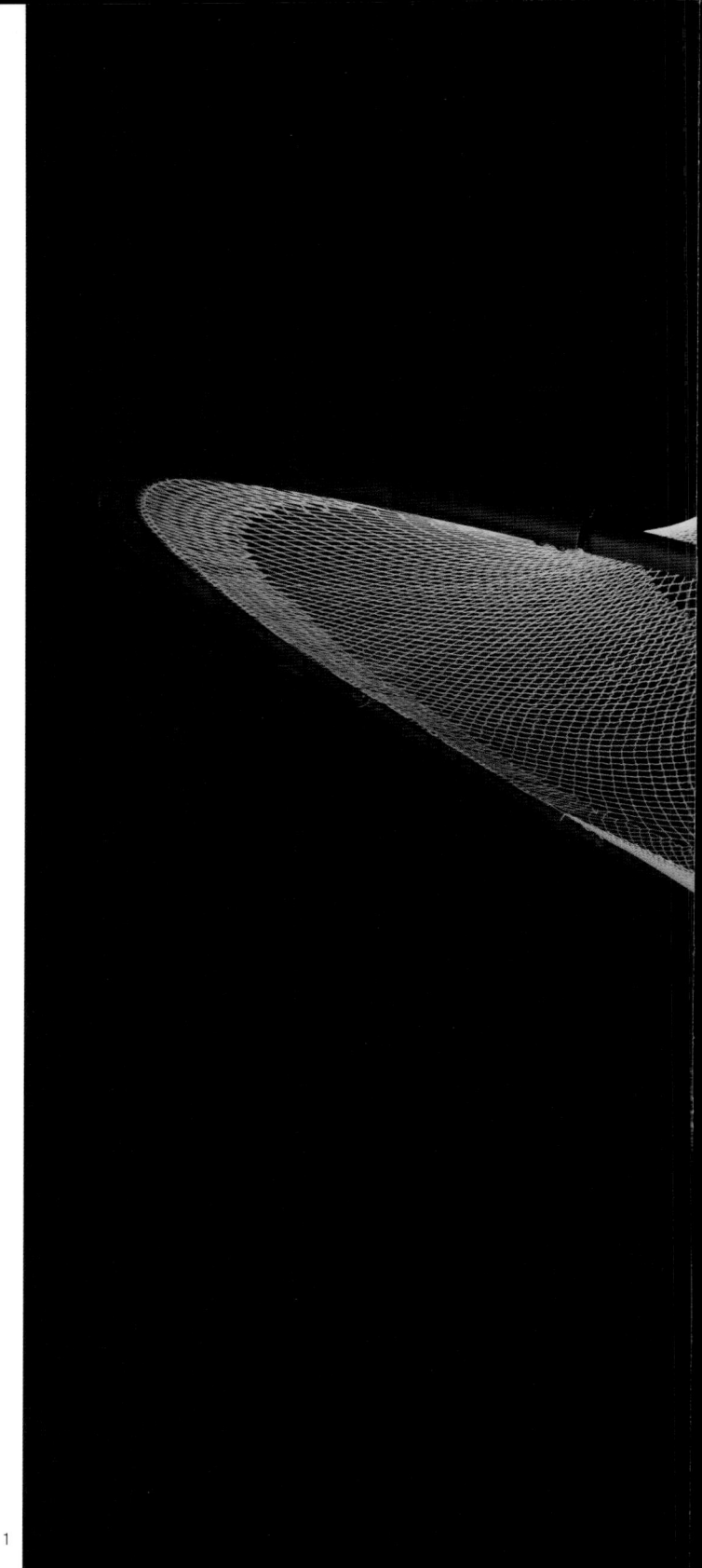

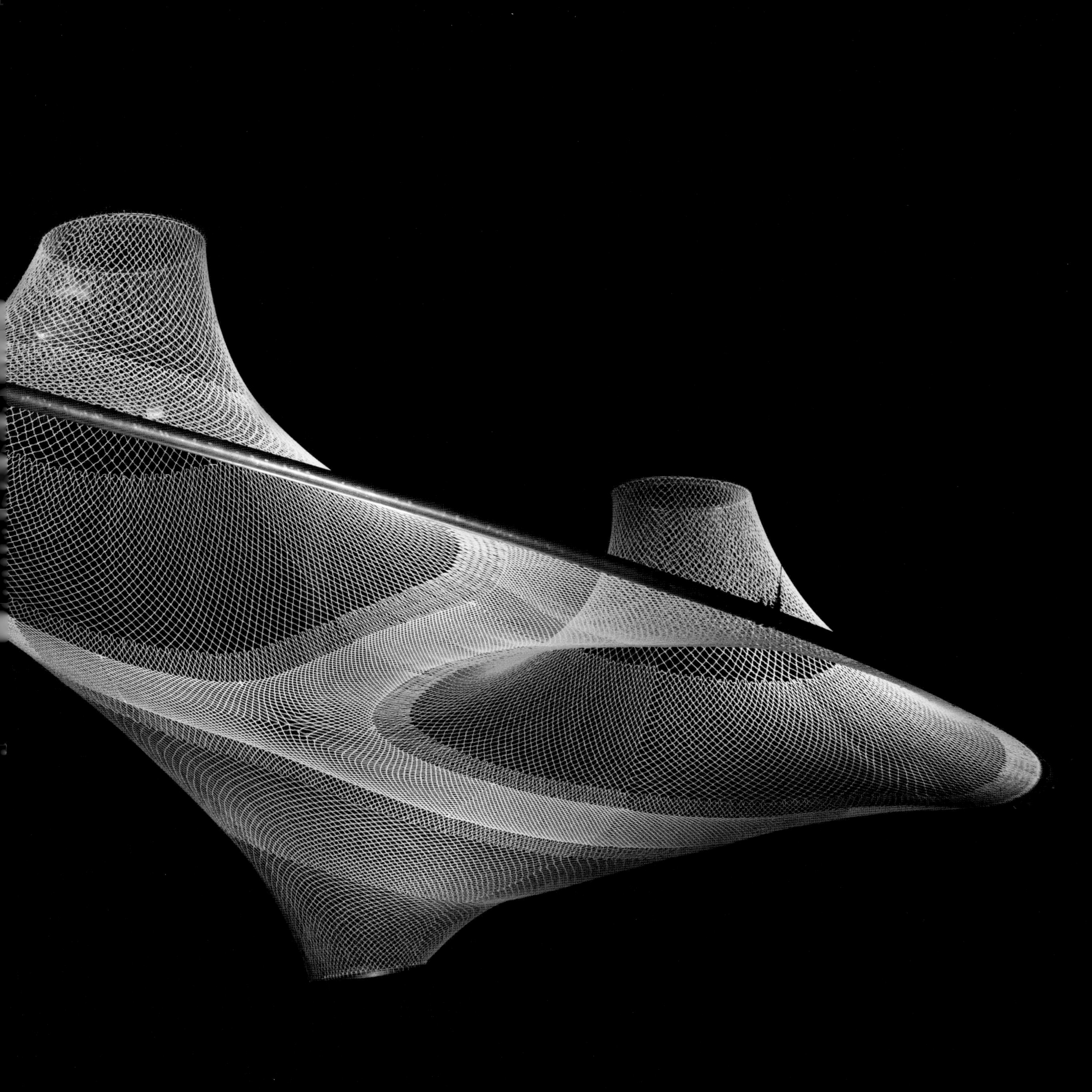

- Advancement in tools and methods for robotic additive manufacturing to enable the 3D printing of thermoplastic elastomers for tensile surfaces.

- A geometric methodology for developing cable nets that can be loaded to produce tailored catenoid forms using spring-based simulation methods.

Central to the work of this project is the development of integrated 1:1 prototyping to assess the interaction among research streams and the fidelity of physical performance relative to the simulations (Fig. 1).

The *Infundibuliforms* project spans three distinct research contexts that have been brought together within the scope of the work: kinetic systems, elastic cable net surface development and control and additive manufacturing. Each has a specific history and body of research literature, and part of the innovation within this project has been to combine these areas in the integrated development of the project. Each also advances specific aspects of the authors' previous research.

Kinetic systems
Traditionally, architecture has been primarily concerned with static structures that aim to attain stability and equilibrium states. Since the middle of the last century, however, there has been increased interest in adaptable structures that would involve motion systems, or what Frei Otto and his researchers at the Stuttgart Institute for Lightweight Structures preferred to call 'convertible' structures (Otto, 1972). Lightweight systems such as tensile structures and inflatables are appropriate for convertible or kinetic structures, since they can be readily deformed at a low energetic cost. More recently, there has been an increased interest in kinetic architectures. However, the majority of this research and development has been at the scale of individual kinetic components, such as actuated surface elements or tessellations as components of responsive envelope systems (Lienhart et al., 2011, Khoo et al., 2011, Thün et al., 2012, Adriannsens & Rhode-Barbarigos, 2013). At the larger spatial scale, while the industry has advanced technologies and techniques for convertible textile and membrane roofs, there still remains relatively little exploration into actively deformable dynamic surfaces. Kinetic, deployable and reconfigurable architectures are areas of research still very much in their infancy, and the advancement of these areas requires simultaneous experimentation with materials, formal exploration, design tools and methods of manufacture, especially in cases where these systems are automated with mechatronics, robotics, communication protocols and control systems.

Additive manufacturing
Additive manufacturing, commonly referred to as 3D printing, has been rapidly advancing the capability of designers across numerous fields to synthesise and manufacture materials and surfaces with complex geometries and materially programmed performances, and is quickly opening up novel formal, structural and performative possibilities for architecture (McGee & Pigram, 2011, Keating & Oxman, 2013, Helm et al., 2015). In the case of tensile and membrane structures, additive manufacturing offers the possibility of developing composite surfaces with programmable material behaviours while also exploring novel performative geometries for net structures (Coulter & Ianakiev, 2015).

Tensile cable net surface design and control
Textiles and cable nets are referred to as 'form-active' structures and are comprised of flexible and non-rigid materials. As their final shape is not known a priori but depends on factors such as boundary conditions, stress distribution, material stiffness and grid topology, there has been extensive research into methods for form-finding, or the process of generating the optimal structural and visual configuration (Bletzinger et al, 2005, Wagner, 2005, Veenendal & Block, 2012). While there are several analysis tools for refining form-active structures, there are few tools available to designers for the creation and rapid assessment of novel structural forms. In this area, particle spring-based systems, which simulate live physical forces, have been explored by a number of designers (Killian & Oschendorf, 2005, Ahlquist & Menges, 2010).

Live-interaction 3D modelling tools, such as the Kangaroo plug-in for the Grasshopper extension of Rhinoceros, use particle spring systems to simulate physical forces and constraints. These tools allow designers to experiment with far more agility in form generation for the development of non-linear structural systems. Additionally, open source tools, such as the ShapeOp library, enable the extension of these form-finding techniques into larger, more complicated and more robust computational models (Deuss et al., 2015). This ability for the model to update at a relatively high rate is essential for work in kinetic systems, and especially so when sensing and interactive behaviours are implemented into the control framework.

1. Second iteration of the installed *Infundibuliforms* prototype, October 2016.

2. Screenshots of concurrently running simulation and control software interfaces (left) paired with a drawing of the extents of the position possibilities and in-between states of the *Infundibuliforms* prototype installation (right).

Images: © RVTR / Matter Design 2016.

Research questions

The primary research question of this project was how to advance the integrated design and control of kinetic lightweight architectures actuated through 3D cable robotics. Within this larger framework, several sub-questions have been advanced.

- Digital design environment: what new hardware and software interfaces can be developed that would enable both the design and the physical motion control of a kinetic tensile system directly from the physics engine-based simulation model?

- Robotic 3D printed cable nets: is it possible to reliably 3D print lightweight and deformable cable net surfaces using materials such as thermoplastic elastomers? How would these surfaces perform under load and what are their formal potentials?

- Flat-to-form geometric method: given the 3D printing manufacturing technique chosen, the geometric challenge of this project was to create an elastic net geometry that could be extruded onto flat surfaces and that could then be stretched into anticlastic forms. Another research question was whether the surface topology could be manipulated to achieve variable elastic behaviour in order to attain more specifically tailored forms which diverge from purely (mathematically) minimal surfaces.

Research methods

Digital design environment

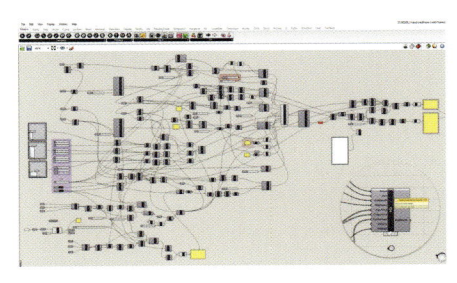

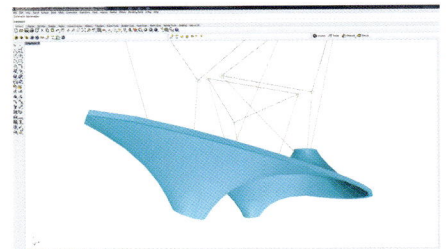

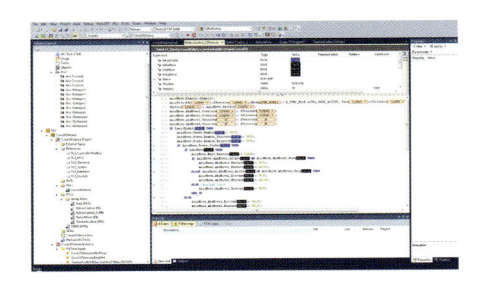

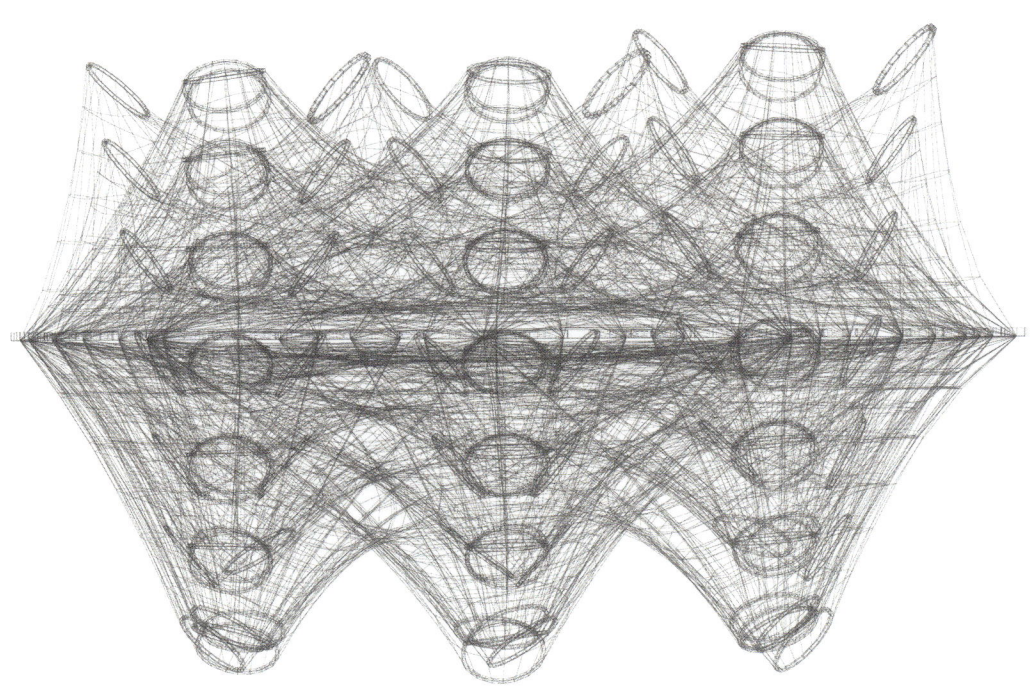

One of the goals of this research has been to develop a physics-based simulation model that could control the operation of the dynamic system in real time. Typical machine control systems only consider kinematics, loosely defined as 'motion parameters'. In contrast, dynamic simulations consider the time-varying phenomena and interactions between motions, forces and material properties. Dynamic simulations are increasingly being applied in robotic control applications, and they have the potential to more accurately predict the true state of the system, creating a 1:1 relationship between the physical system and the simulation. The work developed a custom code written for the Grasshopper extension of Rhinoceros which utilises the Kangaroo physics engine to actively visualise possible forms of the tensile surface. Additionally, a Grasshopper code was developed to allow Ethernet-based communication between the physics simulation/design environment and the TwinCAT motion control software which governed the operation of the motors.

Using the TwinCAT industrial platform has a number of advantages over more commonly used microcontroller-based systems. TwinCAT uses IEC 61131-3 programming, which is standardised across all industrial control systems, as well as the PLCOpen standard, which supplies hundreds of typical motion blocks. This makes the system inherently scalable and portable to other platforms that adhere to the standard (virtually all industrial control manufacturers do so). Additionally, it provides a high performance, allowing cycle times for the system to be in the sub-millisecond range, giving smooth control and instantaneous update to new inputs. TwinCAT also supplies an API (application programming interface), which allows for the easy development of a Grasshopper plug-in to allow Ethernet-based communication between the physics simulation/design environment and the controller, enabling the possibility of fluidly exploring virtual and physical prototyping and operation of deformable structures (Fig. 2). These advancements in the computational simulation and control environment greatly expand the options available for designers.

Robotic 3D printed cable nets

In order to produce the tensile surface, the team pursued the fabrication of a monolithic elastic net fabricated through robotic extrusion of thermoplastic elastomer (TPE). To enable the manufacture of the surface, an existing polyolefin extruder was modified to be servo-driven and mounted to a 7-axis robot (Fig. 3). The design included a specialised hopper for pelletised thermoplastics to feed the pellets into a screw-driven extruder. The SuperMatterTools robotic control software, co-developed by author Wes McGee (Mcgee & Pigram, 2011), was used to direct the toolpath of the robot. The liquid TPE is deposited onto a heated 1,200 x 2,400mm aluminium bed to facilitate joint fusion and then allow for controlled cooling into a monolithic textile surface (Fig. 4).

For the purposes of this project, research had to be undertaken into materials that had high elongation and that could be processed using thermal extrusion. TPEs are a physical mix of thermoplastic and rubber. Like most thermoplastics, they have the potential to be recycled and reused. They are available in a broad range of durometers and melt flows. The melt flow index is a measure of the viscosity of a polymer at specific temperatures in the melting region. TPEs were tested across a range of melt flows in an attempt to balance the characteristics of the extruded bead with the ability to produce a void-free crossing joint in the mesh. A material with low melt flow coupled with a low durometer was chosen for its consistency in processing while maximising the deformation potential of the surface. The use of extruded TPE as the surface material allowed for unique discoveries and a novel geometric approach.

3. Custom robotic thermoplastic extruder end effector.

4. Robotic extrusion of TPE on 4 x 8ft thermally controlled aluminium surface.

5. Geometric methodology to derive a flat printed tailored cable net mesh (above) and 3D printing a specific section of the mesh in TPE (below).

Images: © RVTR / Matter Design 2016.

Flat-to-form geometric method

This project sought to create an elastic cable net geometry that could be extruded onto flat surfaces and that could then be stretched into tailored anticlastic forms. Working with the elastic stiffness parameters for the mesh springs, and based on the actual behaviour of the extruded TPE strands, iterative dynamic relaxation in Kangaroo was used to simulate how the final mesh would deform when loaded. As has been noted previously, mesh topology has a profound influence on the behaviour of the loaded form (Hernandez et al., 2013). Physics-based simulations between a base quadrilateral mesh versus a diagrid mesh indicated that the diagrid mesh would achieve the more acute funnel-shaped forms that were desired for the project. The difference in performance between the quad and diagrid meshes is due to the fact that, in the case of the diagrid, the mesh edges carry loads from ring to frame in both directions. The load paths spiral around the ring, producing more dramatically curved catenoid forms upon stretching, as compared to the conical forms created with the quad mesh's simple point-to-point load transfer.

The question of how to achieve further tailoring of the stretched forms was approached through the basic principle of introducing a curve instead of a line between two nodal points, so that a greater working length under loading could be achieved. By setting a target length derived from the loaded mesh form, it would therefore be possible to control the resultant length of every individual mesh edge, making it possible to programme a range of naturally stressed three-dimensional forms into the flat pattern. Using Kangaroo's spring-based physics, the desired edge length of each vertex of the mesh is 'grown' (the edge difference was slowly stepped up from the 2D length to the 3D length by iteratively increasing a multiplier value from 0 to 1 by 0.01 each time while dynamically solving the physics simulation) into the geometry of the line. The scripted equation used was: edge length = $(i * \Delta L) + 2D$, where 'i' is the iterative multiplier from 0 to 1. Limitations were programmed into the model to more reliably approximate the physical behaviour of the welded connections produced by the robotic extrusion process, and a collision avoidance component was integrated to maintain the mesh topology and account for the physical properties and dimensions of the extruded TPE bead (Fig. 5).

As test prototypes showed, the process works in concert with the material's inherent flexibility, so that when loaded the programmed curves straighten into lines and then relax back into curves when the load is removed, helping the mesh to maintain tension across a number of different states.

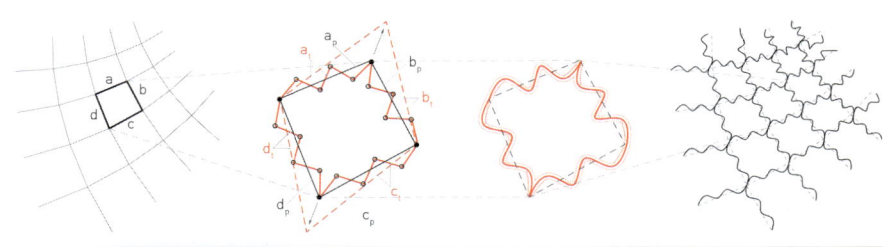

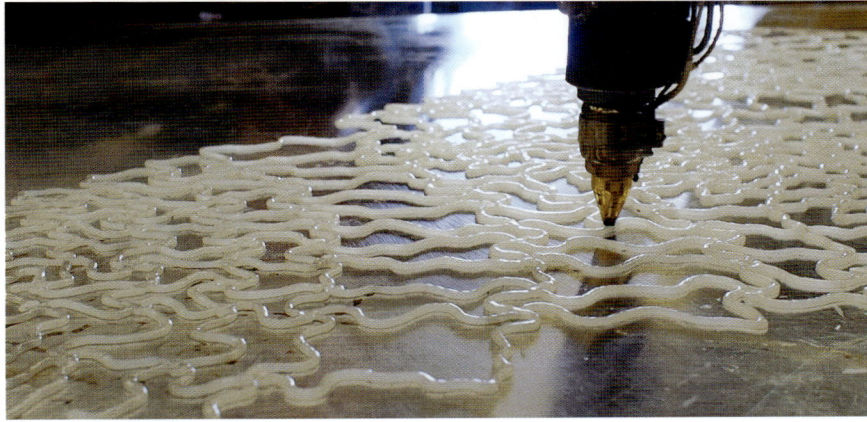

Research evaluation

The work of this project identifies the prototype as the primary mode of evaluation within the 'knowledge-based design' (Coyne, 1990) methodology of this project. The prototype installation of the *Infundibuliforms* project not only allowed the team to test the complex interactions between mesh geometry, its fabrication, the cable robot operation and the design and control environment, but also introduced additional design and implementation frameworks that became productive feedback for the iterative development of the research.

The prototype installation consists of a 28m² 3D printed TPE cable net surface spanning between an outer ovoid ring and three weighted inner rings. Due to the scale of the installation and the manufacturing constraints of the extrusion bed, the surface was subdivided into panels that could be individually fabricated and would then be mated at the seams. To increase the tension at the perimeter, as well as between the individual catenoids, the mesh was subdivided into smaller cells with less curvature between nodes, decreasing the amount of deformation in these areas and increasing the curvature angle of the catenoid. Conversely, cell size and internodal length were increased closer to the inner rings, enabling a greater elongation of that portion of the catenoid. This method allows for formal refinement of the surface through the manipulation of the surface geometry

6

aspects of the mesh geometry, fabrication method, motor design and control were refined and reworked, making the prototype a live testbed for the project. One of the most significant changes between the first and second prototypes was the redevelopment of the mesh geometry to produce more dramatic funnel-shaped forms, as well as the introduction of a reinforced saddle zone between the control ring zones. This zone is extruded with a harder, stiffer TPE material (Shore A hardness of 68, compared to 35 for the rest of the piece) and is intended to maintain tension through the middle of the piece to limit self-load deflections (Fig. 7).

Future directions

The increasing capacity for designers to use robotic manufacturing techniques in order to programme performative capabilities into materials has the potential to expand new possibilities for architectural geometries and physical forms. This paper has presented a novel fabrication and geometric development coupled with a form-control method for elastic net surfaces that is informed by material feedback and the advancement of additive manufacturing techniques for such surfaces. In addition, this work has begun to advance the capabilities of real-time simulation models that can communicate with industrial control systems for dynamic architectures.

Within this specific project, it is also possible to identify a number of areas, both immediate and more distant, that may be pursued with further work. These include:

- Empirically verifying the fidelity of the physical prototype to the computational model by 3D scanning the prototype (such as through LiDAR technology). Since the dynamic computational model is also used to directly control the positions of the individual motors, there is confidence that the spatial location of the inner and outer boundary rings is true. What is less apparent is the geometric accuracy of the surface, relative to the simulation.

- The inclusion of sensors to enable closed loop interactivity with the system. The industrial PLC platform provides an ideal low-level framework for discrete sensors, but higher-level systems (such as depth map cameras like the Kinect) which integrate with the computational model though an API are also a possibility for future exploration.

subdivision and internodal length development. The entire mesh was pre-tensioned by being scaled to 92% of the target surface area, so that it would remain in a tense position when under gravity loads.

Three servo motors control the tilt and position of the outer ring, and three triangulated sets of custom-built stepper motors position the weighted rings located at the end of each infundibulum. These can dynamically reconfigure the surface between vaulted and chimney forms within a broad range of positions (Fig. 6). The full-scale installation also included a pair of custom-designed and built control cabinets, with 12 interpolated axes driven by an industrial embedded PC.

The first *Infundibuliforms* prototype was installed at the University of Michigan Taubman College Liberty Annex in March 2016. A second prototype was reinstalled in the gallery for the ACADIA 2016 exhibition in October 2016. Between the first and second installations, a number of

With this framework, this research has attempted to move toward the integrated development of large-scale kinetic material systems for architecture.

7

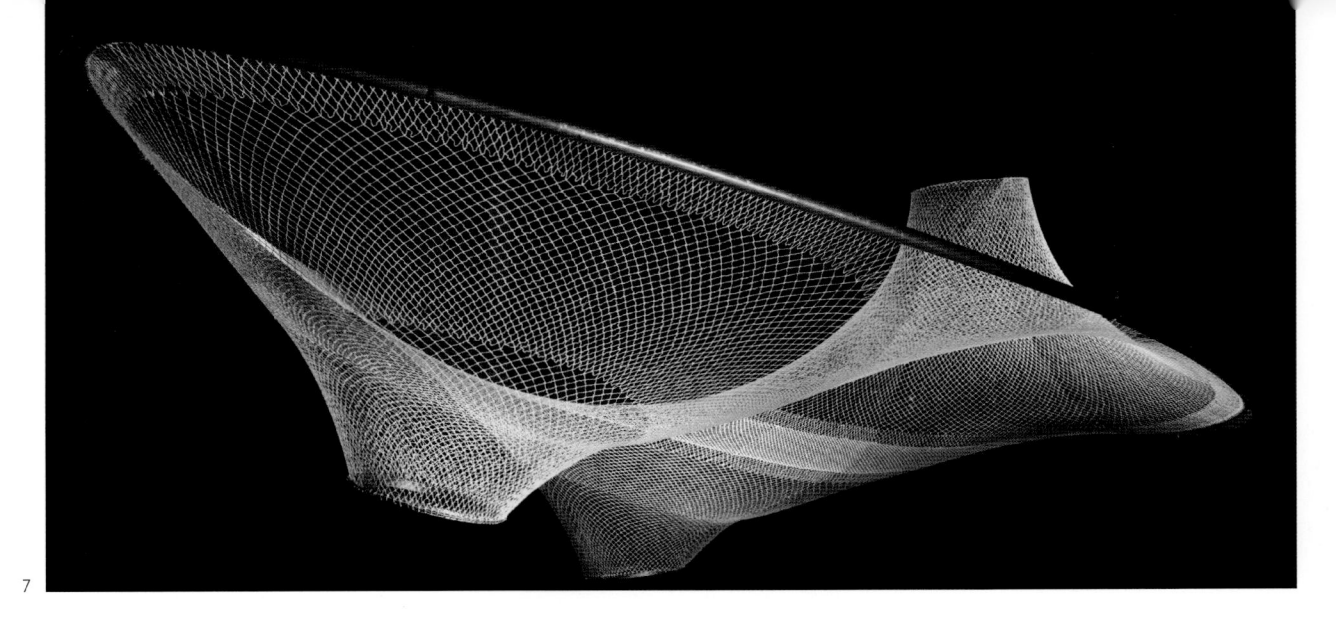

Project Video: https://vimeo.com/173666490

Acknowledgements

Fabrication assistants: Asa Peller, Dustin Brugman, Andrew Kremers, Andrew Wald, Iram Moreno Pinon, Isabelle Leysens.
Technical partners: Buckeye Polymers, Industrial Fabricating Systems, Beckhoff.

Funding: Taubman College of Architecture and Urban Planning: 2016 Research Through Making Programme, University of Michigan Office of Research: Small Projects Grant.

References

Adrianssens, S. and Rhode-Barbarigos, L., 2013, 'Form-Finding Analysis of Bending-Active Systems Using Dynamic Relaxation', in Bletzinger, K-U., Kröplin, B. and Oñate, E. (eds.), *VI International Conference on Textile Composites and Inflatable Structures: STRUCTURAL MEMBRANES 2013*, p.51-61.

Ahlquist, S. and Menges, A., 2010, 'Realizing Formal and Functional Complexity for Structurally Dynamic Systems in Rapid Computational Means: Computational Methodology based on Particle Systems for Complex Tension-Active Form Generation', in Ceccato, C., Hesselgren, L., Pauly, M., Pottman, H. and Wallner, J. (eds.), *Advances in Architectural Geometry 2010*, New York, SpringerWein, p.205-220.

Bletzinger, K-U., Wüchner, R., Daoud, F. and Camprubi, N., 2005, 'Computational Methods for Form-Finding and Optimization of Shells and Membranes' in *Computational Methods in Applied Mechanics and Engineering*, 194 (30-33), p.3438-3452.

Coulter, F.B. and Ianakiev, A., 2015, '4D Printing Inflatable Silicone Structure' in *3D Printing and Additive Manufacturing*, 2(3), p.140-144.

Coyne, R.D., Rosenman, M.A., Radford, A.D., Balachandran, M. and Gero, J.S., 1990, *Knowledge-Based Design Systems*, Reading, Addison-Wesley Publishing Company.

Deuss, M., Holden Deleuran A., Bouaziz, S., Deng, B., Piker, D. and Pauly, M., 2015, 'ShapeOp – A Robust and Extensible Geometric Modelling Paradigm' in Ramsgaard Tomsen, M., Tamke, M., Gengnagel, C., Faircloth, B. and Scheurer, F. (eds.), *Modelling Behavior*, Springer, p.505-515.

Helm, V., Willmann, J., Thoma, A., Piškorec, L., Hack, N., Gramazio, F. and Kohler, M., 2015, 'Iridescence Print: Robotically Printed Lightweight Mesh Structures' in *3D Printing and Additive Manufacturing*, 2(3), p.117-122.

Hernandez, E.L., Sechelmann, S., Rörig, T. and Gengnagel, C., 2013, 'Topology Optimization of Regular and Irregular Elastic Gridshells by Means of a Non-linear Variational Method' in Hesselgren, L., Sharma, S., Wallner, J., Baldassini, N., Bompas, P. and Raynaud, J. (eds.), *Advances in Architectural Geometry 2012*, New York, SpringerWeinNewYork, p.147-160.

Keating, S. and Oxman, N., 2013, 'Compound Fabrication: A Multi-Functional Robotic Platform for Digital Design and Fabrication' in *Robotics and Computer-Integrated Manufacturing* 29, p.439-448.

Khoo, C.K., Salim, F. and Burry, J., 2011, 'Designing Architectural Morphing Skins with Elastic Modular Systems' in *IJAC* 4(9), p.397-419.

Killian, A. and Ochsendorf, J., 2005, 'Particle-Spring Systems for Structural Form Finding' in *Journal of the IASS*, 45 (147).

Lienhard, J., Schleicher, S., Proppinga, T., Masselter, M., Speck, T. and Knippers, J., 2011, 'Flectofin: a Hingeless Flapping Mechanism Inspired by Nature' in *Bioinspiration and Biomimetics* 6, p.1-7.

McGee, W. and Pigram, D., 2011, 'Formation Embedded Design: A Method for the Integration of Fabrication Constraints into Architectural Design' in *Proceedings of ACADIA 2011*, Banff, Canada, p.122-131.

Otto, F. (ed.), 1972, *IL 5 – Wandelbare Dächer/Convertible Roofs*, Mitteilungen des Instituts für Leichte Flächentragwerke (IL), Universität Stuttgart.

Veenendaal, D. and Block, P., 2012, 'An Overview and Comparison of Structural Form Finding Methods for General Networks' in *International Journal of Solids and Structures* 49 (26), p.3741-3753.

Velikov, K. and Thün, G., 2014, 'Towards an Architecture of Cyber-Physical Systems' in Gerber, D. and Ibanez, M. (eds.), *Paradigms in Computing: Making, Machines and Models for Design Agency in Architecture*, Los Angeles, eVolo/ActarD, p.330-341.

Wagner, R., 2005, 'On the Design Process of Tensile Structures' in Oñate, E. and Kröplin, B. (eds.), *Textile Composites and Inflatable Structures I*, Netherlands, Springer, p.1-16.

Note

A previous version of this introductory text was published in Velikov, K., Manninger, S., del Campo, M., Ahlquist, S. and Thün, G., 2016, *Posthuman Frontiers: Projects Catalog of the 36th Annual Conference of the Association for Computer Aided Design in Architecture*.

6. Photograph of *Infundibuliforms* prototype installation taken through one of the control rings.

7. *Infundibuliforms* prototype, second iteration in position with two infundibula down and one up, October 2016.

Images: © RVTR / Matter Design 2016.

ROBOTIC INTEGRAL ATTACHMENT

CHRISTOPHER ROBELLER / YVES WEINAND
Laboratory for Timber Construction IBOIS – EPFL
VOLKER HELM / ANDREAS THOMA / FABIO GRAMAZIO / MATTHIAS KOHLER
Gramazio Kohler Research, ETH Zurich

Integral joints provide a rapid, simple and mechanically strong connection between parts. Our investigation focuses on the assembly of cross-laminated wood veneer plates, where previous studies have shown that the strength of through-tenons is equivalent or superior to state-of-the-art fasteners such as screws or nails. This mechanical behaviour is highly dependent on a precise fit of the joints, where no gaps are left between the parts.

However, the manual assembly of such tight-fitting joints can be complicated. Thanks to its rectangular cross-section profile, a single through-tenon joint is a sufficient assembly guide for an entire plate, but multiple through-tenons are required to establish a mechanically strong connection. This results in a kinematically over-constrained assembly motion (Mantripragada et al., 1996). Additionally, due to fabrication- or material-related tolerances, the joints can be too tight-fitting and manual assembly motions deviate from the precise insertion path. So-called 'wedging' occurs during the assembly of tight-fitting joints, especially with larger parts at a building scale (Fig. 1). This requires high forces to be overcome.

Rather than leaving gaps between the parts, which presents one solution for the manual assembly of such systems, we investigate the idea of assembly using an industrial robot. The robot allows for a more precise assembly motion and the application of higher forces in the direction of assembly. The aim of this research is to use these benefits along with the compressibility of wood for the assembly of oversized tenons. While in regular through-tenon joints the width of the tenon is equal to the width of the slot, the oversized tenons in this paper are slightly wider than their slot parts. This assembly will require a certain insertion force, squeezing the tenons into the holes, but the resulting connection will be tight-fitting without any gaps.

Robotic assembly

Robotic integral attachment demonstrates the advantages of combining robotic assembly (Helm et al., 2016) and integral mechanical attachment, such as through-tenon joints. Both methods are used to facilitate the assembly of complex architectural designs, such as freeform shells and space frames. While integral

1

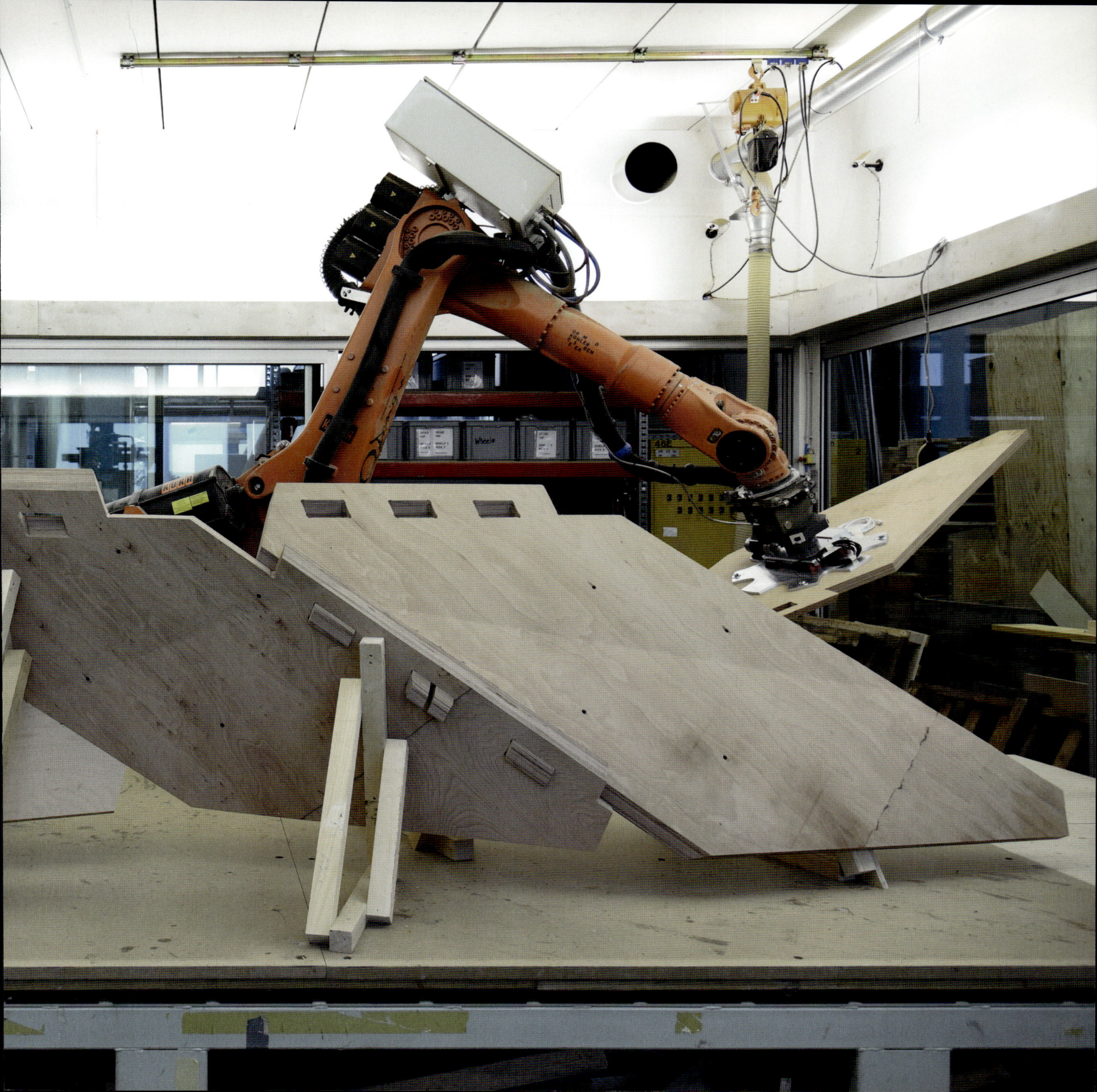

Integral mechanical attachment allows for a simple, fast and precise onsite joining process. It transfers the complex and laborious aspect of assemblies into the prefabrication of the plates. This is made possible through computational design and automatic prefabrication technology. As a consequence of such improved joining strategies, more complex shapes can be efficiently produced and assembled. Robeller and Weinand (2015) provided an example in which a singly-curved folded surface structure using equally shaped parts and regular joints was compared with a doubly-curved folded surface structure with individually shaped plates and integral joints. In a simulation, deflections were 39 percent lower on the double-curved structure, due to the integral joints.

At the same time, integral joints improve structures through the direct transfer of forces through the form of connectors. Roche et al. (2015a) showed that the shear strength of finger- and dovetail-jointed plywood plates is similar to the shear strength of screwed connections. Li et al. showed that the connectors can be combined with metal fasteners, and further research by Roche et al. (2015b) compared the rotational stiffness of joints at ridges, demonstrating the particular strength of through-tenon joints.

Aiming at the automated assembly of timber plates and the elimination of any gaps that would reduce the stiffness of the joints, the main question was to ascertain what forces would occur during the insertion of through-tenon joints, both with and without oversized tenons.

Further questions arose due to the fact that, during the assembly of timber plate shell or folded plate structures, multiple joints must often be inserted simultaneously. These were: how the insertion forces on individual joints could be reduced through modifications in their form; how the forces would add up during such multi-joint assemblies; and how insertion forces and possible wedging could be reduced through custom-built vibration-inducing robot end effectors.

- What force is required for the insertion of a through-tenon joint?
- What force is required for the insertion of a through-tenon joint with oversized tenon?
- Can the effect of wedging be reduced through optimisations in the form of the joints?
- Can the effect of wedging be reduced through automated robotic assembly?

attachment embeds the instructions for manual assembly into the form of prefabricated components (Fig. 2) (Robeller, 2015), robotic assembly integrates the assembly logic into the robotic positioning procedure (Gramazio et al., 2014). The aim of this research is to investigate the combination of these seemingly contrary methods.

The two main benefits of integral joints for the design of timber plate shell and spatial structures are their so-called locator and connector features. Locator features, in the form of the joints, reduce their mechanical degrees of freedom and therefore also the relative motions between the connected parts. This allows for the indication of the correct alignment and position of parts to one another. While some joint shapes, such as finger joints, will reduce the mobility of parts to three degrees of freedom and perform as partial assembly guides, other joints, such as through-tenons, will reduce the mobility of parts to only one insertion direction and perform as fully integral assembly guides.

1. Large-scale robotic positioning of a timber plate.

2. Timber folded plate built from 21mm LVL panels, assembled manually, IBOIS, EPFL, 2014.

3. Simulation of the assembly sequence and plate insertion paths.

Experimental set-up

The robotic assembly of elastic and plastic through-tenon joints for cross-laminated wood veneer plates was examined through physical assembly experiments. Using 40mm-thick beech laminated veneer lumber (LVL) plates and a tenon width of 120mm for all specimens, different joint shapes and parameters were tested.

Elastic joining techniques like cantilever snap-joints are commonly used in other industry sectors, such as consumer electronics or the automotive industry (Messler, 2016). They can be generally applied to elastic materials. The application to cross-laminated wood panels has also been previously investigated (Robeller et al., 2014).

Plastic joining techniques are also commonly used in various industrial applications, especially in the form of press-fit or friction-fit joints. A well-known example using metal materials is staking, where an undersized boss in a regular-sized hole is expanded through a staking punch. The resulting radial expansion will cause a physical interference fit between the two pieces. Metal screws work in a similar way: the thread of the screw creates a large friction surface, while pressure is applied through its inclination and rotation. In addition to the friction interference, the elasticity of the material plays an important role in plastic interlocks, too. Through the press-fit, the parts of the joint are squeezed. The elastic recovery force will apply pressure on the contact surfaces, which further increases the friction interference.

The primary purpose of the plastic and elastic timber plate joints in this investigation is to eliminate gaps, which may be required for the assembly of joints. The regular rigid joints were added as a reference for comparison. This elimination of gaps should be achieved through a press-fit assembly of tenons that are slightly wider than their slots. Multiple series of specimens were tested, where the tenon oversize was increased in small steps: 0.05mm, 0.10mm, 0.15mm, 0.20mm and 0.25mm. During the insertion of the joints, the oversized tenons should be able to fit into the slots primarily due to the material compressibility on the rigid-type through-tenons, and predominantly due to the material elasticity on the elastic-type through-tenons. Here, cuts along the centre line of the tenon allow for lateral deflections during the joint assembly.

Due to the centre line cut on the elastic through-tenons, their shear strength will be greatly reduced in comparison to rigid versions. However, the elasticity was also expected to greatly reduce the required insertion force. Such elastic joints may be particularly interesting in combination with rigid or plastic interlocks (see Fig. 3, the plate held by the robot), providing ideal locator features while requiring reduced insertion forces.

The primary challenge in the assembly of the oversized joints is the so-called effect of wedging, where a friction interlock is established between the two parts during the insertion before the final position is reached. This occurs due to tolerances in the size of the parts, resulting from fabrication imprecision or dimensional changes due to changing environmental conditions, as well as imprecisions in the assembly motion, which must follow one precise path in the case of single-degree-of-freedom joints, such as the through-tenons.

It was expected that the wedging could be reduced through a small inclination of 1° on the small contact faces across the edge on the through-tenons and on the slots. We can achieve an inclination on these faces using a 5-axis CNC router, where the tool is inclined at 1° for the cutting of the slot part. However, the other two contact faces along the edge of the joint cannot be inclined, as those lie on the top and bottom of the cross-laminated wood plate and cannot be easily cut without turning and re-clamping the work pieces.

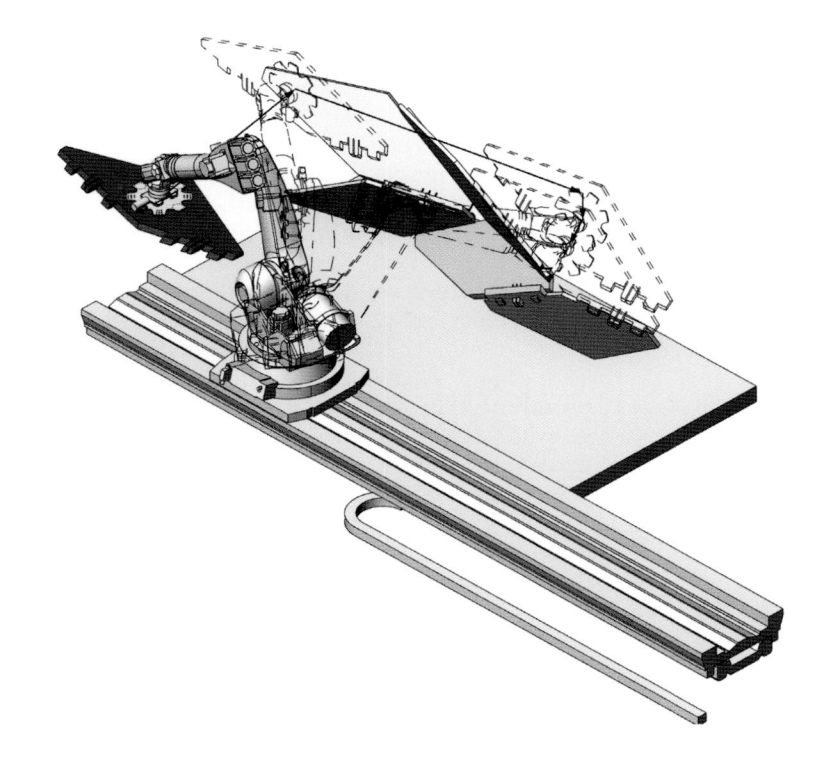

For all assembly tests, a 6-axis industrial robot with a maximum payload of 150kg and an additional seventh linear axis was used to insert the through-tenons (Fig. 3). In the first series of single-joint assembly tests, the slot plates were fixed to a concrete block. The insertion motion was then carried out parallel to the robot's additional linear axis for the single-joint assembly tests. A custom end effector was built with an integrated force measurement device, from which the pressure values were recorded during the assembly motion.

Following the first series of single-joint assembly tests, the assembly of multiple joints was tested on six full-scale plates out of a folded roof structure[1], which was generated with the computational tools presented by Robeller and Weinand (2016).

The multi-plate robotic assembly experiment investigates the offsite robotic assembly of prefabricated segments, which would fit on standard-size trucks for transport to the construction site. With such a prefabricated assembly, 85 percent of the total edge joints in the case study roof would be assembled automatically with robots, while the remaining 15 percent of the edges must be joined onsite with state-of-the-art connectors.

The main challenge in this multi-plate assembly experiment was the cumulative force required for the insertion of entire plates, as well as an increased effect of wedging due to the simultaneous assembly of six through-tenons per plate. A custom end effector was built to hold the plates with an integrated device for the measurement of forces (Fig. 5). This effector was also equipped with an integrated 'vibration-assisted assembly' device for the introduction of vibrations into the plates, in order to reduce the effect of wedging.

The first series of single-joint assembly tests showed the expected increase of insertion forces, along with an increasing oversize of the plastic through-tenon joints. The smallest oversize of 0.05mm would result in a required insertion force of 0.7kN. At an oversize of 0.15mm, we recorded 0.8kN, while the two largest oversizes of 0.20mm and 0.25mm required much larger forces of 1.08kN and 1.57kN. Additionally, the effect of wedging increased along with the oversize value. The inclination of the joint faces across the edge at an angle of 1° resulted in a greatly reduced effect of wedging.

The multi-plate assembly test showed that the simultaneous assembly of six through-tenon joints requires the vibration device to be activated in order to avoid a premature friction-based interlock. Furthermore, the test showed that an additional 'pulse' force in the joints' assembly direction is beneficial in combination with the vibration device. During the tests, this force was applied manually with a mallet. The plates used featured two centre elastic locator tenons and four outer plastic connector tenons.

4. Assembly sequence of the large-scale prototype.

5. A custom-built end effector equipped with a load cell for force measurement and a vibration device.

Boosting the benefits

With their locator and connector features, integral timber plate joints offer considerable benefits for the design of timber plate structures, such as segmented plate shells or folded plates. While the mechanical strength of the joints requires them to be tightly fitted, this can be problematic for the assembly of such kinematically overconstrained joints.

The elastic and plastic interlocks presented in this paper demonstrate how the material properties of compressibility and elasticity can be exploited for an assembly technique that fully eliminates any gaps, through the insertion of oversized tenons. While this basic concept of plastic interlocks is commonly used in mechanical fastening techniques, such as screws and bolts, this paper first applies this concept to the integral attachment of through-tenon-jointed timber plates. This is made possible through the precise assembly motion of an industrial robot, as well as the possibility of it applying an insertion force. Here, the single-joint assembly test series first provides values on the required forces.

Since the assembly of structures such as timber plate shells requires the simultaneous assembly of multiple edges and therefore also multiple joints, it is crucial to estimate the total required insertion force per plate. The multi-plate assembly tests have shown that the assembly of building-scale plates from our case study project is possible with an additional vibration-inducing device. Further research is required into the addition of a pulse force, similar to a jackhammer, which can be induced in the plate's direction of assembly.

Acknowledgements

This research was supported by the NCCR Digital Fabrication, funded by the Swiss National Science Foundation (NCCR Digital Fabrication Agreement #51NF40-141853).

The authors thank their team for their pioneering efforts, particularly Michael Lyrenmann, Luka Piškorec, Laszlo Blaser, Victor Stolbovoy, Stéphane de Weck and Francois Perrin.

References

Gramazio, F., Kohler, M. and Willmann, J., 2014, *The Robotic Touch: How Robots Change Architecture*, Zürich, Park Books.

Mantripragada, R., Cunningham, T.W. and Whitney, D.E., 1996, 'Assembly oriented Design: A New Approach to Designing Assemblies', *Proceedings, IFIP WG5.2 Workshop on Geometric Modeling in CAD*, 19-23 May 1996.

Helm, Volker et al., 2016, 'Additive Robotic Fabrication of Complex Timber Structures' in Menges, A., Schwinn, T. and Krieg, O.D. (eds.), *Advancing Wood Architecture: a Computational Approach*, New York, Routledge, p.29-43.

Messler, R., 2006, *Integral Mechanical Attachment: A Resurgence of the Oldest Method of Joining*, Butterworth Heinemann.

Robeller, C., Mayencourt, P. and Weinand, Y., 2014, 'Snap-fit Joints – CNC fabricated, Integrated Mechanical Attachment for Structural Wood Panels', ACADIA 2014: 34th Annual Conference of the Association for Computer Aided Design in Architecture, Los Angeles, California, USA.

Robeller, C., 2015a, 'Integral Mechanical Attachment for Timber Folded Plate Structures', EPFL PhD Thesis, Lausanne.

Robeller C. and Weinand, Y., 2015b, 'Interlocking Folded Plate – Integral Mechanical Attachment for Structural Wood Panels', *International Journal of Space Structures*, Vol. 30, No. 2, p.111-122.

Robeller, C. and Weinand, Y., 2016, 'Fabrication-Aware Design of Timber Folded Plate Shells with Double Through-Tenon Joints', *Robotic Fabrication in Architecture, Art and Design 2016*, p.166-177.

Roche, S. et al., 2015a, 'On the semi-rigidity of dovetail joints for the joinery of LVL panels', *European Journal of Wood and Wood Products*.

Roche, S. et al., 2015b, 'Rotational stiffness at ridges in folded plate structures', in *Elegance of Structures: IABSE-IASS Symposium*, Nara.

Notes

1. The case study roof structure covers an area of 700m^2. Its doubly-curved surface spans over 20m between two 35m line supports, with a span-to-rise ratio of 4 in the centre point and 8 at the front and end. The surface was segmented with a Miura-Ori fold pattern of 16 plates in the direction of span, 40 plates along the supports and a static height of 650mm. The average dihedral angle between the 640 plates is 130°, with a maximum of 160°, and the average edge length is 1.5m, with a maximum of 1.8m.

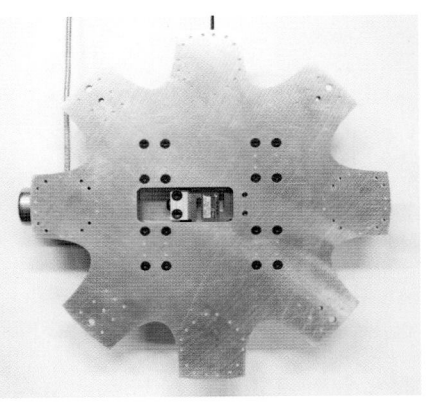

LACE WALL
EXTENDING DESIGN INTUITION THROUGH MACHINE LEARNING

MARTIN TAMKE / MATEUSZ ZWIERZYCKI / ANDERS HOLDEN DELEURAN / YULIYA SINKE BARANOVSKAYA / IDA FRIIS TINNING / METTE RAMSGAARD THOMSEN
CITA | Centre for Information Technology and Architecture, The Royal Danish Academy of Fine Arts, Copenhagen

Lace Wall explores how design-integrated simulations of real world behaviour of building elements and machine learning allow the design and manufacture of large-scale resilient material systems from a minimal inventory of elements: 8mm glass fibre-reinforced plastic rods, textile cables and custom-designed HDPE elements to join cables and rods together. The rods are bent and joined into discrete units stabilised by an internal three-dimensional cable network. 80 units are connected into a 12m-long, 7m-high wall (Fig. 2). While the geometry of the rods is identical, it is the differentiation of cable networks which allows the single units to stand the divergent local strains in the structure and to constrain each individual unit into individual geometries that fit into a desired overall macro shape. The high interdependence of elements and scales in the structure permits the use of established design optimisation strategies to find the specification for the cable networks. In order to explore this, we developed methods that combined lightweight simulation, physical models and machine learning in order to evaluate multiple interdependent design parameters and finally establish a machine-enhanced intuition which is good enough to specify structures that behave as expected.

Building complex geometries

Lace Wall belongs to the family of form-active hybrid structures (FAHS), which allow for the building of complex geometries with hardly any machining effort or waste material (Tamke, 2013, Lienhard, 2014, Holden, 2016). The integration of restraining tensile elements, such as membranes or cables, increases their structural performance (Alpermann 2012). In the case of the units of *Lace Wall*, this increases (in comparison to bending-active-only structures) the possible design space in a dramatic way, as it introduces an added dimension to stabilise, constrain and join elements.

Approaches towards supporting form-active hybrid structure design

The recent efforts of the research community towards approaches that support design, form-finding and structural analysis of form-active structures had a predominant focus on systems of either tensile or compressive members (Menges, 2012, Tamke, 2013). Hybrid systems of interdependent tensioning (rods)

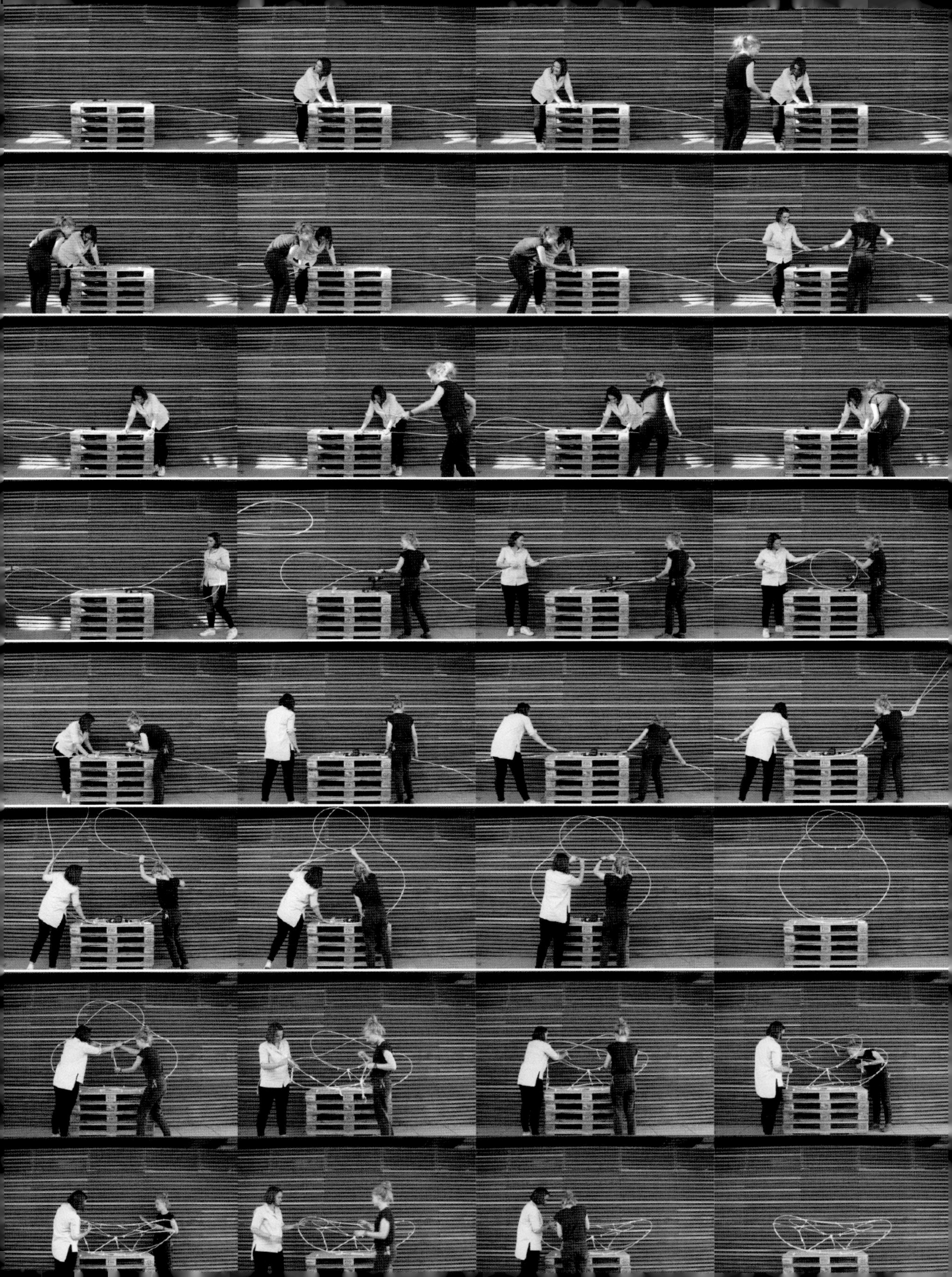

2

1. Assembly process of a single unit and its bespoke cable networks.

2. *Lace Wall* demonstrator during the Complex Modelling exhibition, Copenhagen, Sept-Dec 2016.

3. *Lace Wall*, form-found solution of 80 units with cable networks.

4. Single unit FAHS unit. Initial polyline model with box defining the tiling target (blue), form-found solution (K2), display of forces in the form-found solution (K2E).

and restraining (cables) elements present challenges on all levels from conception to fabrication:

- Existing design approaches are based on explorations with small-scale physical models (Lienhard, 2014, Holden, 2016), which build up an intuition and design repertoire on the part of the designer. However, even small changes in the topology, the length of rods and cables or the position of restraining elements immediately affect the resulting shapes. The sheer number of combinations and interrelations soon creates fatigue on the part of the designer, and the scaling up to large-scale arrays of interacting bending-active units becomes a problem.

- Methods of designing bending-active structures in digital design environments have been a focus of research for several years (Lienhard, 2014). Computational design and analysis tools have only recently emerged that are fast and stable enough to calculate the interaction of many bending-active and restraining tension elements (Ramsgaard Thomsen, 2015). These advances are based on a shift in the underlying computational approach from dynamic relaxation (Day, 1965) to projection-based methods (Bender, 2014), which can include physical dynamics and elastic materials as well (Bouaziz, 2014). The Kangaroo 2 plug-in (Piker, 2016) is based on these methods and has a solver that is stable for arbitrarily high stiffness values, unlike the explicit integration methods used in the earlier versions of Kangaroo. It can be extended through the definition of new goals in a straightforward way, allowing the use of any function which returns a target position even during the run-time of the solver. This allows us to integrate simulation with a sufficient prediction of stress and strain in the form-found structure into the design environment (Quinn, 2016) and most importantly to change the topology of the structure constantly (Deleuran, 2016). The ability to work with open topologies, to continuously add and remove bending-active and tension elements and to find physically correct solutions allows a systematic exploration of options through the designer. Similar to the work with physical models, the designer can build up an intuition about promising design directions and explore them quickly. These explorations can also take place through the automated generation of design options and a subsequent evaluation and reiteration of the form-found solutions according to given aims, such as maximum amount of cables, lengths and relations between elements.

- However, while the methods developed here for design-integrated simulation and the work with open topologies save time compared to the work with physical models, they do not remove the underlying challenge of form-active hybrid structures: a combinatorial explosion of parameters. The sheer number of ways to combine quite simple ingredients (e.g. the units in *Lace Wall*) prohibit established means (such as brute force, evolutionary solvers or other design-related optimisation strategies) of automatically exploring the emerging design spaces here and of finding locally – or even globally – optimal solutions (Rutten, 2014). In the case of *Lace Wall*, it took around two hours to optimise only a single unit described by eight parameters. The optimisation of 80 units with 640 parameters in total and a fitness evaluation seems unviable. An underlying problem can be identified in the fact that these approaches rely on a simplified model and optimisation towards identified key parameters. These approaches require a good understanding of the system – based either on knowledge, a state which might in fact render the whole iterative search superfluous, or more probably on intuition regarding what to search for and where to do so. The question is, however, how to establish this intuition with highly interdependent, complex systems, such as the hybrid one underlying *Lace Wall*.

Machine learning as a means for design search

Machine learning has been introduced in engineering to accelerate complex simulations, as in the case of CFD, and to predict plausible complex wind interference patterns by utilising methods of supervised learning (Wilkinson, 2014). This prediction provides a quick, reliable and precise approximation of the wind interference to inform the designer and optimisation loops 200 to 500 times more quickly about the CFD characteristics of the building than traditional CFD methods.

Queries for structures in large and patchy design spaces have recently used methods of unsupervised machine learning. In Thomsen and Stasiuk (2014), the authors use k-means clustering to analyse the outcomes of the design space exploration – 60,000 bending-active structures have been sorted into clusters of high similarity based on 18 parameters. Solution space exploration is also the focus of Harding (2016), who demonstrates how Kohonen's self-organising maps (SOMs) can be used for dimensionality reduction. A use of k-means for geometry rationalisation is presented by Peña (2012), where the algorithm was used to limit the variation of 15,000 facade panels to conform in 49 families.

While the use of k-means clustering in the case of *Lace Wall* might seem a viable option, it can lead to the wrong classification of unit load cases. This is caused by the distance metric used to decide on similarity between two data points, which doesn't take into account the relationships and characteristics of their values. This makes artificial neural networks (ANNs) more suitable for the task, as these are able to account for both the variance and, more importantly in this case, the relationships between the values.

The simple plots that resulted from calculating this relationship demonstrate how a wrong classification will ultimately result in the wrong association of optimised solutions with a load case. Intuitively, the k-means clustering would categorise them as the same two units with a similar force applied in two different directions, while the neural network can distinguish both direction and amplitude of the load. This problem can be found in every construction system (including *Lace Wall*) where a change in load amplitude is not as crucial as the change in load distribution/direction.

The complexity of the units in *Lace Wall* and their hardly predictable structural behaviour (their reaction to the particular load scenario) was the main problem in developing an overall intuition about the cable networks for the rod topology used. It was this which made the definition of an edge case difficult. The linkage of machine learning with a database enables the memorisation of solutions, in order to build up a kind of experience over time. This is used in *Lace Wall* to identify the cases which are most different to known ones (edge cases), develop solutions for these and reuse/adapt solutions for the cases in between.

Development of Lace Wall

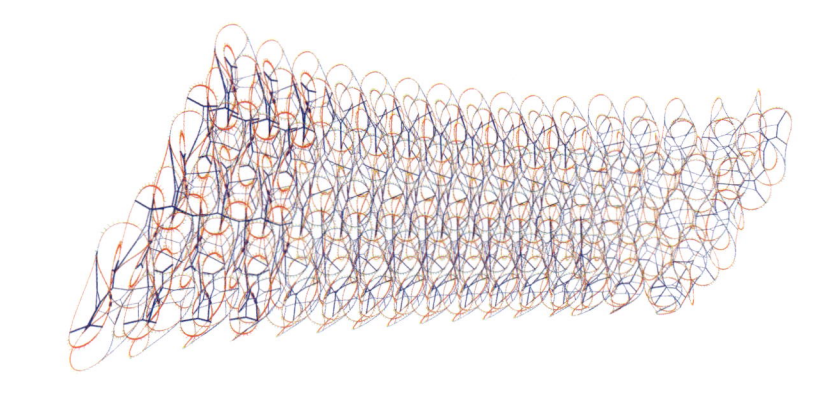

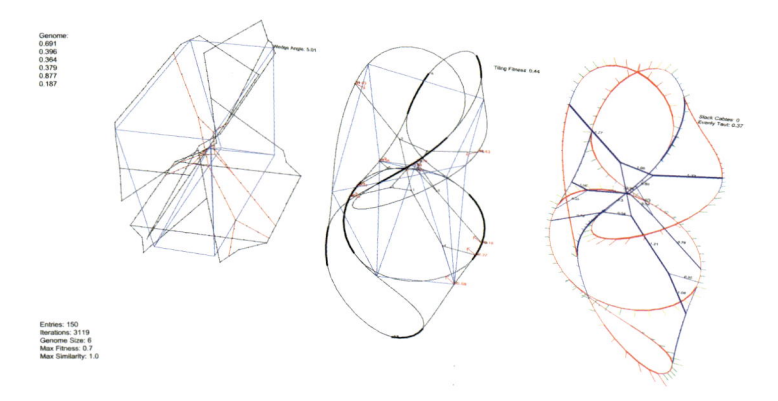

The development of *Lace Wall* started with an exploration of the ways we could create stable spatial units of rods and cables with an equilibrium of forces in the bending-active and tension elements. A further aim was to maintain a minimal and light inventory of parts – for example, in the evasion of complex mechanical joints. The collected experiences of *Tower* (Thomsen, 2015) showed that such stable and balanced bending-active structures are best achieved when they close on themselves, as the '9'-shaped ones which were finally chosen show. In *Lace Wall*, rods are furthermore only joined in parallel, which circumvents the problems of orthogonal connections that require the transfer of great moment forces.

The exploration used, in parallel, both physical models and the above-described custom-made design-integrated simulation environment. The limitations of each of these needed to be reflected in the other; for example, the digital representation of joints informed the way elements were connected in the physical assembly.

The development process included several instances of verification, where an alignment of the digital and physical models was pursued with the aim of creating a coherence between both. These processes showed that the simulation environment was able to predict the emerging shape of our hybrid units to a degree sufficient for design decision-making and fabrication. A lack of final precision in quantitative terms can be removed through the material tolerances in the system.

The development resulted in a single unit made from a set of mirrored rods fixed into a 'double 9' configuration. This is constrained into shape by a three-dimensional cable network (Fig. 1). Single units are arrayed into a diamond-shaped pattern (Fig. 3). Each unit has eight defined points at its perimeter rods that allow it to tile with equivalent points on neighbouring units (Fig. 4).

The development of the single unit and the array of many of them into a larger assembly was guided by the overall aim to create a wall-like structure. Performance goals on a local level were, however, far more fluid, emerging throughout the design process. The interaction with the digital and physical prototypes gave an intuition, rather than a certainty, about the design direction that it might be worthwhile to explore. Crucial parameters and underlying rules for the set-up of the units and the steering of their behaviour were found over many design iterations. However, the knowledge gathered therein is patchy and cannot be considered to apply across the board, as it is based on observed relationships between an introduced means and a unit's overall behaviour – for instance, the idea that a direct connection of the bend rods on each side with tension cables is beneficial for its overall behaviour (Fig. 5). A reflection on the structural set-up of the units links these observations to overarching structural rules, but the complexity of the structural elements impedes a direct linking of the single unit's behaviour to structural first principles.

5. Detailed view at assembly of single units with cable networks in *Lace Wall*.

6. Part of the generated cable networks which fulfil the developed fitness criteria.

The development process, supported by constant feedback from computational and physical models, built up an intuition on the part of the designers and helped to direct the design search. However, this intuition was weak when it came to more fine-tuned decisions. One of these was the distribution of different cable networks across the array of identical units in order to obtain the desired macro shape. A computational global optimisation, where the parameters of every element in every unit are tested and subsequently optimised, was not possible due to the aforementioned combinatorial explosion. Other means of specifying the cable networks in the 80 units according to local force conditions had to be found instead.

Our approach to fabrication using machine learning

The project followed an approach of using machine learning to identify units in the macro shape which prevent the emergence of the desired overall macro shape that we consider structurally sound. The overall stability is hence dependent on the preservation of the macro shape through preventing single units deforming too strongly from the initial shape or even collapsing under the incoming loads. A generation and analysis of new cable nets is hence possible on the level of the single unit. It was expected that the repeated picking and improvement of the structural behaviour of singular cells would, over relatively few iterations, generate an overall increase in structural performance. For this approach, a set of techniques had to be linked:

1) A method to analyse the overall structural behaviour of the single unit, as well as its assembly (customised Kangaroo 2 and the development of Kangaroo 2E).

2) A technique to generate the cable networks: an algorithm was developed, based on the findings from physical prototyping, that showed that a maximum of three cables meeting in a junction and a spatial distribution of cables was preferable.

3) A model to represent the wide range of topologies and performances that the cable networks and linked rods can take on. The encoding capitalises on the fact that the above algorithm creates a cable network unique to any set of distribution points on the rods. An effective and easy way to compare data models emerged where only the order and position of points (genome) and the related performance of the unit (e.g. the deviation from the ideal tiling geometry) needed to be stored in order to represent and, where necessary, reproduce units.

4) Modes to evaluate whether and how well a generated unit fulfils the requirements of design, structural behaviour and fabrication. These emerged during the design and prototyping phases and were verified through observations. Two sets of qualifiers were applied that describe the performance of the units:

(a) Binary ones, which any unit has to meet (durability: rod/cable angle above 50°; fabricability: no overlap of cables; structural performance: all cables tensioned; stability: units which need more than 30,000 iterations to solve tend to be unstable).
(b) Numeric ones, which allow the evaluation of the fitness of a unit (tiling fitness and middle joint position, expressed in total distance of all points to target box in range 0-1, with 0.7 as the minimum necessary to pass this filter).

A second stage evaluates the geometry of the units in order to ensure a healthy breadth of solutions: only those with a genome either better than or different to existing ones are saved for further consideration (Fig. 6).

5) A system that can perform the task of picking the units which need to be optimised (the edge conditions): an artificial neural network trained with back-propagation.

(a) The assembly of units is generated, form-found (Kangaroo 2) and structurally analysed (Kangaroo 2E).
(b) Form-finding and analysis reveal the force distribution and behaviour of the units under load. This data is used to initialise the solutions database, with the two naively picked cases (naive pick initialisation: picked by the lowest and highest sum of load values) then being optimised and saved (optimisation of a single unit with Galapagos and K2E).

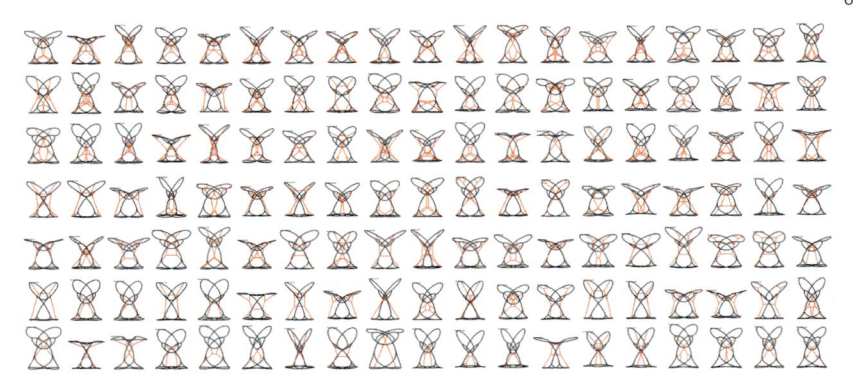

(c) The neural network indicates the cells which it cannot match (smallest output value case) to any of the known solutions (structural analysis data classification). It performs the classification based on the deviation from the 'ideal shape', discretised with ~100 input parameters.
(d) An evolutionary solver (optimisation) is used to find the optimal solution for the indicated units. The result is then added to the database (database solutions) and the cycle repeats.

The general feasibility of the approach was tested on a simplified model, which has the same underlying structural problem and is simple enough to optimise in an exhaustive way using a global optimisation approach which encompasses all variable parameters in the model. A comparison of the results shows that our approach produces results of a similar quality to the 'traditional' methods, but faster and with the advantage that the generated knowledge about the unit's behaviour under load can be applied in other areas.

In order to test the application of the ANN approach on larger structures, it was deployed on several designs of *Lace Wall* (Fig. 7). While the size of these parameter spaces prohibits an application and hence comparison with classical optimisation approaches, we found that the proposed solutions are structurally sound and match the distribution of units which the experienced design and construction team of *Lace Wall* would intuitively use.

What we can learn from the Lace Wall

Lace Wall (Fig. 8) demonstrates how the generation of intuition is crucial for solving complex design problems. We found that the high degree of internal interdependence between these problems and the non-continuous fitness landscapes that result from elements whose performance depends on behaviour prohibit traditional computational design optimisation approaches. *Lace Wall* suggests that the intuition that a designer builds upon to make design decisions for both complex structural performance choices and behaviour can be effectively supported by machine learning. It is supervised machine learning with artificial neural networks which provides a kind of intuition (the means to select) alongside a linked database of previously evaluated solutions, which provides the experience on which the selection is based.

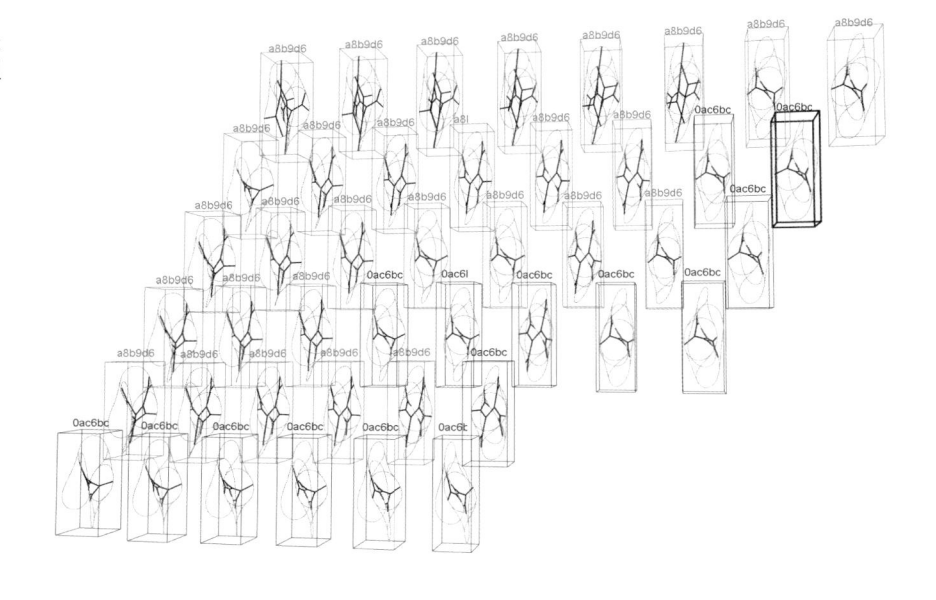

Artificial neural networks, in particular, with their origins in pattern recognition, seem to be well-suited to the investigation of load distribution recognition problems. The machine learning-based approach presented here demonstrates how neural networks can categorise the shape of complex geometries based on high-dimensional discretisations with up to a hundred input parameters. The neural network is able to learn based on an atypical number of parameters compared with other classification methods, which, in our case, ensured that it was able to precisely describe the load distribution. This approach offered flexibility and precision when it came to the classification of previously unseen data. This opened up the possibility of reusing the optimised solutions database and the trained network in multiple iterations of the design.

7. Force distribution in a *Lace Wall* design with cantilever.

8. Detailed view of *Lace Wall*.

Acknowledgments

Lace Wall would not have been possible without the work of Amelie Unger.

References

Alpermann, H. and Gengnagel, C., 2012, 'Shaping Actively-bent Elements by Restraining Systems' in *Proceedings of IASS-APCS 2012: From Spatial Structures to Space Structures*, Seoul.

Bouaziz, S., Martin, S., Liu, T., Kavan, L. and Pauly, M., 2014, 'Projective Dynamics: Fusing Constraint Projections for Fast Simulation' in *ACM Transactions on Graphics 33* (4), Article 154.

Day, A., 1965, 'An Introduction to Dynamic Relaxation' in *The Engineer*, p.218-221.

Dreyfus, S., 1990, 'Artificial Neural Networks, Back-propagation and the Kelley-Bryson Gradient Procedure' in *Journal of Guidance, Control, and Dynamics*, 13(5), p.926-928, available at http://dx.doi.org/10.2514/3.25422.

Harding, J., 2016, 'Dimensionality Reduction for Parametric Design Exploration' in Adriaenssens, S. et al. (eds.), *Advances in Architectural Geometry 2016*, Zurich, vdf Hochschulverlag AG an der ETH Zürich, p.274-287.

Deleuran, A.H., Pauly, M., Tamke, M., Friis Tinning, I. and Ramsgaard Thomsen, M., 2016, 'Exploratory Topology Modelling of Form-Active Hybrid Structures' in *Procedia Engineering*, 0, p.1-10.

Lienhard, J., 2014, *Bending-Active Structures Form-finding Strategies Using Elastic Deformation in Static and Kinetic Systems and the Structural Potentials Therein*, Stuttgart, Universität Stuttgart Inst. F. Tragkonstr.

Menges, A., 2012, 'Material Computation: Higher Integration in Morphogenetic Design' in *Architectural Design*, 82(2), p.14-21.

Nicholas, P. and Tamke, M., 2013, 'Computational Strategies for the Architectural Design of Bending Active Structures' in *International Journal of Space Structures*, 28(3-4), p.215-228, doi:10.1260/0266-3511.28.3-4.215.

Pena, A., 2012, 'Two Case Studies of Freeform Facade Rationalisation' in Achten, H. et al. (eds.), *Physical Digitality – Proceedings of the 30th International Conference on Education and Research in Computer-Aided Architectural Design in Europe*, eCAADe and ČVUT, p.501-510, available at http://cumincad.architexturez.net/system/files/pdf/ecaade2012_316.content.pdf.

Quinn, G., Deleuran, A.H., Piker, D., Brandt-Olsen, C., Tamke, M., Ramsgaard Thomsen, M. and Gengnagel, C., 2016, 'Calibrated and Interactive Modelling of Form-Active Hybrid Structures' in Kawaguchi, K., Ohsaki, M. and Takeuchi, T. (eds.), *Proceedings of the IASS Annual Symposium 2016*, Tokyo.

Rutten, D., 2014, 'Navigating Multi-Dimensional Landscapes in Foggy Weather as an Analogy for Generic Problem Solving' in *Proceedings of the 16th International Conference on Geometry and Graphics*, Innsbruck.

Stasiuk, D. and Thomsen, M.R., 2014, 'Learning to be a Vault – Implementing Learning Strategies for Design Exploration in Inter-Scalar Systems' in *Fusion, Proceedings of the 32nd International Conference on Education and research in Computer Aided Architectural Design in Europe*, 1, p.381-390.

Tamke, M. and Nicholas, P., 2013, 'Computational Strategies for the Architectural Design of Bending Active Structures, I' in *International Journal of Space Structures*, special issue: Active Bending, Vol. 28, No. 3 & 4, s.215-228.

Ramsgaard Thomsen, M., Tamke, M., Holden Deleuran, A., Schmeck, M., Quinn, G. and Gengnagel, C., 2015, 'The Tower: Modelling, Analysis and Construction of Bending Active Tensile Membrane Hybrid Structures' in *Proceedings of the International Association for Shell and Spatial Structures (IASS) Symposium 2015*, Amsterdam, Future Visions.

Wilkinson, S. and Hanna, S., 2014, 'Approximating Computational Fluid Dynamics for Generative Design,' in *International Journal of Architectural Computing*, June 2014, Vol. 12, No. 2, p.155-177.

ROBOTIC FABRICATION OF STONE ASSEMBLY DETAILS

INÉS ARIZA[1,3] / T. SHAN SUTHERLAND[2,3] / JAMES B. DURHAM[3] / CAITLIN T. MUELLER[1] / WES MCGEE[2,4] / BRANDON CLIFFORD[1,4]
[1] Massachusetts Institute of Technology
[2] University of Michigan
[3] Quarra Stone
[4] Matter Design

This research follows an important body of work from the past decade, which focuses on the design of global surface geometries for compression-only structural behaviour. For example, studies in thrust network analysis have made possible the design and computation of complex unreinforced freeform shell structures that work purely under compressive forces once they are completely assembled (Block, 2009). Recent built projects have shown that while it is possible to construct these structures with standard CNC fabrication tools and for them to demonstrate efficient structural behaviour with minimal bending as expected, a major challenge of building these structures is the development of effective assembly strategies during construction to handle tolerance (Rippmann et al., 2016). A second key challenge is the management of falsework, which is structurally necessary to hold individual voussoirs, or compression blocks, in place until the structure is stable, which is sometimes not until the final stone is placed.

These challenges are important to address in order for efficient, geometrically expressive masonry shell structures to play a larger role in the contemporary architectural fabrication landscape alongside conventional steel, concrete and timber structures. In response, the research presented here offers a new approach for the fabrication and assembly of freeform masonry shell structures that can be built with less error and less falsework. Made possible through a computational workflow that simulates structural behaviour during assembly instead of only after a structure is completed, the approach employs cast-metal joining details that bring ancient stonework techniques into the digital age with customised, mechanically responsive geometries.

New agendas for stone carving

Correlating forces (physics) and form (geometry) in 3D, thrust network analysis and accessible physics simulation environments based on dynamic relaxation have extended historical structural form-finding methods into new versatile digital design workflows (Block, 2009, Rippmann et al., 2011, Piker, 2013). One of the results of the availability of these new geometrical exploration approaches has been a renewed interest from designers

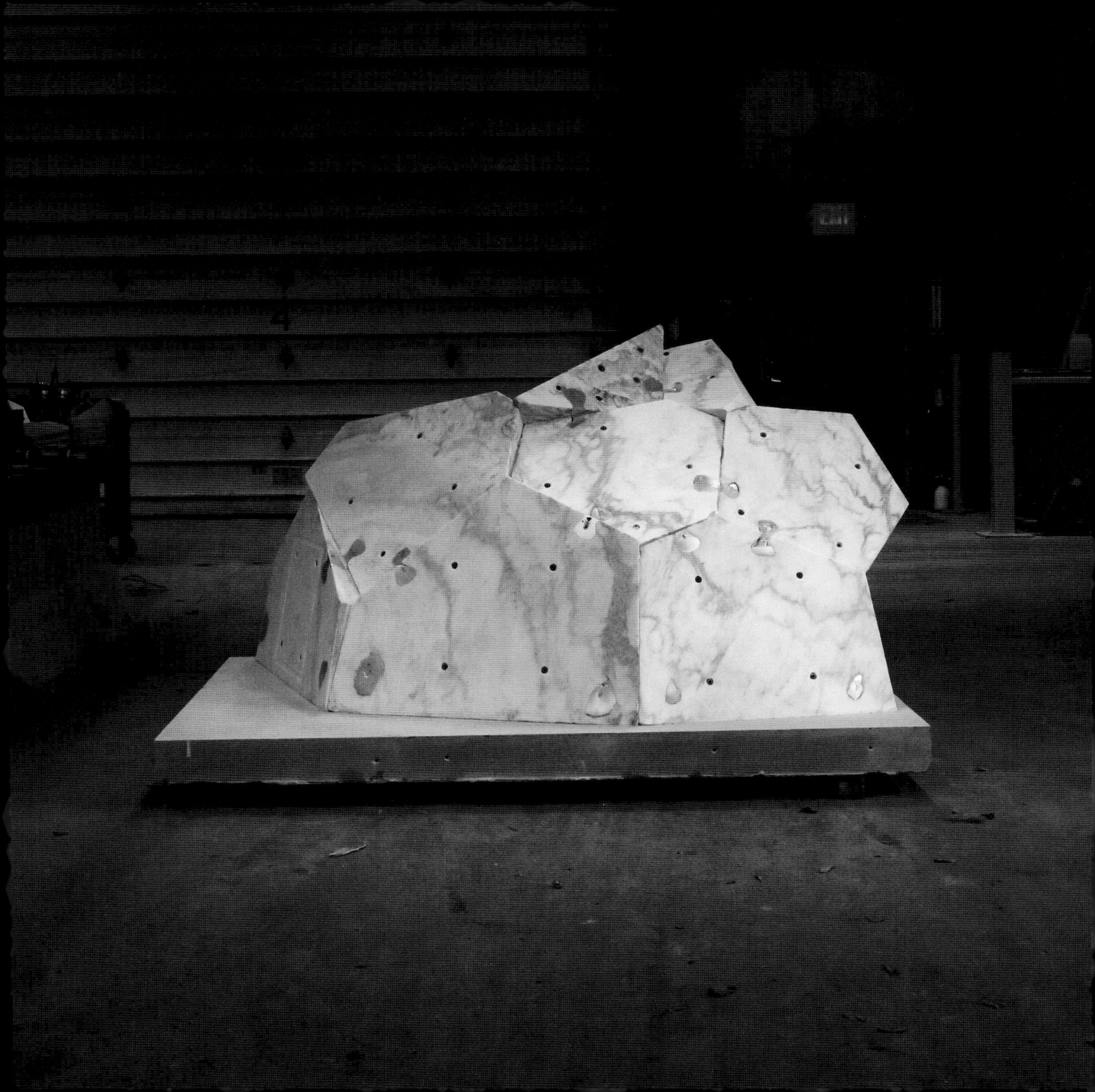

in historical techniques such as stone carving (Lachauer et al., 2011, Rippmann et al., 2016, Clifford et al., 2015, Kaczynski et al., 2011).

Construction of discrete element structures

Most of the current research efforts in discrete element structures have focused on the production of geometrically challenging thin structures that perform efficiently once they are finally assembled. These efforts have not emphasised the forces arising during assembly, or have solved this problem through external means such as scaffolding, chains or ropes (as in Deuss et al., 2014). In contrast, this research approaches the problem of stability during assembly through integrated details.

Stone detail precedents and methods

Two types of detail precedent inform this research. The first is the historic process of carving a detail geometry into stone and direct casting metal into that geometry. This detail is often embedded inside the thickness of stone and is not visible. The motivation of this detail is to resist a possible future force, such as settling or an earthquake. These details are not constrained by the mass of stone, but rather by the properties of metal shaping or casting and the carving tools used (Leroy et al., 2015). The second detail precedent is a procedural one. For instance, Inca stonework carries vestigial details that hint at the sequence in which a wall was constructed. Each detail refers to a particular moment of assembly and its relation to previously placed stones. This concept can be seen not only in the way the stones notch into each other, but also in the nubs used to place the stones (Protzen, 1993). This research seeks to conflate these two detail concepts in order to incorporate procedural and sequential structural analysis to inform detail locations. These locations are responsive not only to the global conditions, but also to the discrete conditions of the in-progress assembly (Fig. 2).

This project examines the problem of assembling masonry structures through the integration of computation, analysis and simulation during the design phases. The motivation of the research is to develop a streamlined workflow which encompasses design, fabrication and assembly of discrete element structures by leveraging the possibilities of digital fabrication methods. Through a focus on historically inspired details, this paper seeks a new approach that can expand the possibilities for designing and building expressive, efficient structural forms.

The assembly method in this research comprises six steps from design to assembly: base geometry, discretisation, physics analysis, detail design, fabrication and assembly. The method is exemplified by an eight-piece masonry structure case study shown in Fig. 5, manufactured at Quarra Stone in Madison, Wisconsin.

Base geometry

This research employs a method which serves to liberate geometry from the exclusive dedication to structural requirements. Though essential, structural forms rarely align with programmatic, ergonomic, thermal or formal concerns. In order to accommodate a confluence of differing concerns, the potentials of depth and volume are employed, resulting in an anti-isomorphic condition, as described previously (Clifford et al., 2015). This deep condition produces a zone of operation that Wolfgang Meisenheimer describes as the 'work body' (Meisenheimer, 1985) – a space between the visible architectural surfaces which is dedicated to the means and methods of making. This method begins with a base geometry informed by the above extra-structural concerns. This singular surface approaches a structural logic, but does not satisfy it. Through variable depth and detailing strategies, this non-idealised form transforms into a proposal which satisfies a thrust network within the middle third of the material depth (Fig. 3).

Discretisation

Next is the discretisation of the base geometry into voussoirs. Many different discretisation methods are possible – in this case, a Voronoi-based discretisation is created using a particle-spring system, which creates a random gradient distribution of 3D voussoirs that are larger toward the base of the structure (Fig. 4).

1. Six-piece mock-up, exterior.

2. (a) Cavities that were carved into stones and fitted with steel joints during the Angkorian era (Mitch Hendrickson, source: Cambodia Daily) and (b) Inca wall assembly detail (Brandon Clifford).

3. Section of assembly strategy.

4. A 3D diagram showing particles, springs and final voussoirs.

5. A 3D diagram showing the variable volume eight-piece mock-up and the results from the overall analysis showing reaction forces at the base.

Physics analysis

This method proposes an alternative assembly strategy for freeform stone shells that relies on a local understanding of forces at each step of the assembly sequence (Ariza, 2016). The structural analysis includes two steps: a global analysis that evaluates the equilibrium of the structure in its final state and a local analysis that evaluates all intermediate equilibrium states during assembly. The analyses are conducted with Karamba v.1.2.1, a finite element analysis plug-in for Grasshopper (Preisinger, 2013), and directly contribute to the design and distribution of cast tension details. Specifically, the analyses consider reactions generated at boundary conditions between elements and at the interface with the ground to determine the types and locations of necessary details.

Global equilibrium analysis

Because the base geometry is not generated to fulfil one single constraint (i.e. structural performance), global stability is not guaranteed. The results of the overall calculation of reaction forces at the base of the eight-piece section of the structure are shown in Fig. 5.

Local equilibrium analysis

The discrete analysis step comprises assigning an assembly sequence of voussoirs, determining the support location and condition of each voussoir according to the sequence and visualising the reaction forces at each support.

Assembly sequence

The sequence of assembling voussoirs does not affect the global stability of the final assembled structure. However, there is a big impact on stability during the assembly process. While this research does not rigorously address this question, the topic has been studied in Deuss (2014). This research establishes a reasonable assembly sequence using rings of voussoirs, and the most stable unit of each ring is assembled first. As each new voussoir is added, it is not possible to assume that the previous state of equilibrium is still valid. Ultimately, every previous interface between voussoirs needs to be checked, since each is affected by every new addition. As a proxy, in this case study the stability of the global intermediate, or the sum of all previously assembled voussoirs, is checked at the base (Fig. 5).

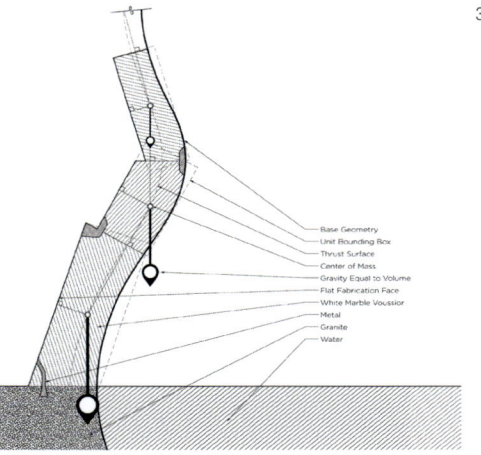

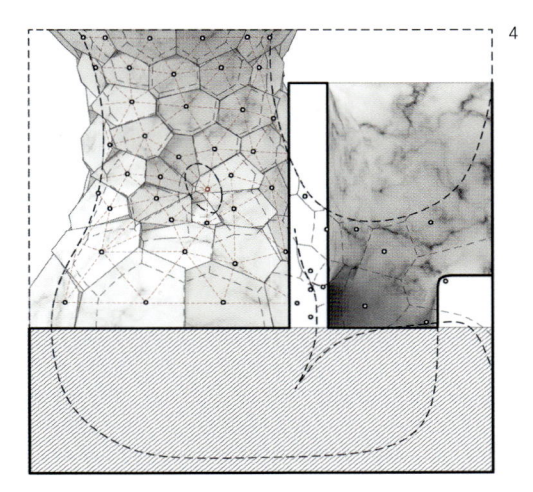

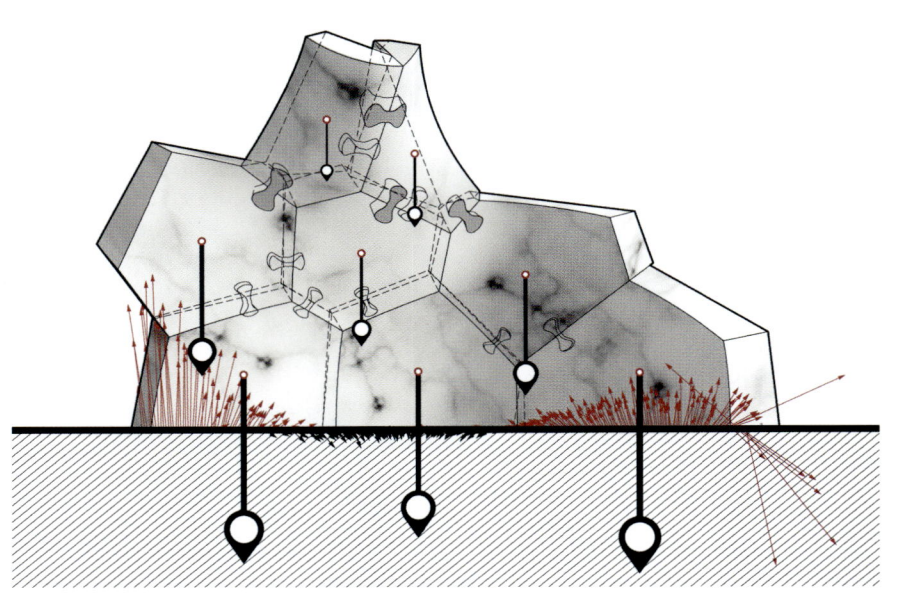

Detail design

Details can be inspired by different motivations. In this project, the role of the details is to coordinate different type of constraints: structural (type, direction and magnitude of reaction forces), fabrication (properties of the carving and casting tools and machines) and assembly (direction and fixing steps of units). This approach takes advantage of the ability of robots to perform custom non-repetitive stone carving and match it with cast metal's ability to be formed with geometric flexibility.

Structural constraints

The reaction forces of the discrete analysis are interpreted one by one, matching type, direction and magnitude with specific geometric detail strategies. Compression forces require surface area, so the planar edges of the voussoirs are left unmodified. Tension forces in the plane require a locking geometry in plane and in the direction of the tension vector to avoid units pulling apart. Out-of-plane tension forces and bending moments are counteracted with couples on opposing faces. In-plane shear forces require a locking geometry perpendicular to the plane of action of the force.

Fabrication constraints

The type of stone, the geometric properties and the performance of tools define the carving constraints. This paper's case study uses Vermont Marble and a blunt electroplated tool. The tool diameter defines the minimum radius of possible carved curvature, and the tool shaft height defines the maximum carving depth. This last parameter is key to specifying possible locations of tension details.

Casting constraints are dictated by the way in which the metal flows through and freezes in the mould when poured. Sharp external corners result in more rapid cooling, leading to increased grain size and brittleness. Sharp internal corners often result in cracking during freezing. Drastic changes in cross-sectional area and volume result in uneven cooling and grain structure. Since traditional clips and butterfly joints in wood or wrought metal do not suffer from such constraints, cross-sectional areas can be varied as much as needed. The translation of this geometry to cast metal requires modification to maintain a constant cross-sectional area throughout the joint.

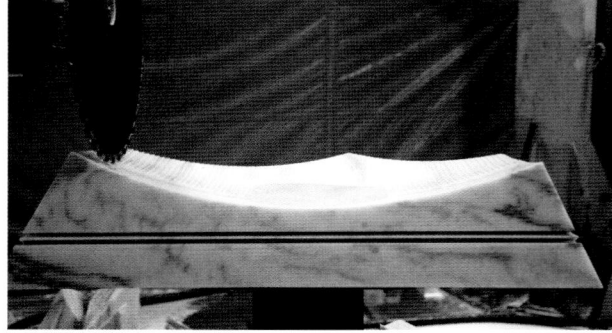

Assembly constraints

The assembly strategy is composed of two steps: registration and fixing. In order to register the pieces that are in place, a precast metal drift-pin is inserted, followed by the cast in-situ final fixing of the unit. This two-step assembly strategy determines the drafted geometry and the material selection of the drift-pins.

Robot control and constraints

Industrial robots are designed to be highly flexible manipulators, but this flexibility results in compromises with respect to overall volumetric accuracy. One technique for minimising positioning error is to utilise an external synchronous positioning axis (rotary table). By allowing the robot pose to be restricted to a smaller range of

6. Cutting operations: (a) edge saw cutting, (b) face side-cutting and (c) detail milling.

7. (a) Casting of specimens, (b) casting of joints in-situ and (c) sample specimen of tension joint.

motion and a reduced range of joint configurations, accuracy can be improved; in addition, the overall work volume of the robot increases significantly. Both of these techniques are employed in the fabrication of the case study. In order to maximise part accuracy, individual voussoirs are processed from a solid blank to the finished part using a single fixturing set-up on a flat back face.

Cutting operations

The production of individual voussoirs benefits from an automatic tool changer set-up and comprises four robotic carving operations (Fig. 6). The majority of the stock is removed with a thick diamond composite blade. The first operation, a saw slab-cutting strategy, is used for cutting the flat-bearing surfaces of the voussoir. This proved to be the most efficient operation, with a higher material removal rate (material removed per minute). Then the internal face is accomplished with a parallel kerf-roughing and a side-cutting finishing, the latter in a motion perpendicular to the previous direction of the blade. Finally, an electroplated diamond tool is used for a pocketing milling operation that produces the joint voids.

Automation of geometry for toolpathing

While the implemented algorithmic design approach generates highly unique geometries with relative ease, it was important to identify production bottlenecks early in the project. While fully automated design-to-machine code strategies have been implemented in certain projects, it was determined that a hybrid approach would integrate better with the fabrication workflow at Quarra Stone. This involved the automated generation and organisation of 3D part files with the needed 'helper' geometry to work smoothly with the production CAM package used by the fabrication team.

Assembly

Several challenges arise in the placement of the individual voussoirs. First, the stones are never set upon a level surface and the centre of mass of the piece is often not directly over the bearing surface, resulting in temporary instability during assembly. Second, while the meeting faces of the voussoirs are drafted in all directions, which facilitates positioning, there are still several degrees of freedom in the movement of the stones as they are individually placed. To counteract this temporary instability, a two-step assembly method is implemented.

Fitting and registration

Using minimal, adjustable tension and compression falsework, each voussoir is fitted in place by hand and registered to its correct location by a precast tapered drift-pin applying tension normal to the adjacent faces of the two stones. This registering operation facilitates the minute adjustment of the voussoirs after placement and temporarily holds them in place during the completion of the ring. The malleable drift-pins also have the capacity to be adjusted to fit in case of fabrication inaccuracies.

Casting and fixing

After the placement of an entire ring of voussoirs, the pre-machined drafted voids of the shear details located between the most vertical faces of the stones are filled with metal in-situ, permanently fixing the ring together. Finally, the precast adjustable pins holding the course in place are cast over in-situ, permanently locking the drift-pin in place. Additionally, any gaps between voussoirs resulting from the tolerances in fabrication are filled during the pouring of the in-situ joints. This series of operations is then repeated for each consecutive ring.

Research evaluation

The validity of the structural analysis and assembly method was assessed through a series of structural tests of specific cast details and prototypes. The former evaluated the material strength and efficiency of the joint geometry throughout a series of controlled specimens. Different mock-ups explored the possibilities and performance of the various available machining methods, the casting and assembly processes and the materials to be used in the precast and in-situ details. The final eight-piece case study served as a final evaluation of the overall detailing and assembly method.

Material tests

Structural tests were performed on details with two different casting alloys: pewter (AC or Brittania), an alloy of tin, copper and antimony; and zamak 3, an industrial die-casting alloy of mostly zinc, copper and magnesium. Despite having a much lower ultimate tensile strength (51.7 MPa) than zamak (284.8 MPa), pewter was selected due to its lower melting point, shrinkage and brittleness, its resistance to work hardening and its higher flow rate (Fig. 7).

Ten geometric variations of tension joint were tested. Controlling variables included the length, depth and thickness of the joint. Three specimens of each geometry were tested to failure under tension. The most successful specimens transferred between 9 and 12.5kN under tension (Fig. 8).

Eight-piece case study

The eight-piece case study made from Vermont Marble served to evaluate the various aspects of the research. In terms of fabrication, inaccuracies (up to 3mm) related to the location of joints were handled with the specific assembly strategies described above. The most critical inaccuracy location was found to be the intrados of the voussoir, for which further fabrication and assembly strategies need to be studied. In terms of assembly, ratchet straps attached to the fixtures of the flat back face were found to be a useful temporary falsework method to support pieces in place until the final fixing of the ring was achieved. Regarding structural performance, units with larger instability were successfully supported by drift-pins in cases of no larger than 3mm inaccuracies. This last test proved the importance of the geometry of the drift-pin as a tolerance-handling method.

Conclusion

This research successfully demonstrates a new method to design, analyse and construct complex geometry shell structures which satisfy a confluence of architectural concerns, without the need for extensive falsework, formwork or templating. Through computation, digital fabrication and the adaptation of ancient detailing strategies, this method points to a possible application of design in synchronous feedback with the constraints of assembly. While the potentials of such a method accommodate an endless number of possible geometries, the analysis points to a series of constraints. These constraints exist primarily in the structural and material properties of stone and metal, the geometric constraints of fabrication and the problematics of compounding errors during assembly.

Future research seeks to further evaluate the capabilities of assembly simulation and sequential fixing in the construction of a full-scale marble caldarium.

8. Geometric variations of joints (from upper left, A to J) and tension testing of specimen F1.

9. Six-piece mock-up, detail showing units 3 and 5 locked with the in-situ casting technique, and unit 6 supported by two drift-pins.

10. Six-piece mock-up, interior.

9
10

Acknowledgements

This research was conducted as part of the 2016 QuarraMatter Fellowship, an industry/academy partnership between Quarra Stone (www.quarrastone.com) and Matter Design (www.matterdesignstudio.com). Each summer, two fellows are embedded in Quarra Stone to produce a research project in advanced fabrication techniques. The form generation employs T-splines (www.tsplines.com) as an organic modeller to inform Grasshopper (www.grasshopper3d.com), a plug-in developed by David Rutten for Rhinoceros (www.rhino3d.com), a programme developed by Robert McNeil. In addition, the analysis computation employs Karamba (www.karamba3d.com) by Clemens Preisinger, and Kangaroo (www.grasshopper3d.com/group/kangaroo), Plankton (Ibid./plankton) and MeshMachine (Ibid./meshmachine) by Daniel Piker. The fabrication computation utilises a custom C# script by Wes McGee to automate toolpath geometries, SUM3D (cap-us.com) for toolpath generation and RoboMOVE™ (www.qdrobotics.com/eng/robomove) by QD Robotics for robot programme simulation. Structural testing was provided by Daren Kneezel and Jeff Scarpelli of Wiss, Janney and Elstner (www.wje.com). The project team includes Brian Smith and Alireza Seyedahmadian. The authors would like to thank Alexander Marshall, Eric Kudrna, Ryan Askew and Edgar Galindo for their fabrication support.

References

Ariza, I., 2016, *Decoding Details: Integrating Physics of Assembly in Discrete Element Structures*, Master's thesis, Massachusetts Institute of Technology, available at http://dspace.mit.edu/handle/1721.1/106365 (accessed 24 January 2017).

Clifford, B. and McGee, W., 2015, 'Digital Inca: an Assembly Method for Free-Form Geometries' in *Modelling Behaviour*, Springer International Publishing, p.173-186.

Clifford, B. and McGee, W., 2014, 'La Voûte de LeFevre: a Variable-volume Compression-only Vault' in Gramazio, F., Kohler, M. and Langenberg, S. (eds.), *Fabricate Negotiating Design and Making*, Verlag.

Deuss, M., Panozzo, D., Whiting, E., Liu, Y., Block, P., Sorkine-Hornung, O. and Pauly, M., 2014, 'Assembling Self-supporting Structures' in *ACM Transactions on Graphics (TOG)*, 33 (6), p.214.

Kaczynski, M.P., McGee, W. and Pigram, D.A., 2011, 'Robotically Fabricated Thin-shell Vaulting: a Methodology for the Integration of Multi-axis Fabrication Processes with Algorithmic Form-finding Techniques', ACADIA 2011.

Leroy, S., Hendrickson, M., Delqué-Kolic, E., Vega, E. and Dillmann, P., 2015, 'First Direct Dating for the Construction and Modification of the Baphuon Temple Mountain in Angkor, Cambodia' in *PloS one*, 10 (11), p.e0141052.

Lachauer, L., Rippmann, M. and Block, P., 2010, 'Form Finding to Fabrication: a Digital Design Process for Masonry Vaults' in *Proceedings of the International Association for Shell and Spatial Structures (IASS) Symposium*.

Meisenheimer, W., 1984, 'Von den Hohlräumen in der Shale des Baukörpers' ['Of the Hollow Spaces in the Skin of the Architectural Body'] in *Daidalos*, 13, Avery Index to Architectural Periodicals, p.103-111.

Piker, D., 2013, 'Kangaroo: Form Finding with Computational Physics' in *Architectural Design*, 83 (2), p.136-137.

Preisinger, C., 2013, 'Linking Structure and Parametric Geometry' in *Architectural Design*, 83 (2), p.110-113.

Protzen, J.P. and Batson, R., 1993, *Inca Architecture and Construction at Ollantaytambo*, Oxford University Press, USA.

Rippmann, M. and Block, P., 2011, 'Digital Stereotomy: Voussoir Geometry for Freeform Masonry-like Vaults Informed by Structural and Fabrication Constraints' in *Proceedings of the IABSE-IASS Symposium*.

Rippmann M., Van Mele T., Popescu M., Augustynowicz, E., Méndez Echenagucia T., Calvo Barentin, C., Frick, U. and Block P., 2016, 'The Armadillo Vault: Computational Design and Digital Fabrication of a Freeform Stone Shell' in *Advances in Architectural Geometry 2016*, p.344-363.

ADAPTIVE ROBOTIC FABRICATION FOR CONDITIONS OF MATERIAL INCONSISTENCY
INCREASING THE GEOMETRIC ACCURACY OF INCREMENTALLY FORMED METAL PANELS

PAUL NICHOLAS / MATEUSZ ZWIERZYCKI / ESBEN CLAUSEN NØRGAARD / DAVID STASIUK / METTE THOMSEN
CITA | Centre for Information Technology and Architecture, The Royal Danish Academy of Fine Arts, Copenhagen
CHRISTOPHER HUTCHINSON
Monash University

This paper describes research that addresses the variable behaviour of industrial quality metals and the extension of computational techniques into the fabrication process. It describes the context of robotic incremental sheet metal forming, a freeform method for imparting 3D form onto a 2D thin metal sheet. The paper focuses on the issue of geometric inaccuracies associated with material springback that are experienced in the making of a research demonstrator. It asks how to fabricate in conditions of material inconsistency, and how might adaptive models negotiate between the design model and the fabrication process? Here, two adaptive methods are presented that aim to increase forming accuracy with only a minimum increase in fabrication time, and that maintain ongoing input from the results of the fabrication process. The first method is an online sensor-based strategy and the second method is an offline predictive strategy based on machine learning.

Rigidisation of thin metal skins

Thin panelised metallic skins play an important role in contemporary architecture, often as a non-structural cladding system. Strategically increasing the structural capacity – particularly the rigidity – of this cladding layer offers a way to integrate enclosure, articulation and structure, but requires a consideration of scale and fabrication that lies outside a typical architectural workflow. Thin sheets can be stiffened via isotropic or anisotropic rigidisation techniques that selectively move local areas of the sheet out of plane, with the effect of increasing structural depth. The use of these techniques marked the early development of metallic aircraft, were pioneered by Junkers and LeRicolais within architecture and are currently applied within the automotive industry.

This research takes inspiration from Junker's proposition, made through the transfer of these techniques into building, of thin-skinned metallic architectures. *A Bridge Too Far* (Fig. 2) presents as an asymmetric bridge. The structure consists of 51 unique planar, hexagonal panels, arranged into an inner and outer skin. The thickness of each panel varies locally, though it is at maximum 1mm thick. Excluding buttresses, the bridge spans 3m and weighs 40kg. Geometric features for resisting local footfall, buckling within each panel and structural

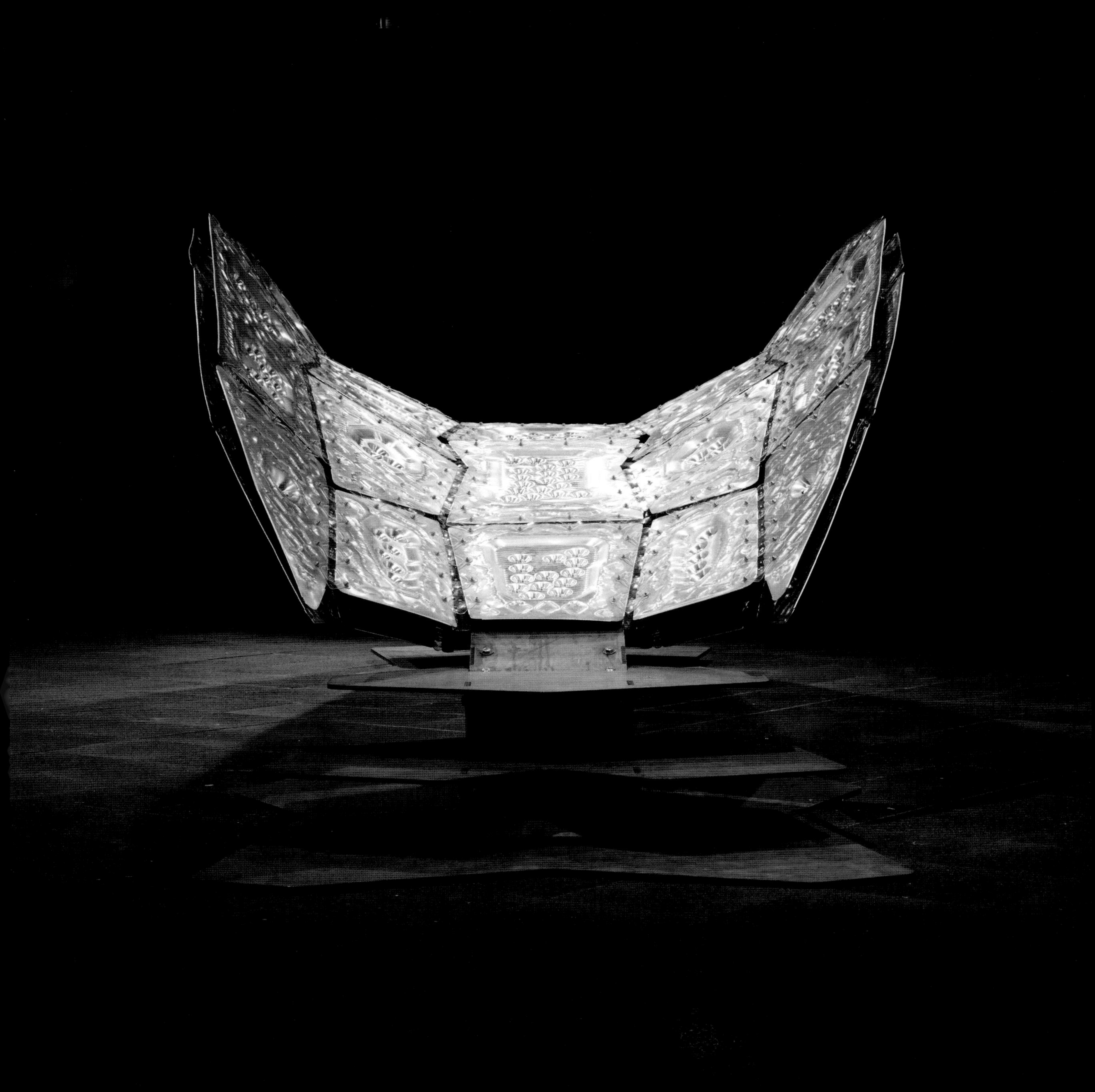

connections – for managing shear forces across inner and outer skins – are produced through the custom robotic forming of individual panels.

Robotic incremental sheet forming

The incremental sheet forming (ISF) method imparts 3D form onto a 2D sheet, directly informed by a 3D CAD model. A simple tool, applied from either one or two sides, facilitates mouldless forming by moving over the surface of a sheet to cause localised plastic deformation (Bramley et al., 2005). A double-sided robotic approach provides further flexibility for forming out of plane in opposing directions (Fig. 3). Moving from SPIF (single point incremental forming) to DPIF (double point incremental forming) removes the need for any supporting jig. This allows for more freedom and complexity in the formed geometry, including features that it would be difficult or impossible to create supports for. A second advantage is the creation of a hydrostatic pressure between the two tools, which has been found to delay the initiation of necking for any strain path.

Transferred into architecture, ISF moves from a prototyping technology to a production technology. Within the context of mass customisation, it provides an alternate technology through which to incorporate, exploit and vary material capacities within the elements that make up a building system. Potential architectural applications have been identified in folded plate thin metal sheet structures (Trautz & Herkrath, 2009) and customised load-adapted architectural designs (Brüninghaus et al., 2012). Recent research has established ISF as structurally feasible at this scale (Bailly et al, 2014) and has explored the utilisation of forming cone geometries as a means to reach from one skin to another (Kalo & Newsum, 2014).

The DPIF set-up used to fabricate panels for *A Bridge Too Far* incorporates two ABB industrial robots working on each side of a moment frame that allows for a working area of approximately 1,000 x 1,000mm. Working with DPIF requires a precise positioning of two tools, one that works as a forming tool and one as the local support. The supporting tool can be positioned in two different ways, following the top perimeter of the feature or following the forming tool down the geometry (Paniti, 2014). Early investigation of both methods showed that, for our set-up, moving the supporting tool only along the feature perimeter quickly led to tearing of the metal, due to the repeated tooling of the same area.

Material considerations

The DPIF process has effects that are both geometrically and materially transformative. Geometric features locally stretch the planar sheet to increase structural depth or to provide architectural opportunities for connection and surface expression. Depending on the geometric transformation, the effects of the material transformation are locally introduced into the material to different degrees according to the depth and angle attained. Calculation in advance to inform generative modelling and fabrication is important, as local thinning of the stretched metal can lead to buckling or tearing when approaching zero thickness (Fig. 4). Work hardening during the forming process also induces different yield strengths, and even strain softening, depending on the base materials.

The choice of material for *A Bridge Too Far* was a negotiation between formability and yield strength to ensure a stable structure but not exceed the force capability of our robotics set-up. Aluminium 5005H14 was chosen, as it provided a good balance between formability, forming speed, initial thickness and initial hardness. In comparison to previous research demonstrators (Nicholas et al., 2016), a higher fixed speed could be used in order to ensure faster production without risking a significantly higher amount of material failures. This choice of material also impacted the design, where the average wall angle of the rigidisation pattern and other geometries was increased from previous prototypes. Because AL5005H14 is pre-hardened, forming at low wall angles softens the metal, while higher wall angles harden it again.

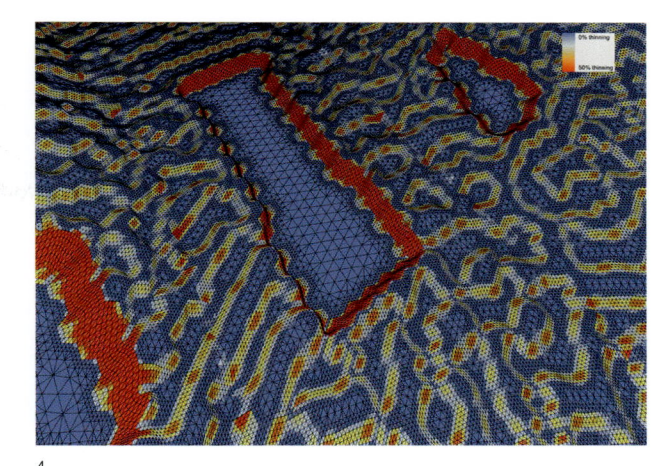

Robotic fabrication

Toolpaths for 51 panels were generated automatically from a 3D mesh using HAL and a custom toolpathing algorithm based on the creation of spirals. The main parameters of this algorithm were the grouping and positioning of features. Because the pattern of rigidising points at which the upper and lower skins connected (Fig. 5) had not yet been designed or located, these geometric features were not included in the initial fabrication pass. However, leaving the formed panels in their frames provided a means to exactly relocate them in the moment frame for continued forming at a later point. Panel fabrication times for 51 panels varied between 4 hours and 8.5 hours. After fabrication, the panels were measured for accuracy, where tolerances of up to 16mm from the digital geometry were found.

The problem of accuracy

Incremental forming is a formative fabrication process, in which mechanical forces are applied to a material so as to form it into a desired shape. A characteristic of formative fabrication processes, particularly mouldless, freeform approaches, is that their positional accuracy is more highly dependent upon a combination of material

1 & 2. *A Bridge Too Far*, at the Royal Danish Academy of Fine Arts, School of Architecture, 2016. Image: © Anders Ingvartsen.

3. Double point incremental forming. Image: © CITA.

4. The pre-calculation of material thinning is materially informed (AL5005-H14 is shown here) and used to prevent tearing during the forming process. Image: © CITA.

5. Points connecting the upper and lower skins provide local rigidisation capable of sustaining significant point loads.
Image: © Anders Ingvartsen.

6. Single point distance sensor mounted to the robot arm.
Image: © CITA.

behaviour and forming parameters than subtractive or additive approaches. Research into resultant incrementally formed geometries has shown significant deviations from the planned geometries (Bambach et al., 2009), and that parameters including the forming velocity, the toolpath, the size of material and distance to supports and particularly the material springback of the sheet during forming all affect the geometric accuracy of the resulting shape. These geometric deviations are a key deterrent from the widespread take-up of the process (Jeswiet et al., 2008).

There are several approaches to improving geometric accuracy, the most direct of which is reforming. This approach simply re-runs the whole, or significant parts, of the original toolpath. It has been shown to achieve considerable improvement, but can potentially double the amount of fabrication time. A second approach is to use a sensor-based measuring strategy, where the deviations are detected and accounted for on the formed shape. After forming, new adjustment lengths for the next forming cycle can be calculated from accurate measurement of the formed shape. This workflow can again lead to considerably longer fabrication times and also requires sophisticated machine vision and path offsetting approaches. A third approach is to use a model-based technique, in which a finite element model of the material and a model of the compliant robot structure are coupled together (Meier et al., 2009). The forces in the tool tip are computed by the FEA, while the path deviations due to these forces can be obtained using the MBS model. Coupling both models gives the true path driven by the robots. While predictive, and therefore minimising the time used to increase fabrication accuracy, this approach is entirely dependent on simulation, which may not accurately represent the reality of fabrication.

In contrast to these approaches, we have investigated two methods that aim to increase forming accuracy with only a minimum increase in fabrication time and that maintain ongoing input from the results of the fabrication process: an online sensor-based strategy and an offline strategy based on machine learning.

Sensor-based strategy to increase accuracy

The first method for increasing forming accuracy during forming was implemented on the rigidising cones that connect the upper and lower skins. A single point laser distance measure was mounted to the robot arm and used to measure, at each cone centrepoint, the local deviation between actual formed depth and ideal geometry. This deviation was then automatically added to the target depth for a given cone, and from this combined target depth an appropriate cone was chosen from a series of toolpaths pre-programmed into the controller, with depths of 20mm, 23mm, 26mm and 30mm.

But because of springback during the forming process, a cone that has the same forming depth as the combined target depth is not the correct choice – the forming depth needs to be larger than the target depth. To determine the correct amount, curve fitting was used to model the relationship between target depth and forming depth. After each cone was formed, the resultant depth was scanned and this data was used to refine the curve-fitting model, allowing a continued improvement in accuracy across the course of fabrication (Fig. 6). After forming and scanning, two further automated correction methods could be triggered. If the formed cone geometry was greater than 5mm off the target geometry, the cone was reformed. If it was between 5 and 2mm off the target geometry, the tip of the cone was extended by 2mm. Tolerances below 2mm were considered acceptable.

Force feedback

While tolerances could be adequately corrected for using the sensor-based strategy outlined above, this method did not provide any deeper understanding of the forming process and the resulting imprecisions. To establish meaningful input parameters for the machine learning algorithm, a load transducer was attached to the forming tool to register changing forces on the tool tip during the fabrication process. A live stream of read-outs (approximately one per 50ms, or every 2cm along the toolpath) was established and the data was stored directly in a binary file. This data was used to identify the right type and amount of data for the training of a neural network as a material behaviour model. Visualising this information revealed relationships between the fabricated shape and forces acting on the sheet, and showed the following parameters to be significant:

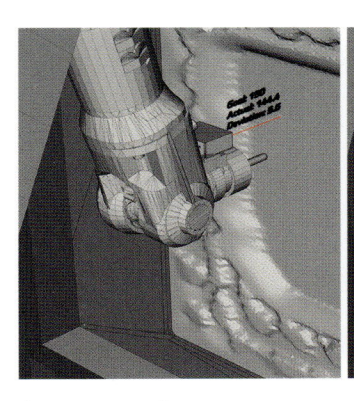

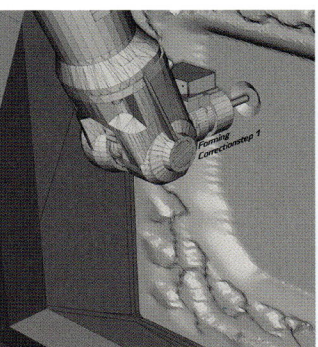

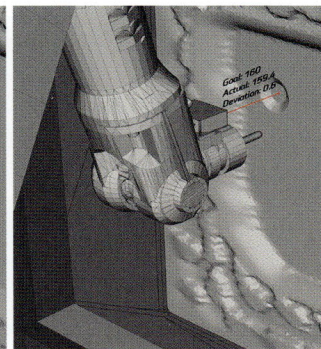

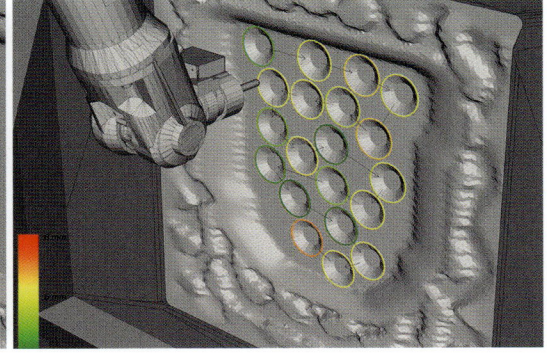

- Local feature.
- Distance to fixed panel edge.
- Current depth of the shape.

A 'local feature' is understood to be a small fragment of the shape being currently formed. It informs the model about edges, ridges and other small-scale geometry of the panel.

Distance to the edge of the panel is the parameter describing the distance to the closest point on the edge of the formed geometry. It is a result of the physical set-up and how the panel was placed in the forming frame (pinned to the underlying MDF board with a panel-specific cut-out). Current depth of the shape is the distance from the initial sheet plane to the current position of the tool tip. It is directly dependent on the material properties and their change over deformation depth. Other parameters – such as the slope angle – are not provided directly to the model. Instead, the local feature is understood as an indirect provider of such information.

Network architecture and learning process

The information gained from the force gauge read-outs was overlaid with a 3D scan of the fabricated panels. This coupling of input and output parameters (local feature, distance to the edge, depth vs. formed shape) constitutes the input and output set for the supervised learning process. Given that the output of the network is the depth of the analysed point after forming, the problem is substantially a regression analysis.

The local feature and current depth are encoded as a heightmap, with a real-world size of 5 x 5cm. With the resolution of 1 pixel per millimetre, without pre-processing the input vector would have to have 2,500 dimensions, making the training process unnecessarily detailed and slow. To reduce its dimensionality, a max pooling technique was applied, resulting in a 9 x 9 pixel – 81 dimensional – heightmap.

The network consists of an input layer with 82 neurons (81 + 1 additional for edge-proximity parameter), a hidden layer with 30 neurons and an output layer with 1 neuron indicating the depth of the resulting point. Back-propagation-based learning was performed on a set of ~1600 samples and took approximately an hour on a regular desktop computer.

Results

The network is able to predict, to some extent and resolution, the resulting geometry based on an input heightmap of the target piece. The authors find the network unexpectedly accurate, given that the training was based only on data gathered from a small number of panels. Additionally, the exploration of the network predictions gave more information on the trained model itself, showing that material behaviour isn't strictly linear – therefore it would be reasonably more challenging to find appropriate functions and ways to encode shape information with a curve-fitting approach (although the neural network is function-fitting as well).

With this neural network-based model, it is possible to predict the forming process result upfront, and with multiple queries the resulting panel surface can resemble the target much more precisely.

The training set is a set of randomly distributed fragments on the surface of the panel. The training set output is a heightmap based on a 3D scan of the formed panel (the ground truth), and is used as the training set output.

As the training process might end up with function overfitting, a comparison is made on another panel to assure the network's versatility.

The values obtained from prediction were used to adjust the fabrication geometry. The method for adjusting the geometry is straightforward: the input mesh heightmap values are increased by the difference between the target and prediction heightmaps. While this method yields a substantial increase in precision, more advanced methods will be a subject of future research.

Conclusion

This paper has addressed the issue of material springback and geometric inaccuracy in the incremental forming process. It has demonstrated the use of sensing and feedback to manage springback and to reduce geometric inaccuracies within the forming process. Two different methods have been presented, the first based on online adaptation and the second based on offline prediction. Both models negotiate between the design model and the fabrication process. The first method changes the design parametrically during the fabrication process, diverging from the desired design, while the second method changes the fabrication geometry prior to fabrication to achieve the desired design. These models are necessary because, for the incremental forming process, the information contained within the design model is not by itself enough to achieve accurate forming. On this basis, the authors believe that machine learning processes could provide new bridges between designing and making, especially where the material behaviour model is a combination of multiple functions.

Acknowledgements

This project was undertaken as part of the Sapere Aude Advanced Grant research project *Complex Modelling*, supported by The Danish Council for Independent Research (DFF). The authors want to acknowledge the collaboration of Bollinger Grohmann consulting engineers, SICK Sensor Intelligence Denmark, Monash University Materials Science and Engineering and the robot command and control software HAL.

References

Bailly, D. et al., 2014, 'Manufacturing of Innovative Self-supporting Sheet-Metal Structures Representing Freeform Surfaces' in *Procedia CIRP*, 18, p.51-56.

Bambach, M., Taleb Araghi, B. and Hirt, G., 2009, 'Strategies to Improve the Geometric Accuracy in Asymmetric Single Point Incremental Forming' in *Production Engineering*, 3(2), p.145-156.

Brüninghaus, J., Krewet, C. and Kuhlenkötter, B., 2013, 'Robot Assisted Asymmetric Incremental Sheet Forming' in Brell-Çokcan, S. and Braumann, J. (eds.), *Rob | Arch 2012: Robotic Fabrication in Architecture, Art, and Design*, Vienna, Springer Vienna, p.155-160. Available at http://dx.doi.org/10.1007/978-3-7091-1465-0_16.

Jeswiet, J. et al., 2005, 'Asymmetric Single Point Incremental Forming of Sheet Metal' in *CIRP Annals – Manufacturing Technology*, 54(2), p.88-114. Available at http://www.sciencedirect.com/science/article/pii/S0007850607600213.

Jeswiet, J. et al., 2008, 'Metal Forming Progress since 2000' in *CIRP Journal of Manufacturing Science and Technology*, 1(1), p.2-17. Available at http://www.sciencedirect.com/science/article/pii/S1755581708000060.

Kalo, A. and Newsum, M.J., 2014, 'An Investigation of Robotic Incremental Sheet Metal Forming as a Method for Prototyping Parametric Architectural Skins' in McGee, W. and de Leon, M. (eds.), *Robotic Fabrication in Architecture, Art and Design 2014*, Cham, Springer International Publishing, p.33-49. Available at http://dx.doi.org/10.1007/978-3-319-04663-1_3.

Meier, H. et al., 2009, 'Increasing the Part Accuracy in Dieless Robot-based Incremental Sheet Metal Forming' in *CIRP Annals – Manufacturing Technology*, 58(1), p.233-238. Available at http://www.sciencedirect.com/science/article/pii/S0007850609000845.

Nicholas, P. et al., 2015, 'A Multiscale Adaptive Mesh Refinement Approach to Architectured Steel Specification in the Design of a Frameless Stressed Skin Structure' in Thomsen, R.M. et al. (eds.), *Modelling Behaviour: Design Modelling Symposium 2015*, Cham, Springer International Publishing, p.17-34. Available at http://dx.doi.org/10.1007/978-3-319-24208-8_2.

Nicholas, P. et al., 2016, 'An Integrated Modelling and Toolpathing Approach for a Frameless Stressed Skin Structure, Fabricated Using Robotic Incremental Sheet Forming' in Reinhardt, D., Saunders, R. and Burry, J. (eds.), *Robotic Fabrication in Architecture, Art and Design 2016*, Cham, Springer International Publishing, p.62-77. Available at http://dx.doi.org/10.1007/978-3-319-26378-6_5.

Paniti, I. and Somló, J., 2014, 'Novel Incremental Sheet Forming System with Tool-Path Calculation Approach' in *ACTA POLYTECHNICA HUNGARICA*, 11(7), p.43-60. Available at http://eprints.sztaki.hu/8132/.

Trautz, M. and Herkrath, R., 2009, 'The Application of Folded Plate Principles on Spatial Structures with Regular, Irregular and Freeform Geometries' in *Symposium of the International Association for Shell and Spatial Structures (50th. 2009. Valencia). Evolution and Trends in Design, Analysis and Construction of Shell and Spatial Structures: Proceedings*. Available at https://riunet.upv.es/handle/10251/6765 (accessed 8 April 2016).

7. *A Bridge Too Far*, at the Royal Danish Academy of Fine Arts, School of Architecture, 2016. Image: © Anders Ingvartsen.

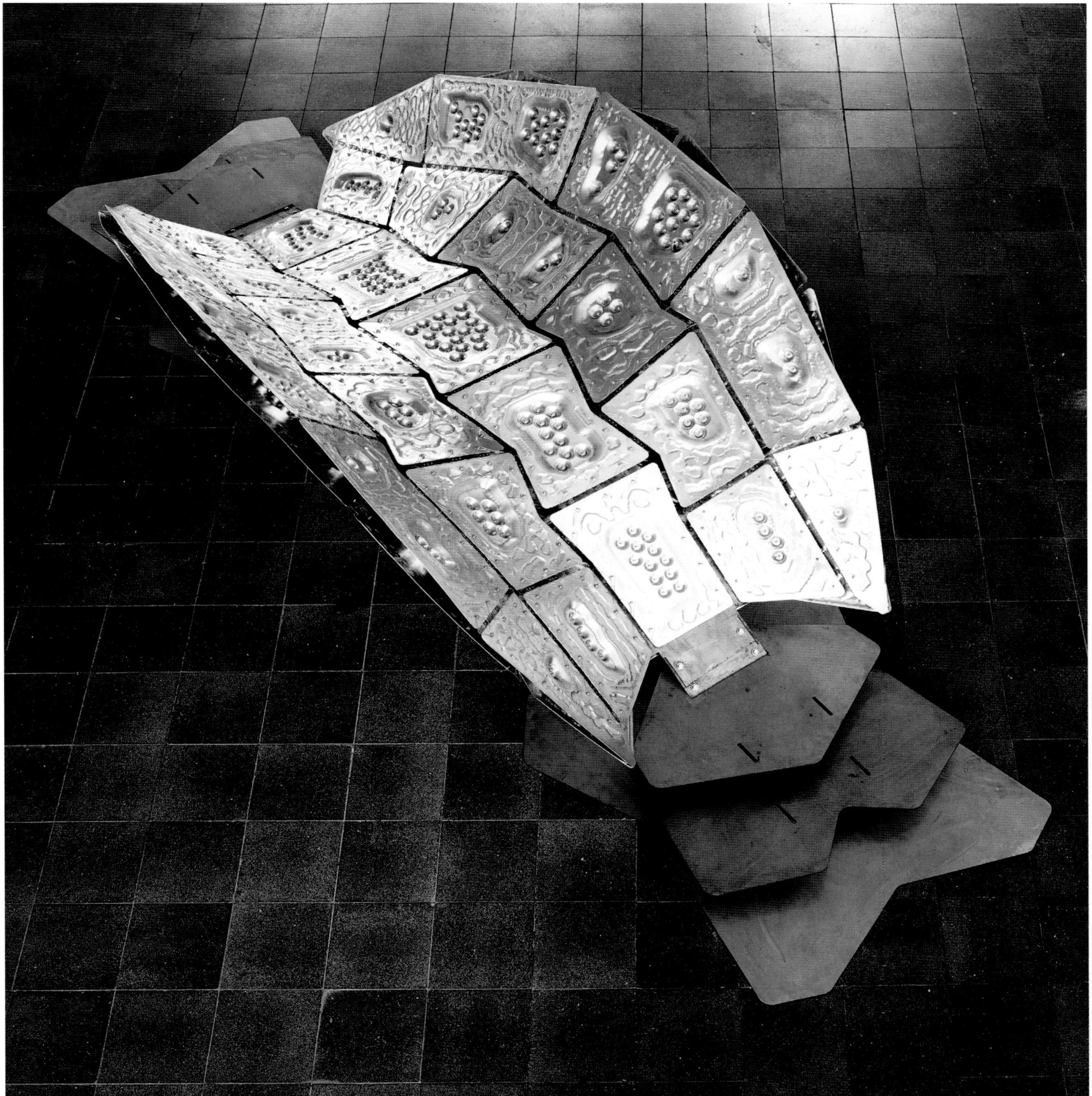

DIGITAL FABRICATION OF NON-STANDARD SOUND-DIFFUSING PANELS IN THE LARGE HALL OF THE ELBPHILHARMONIE

BENJAMIN S. KOREN
One to One, Frankfurt/New York
TOBIAS MÜLLER
Peuckert, Mehring

Classical music and performances have long been considered an exclusive pastime. In the past decade, classical performances have faced a declining number of concert-goers, whose median age is simultaneously on the rise. Despite this negative trend, new concert halls and opera houses are being built around the world by some of the most prestigious architectural offices, resulting in some of the most exciting contemporary architectural projects. This is evidenced by the fact that, for example, three of the last four Mies van der Rohe Award winners were concert or opera hall projects: the Norwegian National Opera & Ballet by Snøhetta in 2009, the Reykjavik Concert Hall by Henning Larsen Architects with Batteríid and Eliasson in 2013 and most recently the Philharmonic Hall in Szczecin by Barozzi/Veiga in 2015. One may argue that this current interest in new concert hall projects does not contradict the aforementioned attendance crisis, but may instead be interpreted as an effort to rectify it.

In this effort to revive interest in classical concerts, contemporary architects play a vital role. They can help to renew interest by making concert hall buildings more open and accessible to wider, younger audiences as well as augmenting the concert experience as a whole. At the centre of the experience, however, is the performance itself, which may be enhanced on an aural, visual or even tactile level. It is therefore not surprising that in current concert hall projects there is a concerted effort to achieve excellent acoustics, which are in harmony with the architectural language of the building as a whole.

New design and fabrication methodologies open up new possibilities, which are a result in part of design software developments over the past decade, an improved understanding of concert hall acoustics towards the end of the last century and a surge in and access to digital fabrication technologies. In order to make these enhancements, it is critical that a close collaboration between the architect, the acoustician and the fabricator exists.

This paper aims to document one such close collaboration: the development and execution of non-standard sound-diffusing acoustic panels in the large concert hall of the Elbphilharmonie in Hamburg.

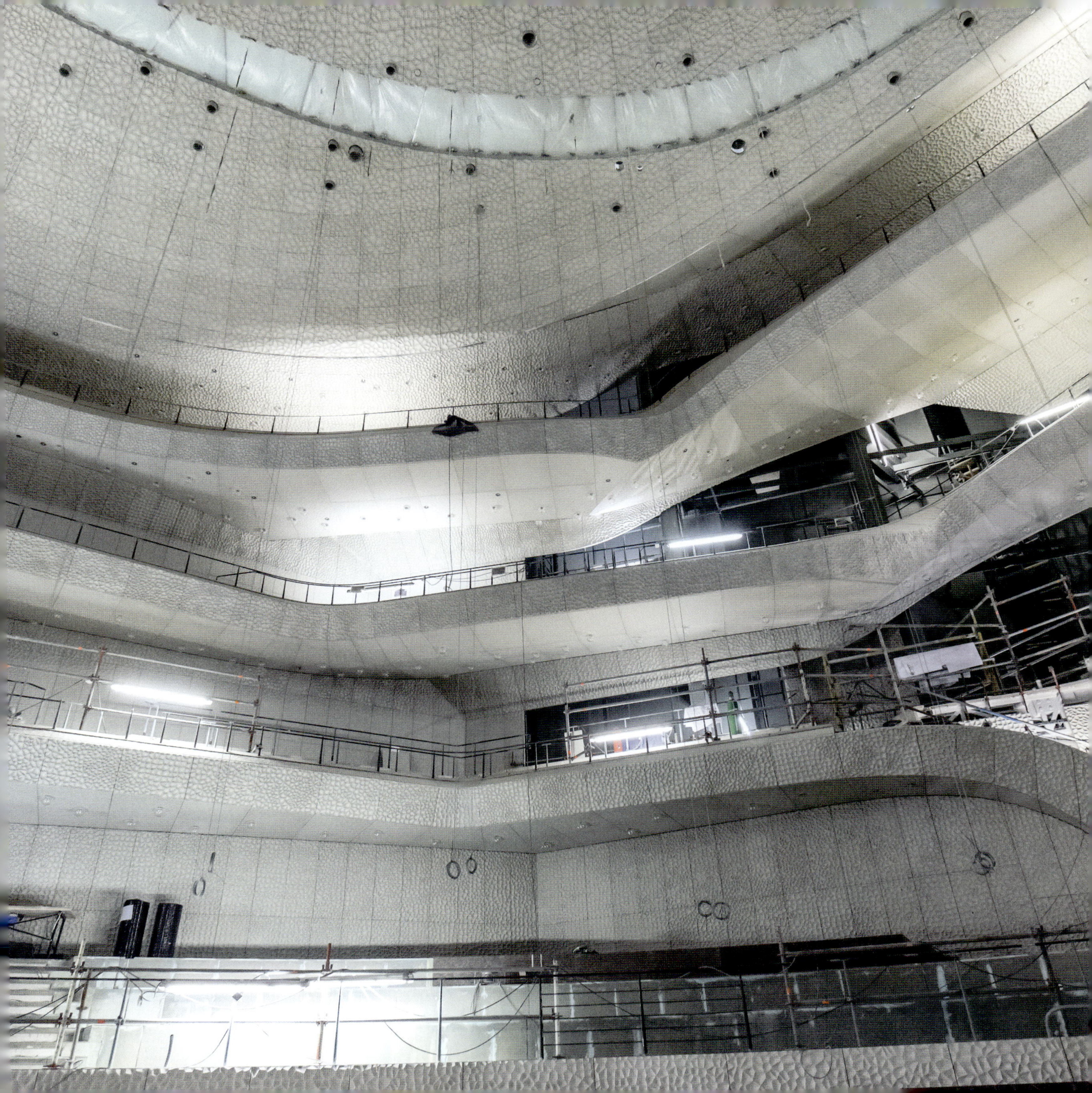

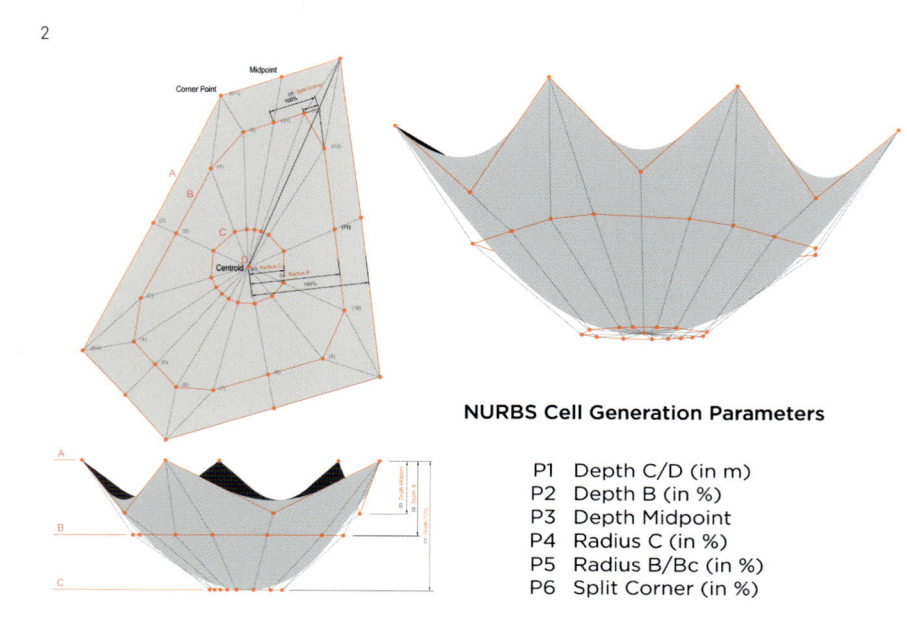

Sound diffusion

Designed by Herzog and de Meuron, the Elbphilharmonie is located in the HafenCity area of Hamburg, Germany. It comprises approximately 120,000m² of space, including three concert halls, a hotel, apartments, restaurants, a parking garage and a public observation platform. The large concert hall lies at the heart of the project, seating 2,150 people. Yasuhisa Toyota, of Nagata Acoustics, was responsible for acoustical engineering. He collaborated with the architects from the early stages of design through to completion. They approached the project from two perspectives: first, in terms of the overall shape of the design – by optimising the orientation of the sound-reflective surfaces – and second, by developing the sound-diffusing surface geometry applied to the individual acoustic panels.

In broad terms, sound diffusion is the even scattering of sound energy in a room. Non-diffusive, reflective surfaces in concert halls can lead to a number of unwanted acoustic properties, which can be rectified, in part, by adding diffusers. A perfectly diffusive space is one where acoustic properties, such as reverberation, are the same, regardless of the location of the listener. Diffusion in some of the best concert halls in the world, such as the Great Hall of the Musikverein, built in 1870 in Vienna, is now understood to be a byproduct of the uneven surfaces of the rich neoclassical ornamentation of its interior. The antipathy to elaborate ornamentation by twentieth-century architects may have come at the expense of good concert hall acoustics. In the past, bad acoustics could be treated at a later stage, by selectively retrofitting absorbers or diffusers, which resulted in a disjunction between the original architectural intent and its modifications.

Commercial diffusers began to appear towards the end of the twentieth century, as engineers studied the science and physics of sound diffusion. This process was started with the seminal work of Manfred R. Schroeder in the 1970s, which led to the development of his 'Schroeder diffusers'. It has been noted that Schroeder's utilitarian approach to diffuser design corresponded well with the architectural styles of his time and were successfully applied to concert hall designs. However, contemporary engineers recognise that the shape of such diffusers is not necessarily in line with contemporary architectural design. Cox (2004), for example, laments: "When Schroeder invented his diffusers, they fitted in with some of the artistic trends of the day. With abstraction at the fore, the fins and wells formed elements in keeping with the style

1. Concert hall
under construction
in January 2016.
Image: Michal Commentz.

2. Single NURBS cell
generation parameters.
Image: ONE TO ONE.

3. One of the panels being
milled on a CNC machine.
Image: PEUCKERT.

4. Final milled panels
at the workshop.
Image: PEUCKERT.

of the day. But in the intervening decades, tastes have moved on. Architecture has been greatly influenced by advances in engineering to allow previously unimaginable shapes to be constructed. Landmark buildings are becoming sculpted with complex geometries and curved forms. To many, Schroeder diffusers no longer match the style required. Fortunately [...] it is possible to design arbitrarily shaped diffusers, echoing the ability of architects to seemingly work with any shaped building. Diffusers can usually be created which have harmony with the architectural style of the building."

Faced with the curvilinear, intricately intertwining wall, balustrade and ceiling surfaces in the Elbphilharmonie's design (Fig. 1), the acousticians had no choice but to develop a bespoke sound-diffusing panel system that was in agreement with the architectural design intention.

Parameters

While an in-depth discussion of the science and physics of sound diffusion is beyond the scope of this paper, one can, however, summarise a few key concepts to provide an understanding of how acousticians derive the specifications for a diffusing pattern, which, in this project, were ultimately translated into code and architectural form.

Sound travels through air in longitudinal waves, meaning that air molecules vibrate back and forth, colliding with each other. Repeated periodic pressure differences are perceived as musical notes. The wavelengths of notes at lower frequencies are longer, whereas notes at higher frequencies are shorter. Because of this property of sound, there is a direct relationship between the spatial dimensions of the diffusing pattern and the musical notes this pattern will have an effect on, based on the wavelengths of the notes. A sound-diffusing pattern will affect a specific frequency range based on its physical dimensions. It is generally understood that the lower cut-off frequency of this range is influenced by the depth of the surface pattern, while the higher cut-off frequency is influenced by its width.

The acousticians therefore defined the specifications of the sound-diffusing pattern, which would ultimately consist of randomly placed, individually shaped cells for specific regions in the concert hall. The shape of these cells ranged from 5mm to 90mm in depth and 40mm to 160mm in width, depending on the region identified by the acousticians.

Parametrically defined surface

In order to interpret the specifications of the acousticians, i.e. the width, depth and randomness of the pattern,

5. Close up view of four assembled panels with a seamless pattern across 5mm gaps.
Image: ONE TO ONE.

6. Transition zone between solid and transparent panels.
Image: PEUCKERT.

7. Concert hall under construction in February 2016.
Image: PEUCKERT.

a bespoke software plug-in was developed for Rhino 3D to generate approximately one million parametrically defined, uniquely shaped NURBS cells. A paper by one of the authors of this paper presented at the Design Modelling Symposium in Berlin in 2009 outlined this development in detail. The pattern itself was initially based on the distortion of a two-dimensional, orthogonal grid of Voronoi seeds. The program allowed for random seed displacements, deletion and insertion in order to control both the degree of randomness and the scale, i.e. the cell width of the pattern. In a subsequent step, each closed 2D polygon of the Voronoi pattern was used as input in the 3D formation of a parametrically defined NURBS cell (Fig. 2), which exhibited a peak-and-trough shape, a motif characteristic to the project as a whole and found in areas such as the roof of the building or the overall shape of the concert hall. The placement of the control points was driven by a total of six parameters, which allowed for the precise definition of the depth of each cell and also its overall shape, which included a range of harder and softer edges. All the parameters were driven using grayscale bitmap images, which mapped XY coordinates from the bitmap space to each of the concert hall's wall and ceiling surfaces' UV coordinates. In a last step, every control point of every NURBS cell was mapped topologically onto either flat, single- or double-curved surfaces in 3D, with special attention given to the continuation of the pattern across the seams between each connecting surface.

Digital fabrication and assembly

As each panel was unique, further software programmes were developed to automate the 3D planning and digital production of approximately 10,000 CNC-milled gypsum fibreboard panels, as well as to optimise the acoustic surface's substructure. For acoustical reasons, the weight per unit area of the panels, up to 150kg/m^2, was fairly large, and had to be achieved by giving the panel thickness a range of between 35 and 200mm. Therefore the highest available density fibreboard panel, with a volumetric density of 1,500kg/m^3, was chosen. Since the material is only produced up to 40mm thickness, most of the panels had to be built up in several layers, glued and mechanically fixed together, in order to achieve the desired weight. The architects defined a precise and intricate network of gap lines, which, not unlike the sound-diffusing pattern itself, was meant to be seamless across the hall's surfaces. Therefore the edges of the panels were made to always align with the edges of the neighbouring panels, resulting in planar, curved and twisted edges, including rabbets in some cases. Because of the varying degrees of complexity in edge conditions, a 5-axis milling machine was used to manufacture the panels (Fig. 3). The curvature of the front surface was achieved by keeping the back of each panel planar, while the front was milled to shape. For each panel, the edges had to be digitally generated, the fixings had to be placed and a groove along the entire perimeter, for the placement of a sealing band, had to be positioned exactly 5mm below the lowest point of the sound-diffusing pattern. In addition, mechanical fixings were placed to secure the glued layers and, most importantly, the previously generated diffusing pattern was assigned to each panel. After this, the panels were ready for manufacturing.

Each raw panel was prepared to size. The panels were CNC-milled in two stages. First, each panel was milled from the back, which included the 5-axis formatting of

8

the edges and the placement of the holes for fixing the substructure and for mechanically securing the glued layers. Then each panel was flipped, repositioned on the machine and milled from the front, which included a stage for 3-axis milling of the sound-diffusing pattern using a ball-end cutter, milling in parallel tracks spaced at fairly large distances. This resulted in a rough, final surface texture that also exhibited the peak-and-trough motif down at the scale of the trails left by the milling head (Fig. 5). Once the panels were milled, they were lacquered on both sides with a clear lacquer. Part of the substructure – profiles standardised in length at 100mm intervals – were prefixed to the back of each panel using a combination of four standard screw lengths. The depth of penetration of the screws was determined computationally beforehand and was controlled by pairing each screw with one of ten washer types, at 1.5mm incremental thicknesses, to allow for a large range of precise screw penetrations. Finally, the panels were quality checked and packed for shipping. Once they arrived onsite, each panel was manually installed. Very simple details of the substructure allowed for panel adjustments with three degrees of freedom, allowing them to be fitted with a 5mm gap between panels, and at the required precision.

Trialling new technology

Apart from the architectural achievement, the final project can be evaluated from both a fabrication and acoustical standpoint. The seamless design and the acoustical specifications both necessitated highly precise computational design and digital fabrication methods. While 3-axis and 5-axis CNC-milling techniques have become the norm in architecture today, one needs to point out that development of the panels for this project started in the mid-2000s, when 5-axis milling machines, for example, were not readily available. The workshops involved in this project had to invest in new machinery and train their staff in order to deal with this new technology, which necessitated a close collaboration between all the parties involved. In addition, no one had previously attempted to mill very dense gypsum fibreboard panels in such a way and in such large quantities. In order to swiftly and precisely produce large quantities of panels without wearing out the machines and tools, several rounds of meticulous tests and trials were conducted, with several mock-ups built. Once parameters had been determined, panels were produced efficiently and to an incredibly high degree of precision, which was necessary in order to assemble the final panels at the desired tolerances. In the end, all the panels fitted together perfectly, keeping the number of faulty panels at a minimum despite each one being non-standard and assembled in a complex manner. Only 20 out of a total of about 10,000 panels had to be replaced due to dulled tools – an error rate of only 0.2%. This extremely small error rate is an achievement in itself given the scale and complexity of this project.

As for the acoustic evaluation, at the time of publication of this paper final measurements have not yet been published. Tests are generally conducted on concert halls before the opening concert, which took place on 11 January 2017.

Pioneering collaboration

The development and execution of the non-standard sound-diffusing panels in the large concert hall of the Elbphilharmonie is a noteworthy collaborative effort between architectural design, acoustic engineering and digital fabrication, which resulted in the intentional application of a sound-diffusing surface treatment in harmony with a contemporary, complex architectural design. Software and manufacturing methodologies, as well as related technologies, have advanced greatly since this project began 10 years ago. Improvements in software and computing power for precise acoustic simulations, as well as readily available access to new fabrication technologies, such as 3D printers and robots, alongside the computational methods outlined in this paper, offer great potential for similar projects in the future. Fromm (2014), for example, investigated the potential use of 3D printed cement-bound elements as an alternative, comparing and contrasting them specifically with the CNC-milled gypsum fibreboard panels of the Elbphilharmonie. This points to one of many exciting new possibilities for the design and application of sound diffusers in future concert hall projects.

8. Concert hall under construction in February 2016. Image: PEUCKERT.

9. Concert hall, under construction in January 2016. Image: Michal Commentz.

References

Audience Insight LLC, 2002, *Classical Music Consumer Segmentation Study*, John S. and James L. Knight Foundation.

Blasi, I., 2015, *European Prize for Contemporary Architecture, Mies Van der Rohe Award 2015*, Barcelona, Fundacio Mies van der Rohe.

Cox, T., 2004, 'Acoustic diffusers: the good, the bad, the ugly' in *Reproduced Sound*, Proceedings of the Institute of Acoustics.

Cox, T. and D'Antonio, P., 2005, 'Thirty years since "Diffuse Sound Reflection by Maximum-Length Sequences": Where Are We Now?', Forum Acusticum, p.2129-2134.

Fromm, A., 2014, *3D Printing zementgebundener Formteile*, Kassel, Kassel University Press.

Gray, D., Cuspinera, M. and Bes, A., 2009, *European Union Prize for Contemporary Architecture*, Barcelona, Actar.

Koren, B., 2009, 'A Parametric, Sound-Diffusive Surface Pattern for the Symphonic Concert Hall of the Elbphilharmonie', *Design Modelling Symposium 2009*, Universität der Künste Berlin, p.237-248.

National Endowment for the Arts, 2015: *A Decade of Arts Engagement: Findings from the Survey of Public Participation in the Arts, 2002-2012*.

Schroeder, M., 1975, 'Diffusive Sound Reflection by Maximum-Length Sequences' in *The Journal of the Acoustical Society of America*, Vol. 57, No.1, p.149-50.

QUALIFYING FRP COMPOSITES FOR HIGH-RISE BUILDING FACADES

WILLIAM KREYSLER
Kreysler & Associates

Using fibre-reinforced polymer on the SFMoMA addition

Fibre-reinforced polymer (FRP), in this case glass fibre-reinforced polyester resin composite with a polymer concrete face coat, was used in the US for the first time as exterior cladding on a Type 1 multi-storey building on the San Francisco Museum of Modern Art (SFMoMA) addition. This 11-storey addition, completed in May 2016, makes SFMoMA the largest museum of modern art in the US, with the largest architectural FRP facade application in the US to date.

FRP was chosen to mimic the rippling water of the nearby San Francisco Bay on the east and west elevations. Although recognised by the IBC (International Building Code) in 2009 as an accepted building material (International Code Council, 2009), any FRP material used must pass the same code requirements as other combustible materials. The most difficult of these requirements is the NFPA 285 test. Until this and other requirements are met, no combustible material, including FRP, is allowed.

The design for the SFMoMA project called for over 700 unique, individual, constantly curving panels (Fig. 1).

Although it is possible to construct such panels with metal, the only practical option was to mould the 710 unique panels, thus suggesting precast concrete or the lighter UHPC or GFRC. The less familiar FRP was listed as an alternative by the façade consultant in part because of its more widespread use in European construction.

Although used sparingly on US buildings for decades, FRP has dominated other industries such as corrosion-resistant ducting and chemical storage tanks, wind energy, marine and heavy truck components. However, it has seen no extensive use on Type 1 buildings. This has been partly because of codes and partly because its primary advantages over other materials are its high strength-to-weight ratio and its ability to be formed into complex shapes. Neither of these characteristics has been very important in construction until recently.

1

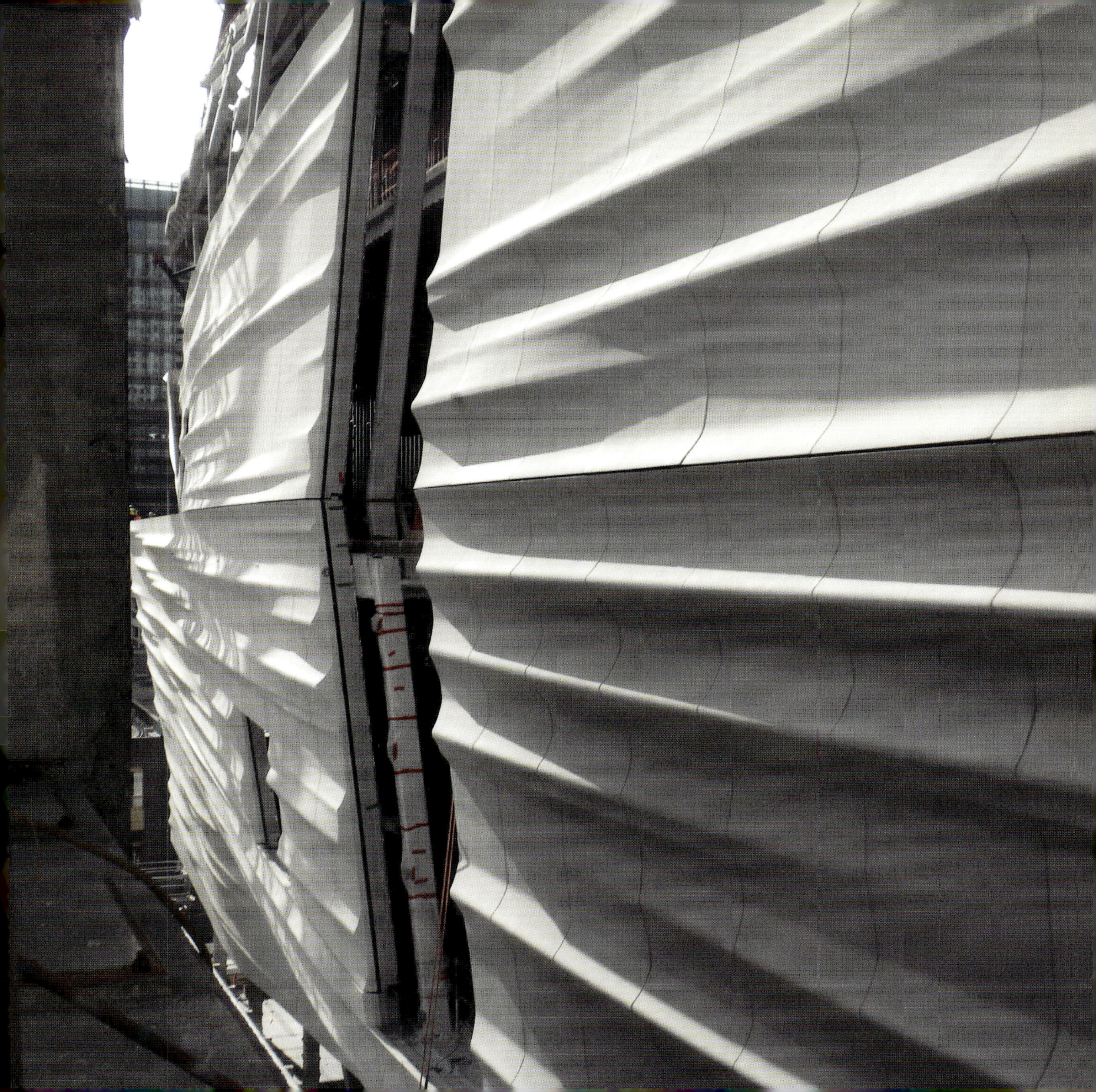

1. SFMoMA east façade during construction. Image: © Enclos.

2. A large CNC machine cuts blocks of EPS foam into the unique geometry of each SFMoMA panel. Image: © Tom Paiva Photography.

3. CNC-machined mold surface being prepped prior to composite lamination. Image: © Tom Paiva Photography.

4. Tower crane lifts 26ft-tall rain screen panel unit into place during construction of the SFMoMA expansion. Image: © Enclos.

After successfully passing a rigorous evaluation process, FRP was chosen because it offered solutions to several problems presented by the use of other systems. Its primary advantages were its light weight and formability, the very features exploited by other industries in the past and now increasingly relevant in contemporary design and construction.

New means, models and materials

Aside from curiosity about something new, several factors are pushing building designers towards sometimes radical departures from traditional means and methods. This shaking up of the status quo, in an otherwise risk-averse industry, is leading to the startlingly rapid deployment of fundamentally new building systems, including to a large degree the building envelope itself. Environmental concern for building materials as well as building operations, health and safety issues relating to building construction and occupancy, rapidly changing regulations and code modifications are driving these new approaches. Additionally, jobsite labour costs, time to delivery and an evolving design ethic brought about by 3D computer modelling are leading designers to consider an array of new ideas, methods and materials. FRP composites represent one of these 'new materials' that offer a fundamentally new approach to building construction. Although still some way off, there is technically no reason why FRP cannot be used to create entire building structures as well as complete envelopes (Lambrych, 2008). Indeed, such composite structures are common in other industries such as aerospace, transportation and marine where monocoque structures are routine.

Meanwhile, FRP composites will find increasing use in non-load-bearing architectural applications in AEC due to their formability, high strength-to-weight ratio, durability and minimal maintenance requirements.

Designers, engineers, builders, owners and even fabricators of FRP products need reliable information about the proper use of FRP in construction. This paper is an attempt to improve the understanding of one of these materials and to address questions, concerns and misconceptions relating to the proper use of FRP on building façades.

Using FRP: context and background

FRP has found limited acceptance in construction despite its proven success in other industries. Its principle advantages are its high strength-to-weight ratio compared to other materials and the ability to consolidate what would otherwise be assemblies of other materials such as wood or metal into a single moulded part. For example, on the Boeing 787, a primary benefit of composites was to significantly reduce the part count.

Recent changes in building codes and design are opening the door to more widespread use of composites in architectural and even structural applications. The American Concrete Institute has adapted a design standard for the use of FRP in concrete structures (American Concrete Institute, 2008, p.440.2R-16), including a design standard for FRP composite rebar in structural concrete (ACI Committee 440, 2015). AASHTO has published a standard for pedestrian bridge designs using composite structures (American Association of State Highway and

Transportation Officials, 2008). DOT initiatives throughout the US and other countries have had experimental bridges and other structures in place for decades and are beginning to publish results indicating successes (American Association of State Highway and Transportation Officials, 2009).

Research questions

- What are the engineering and building code obstacles to overcome to use FRP as an exterior cladding on Type 1 multi-storey buildings in the US?
- How can these obstacles be overcome within the schedule and budget constraints of a project?
- What advantages would an FRP rain screen provide compared to more traditional material alternatives?

The use of FRP cladding on the SFMoMA provides a case study for the use of FRP as cladding on any multi-storey commercial building. Initial prototypes, cost estimation, design assist procedures, code compliance strategies and engineering and installation methods were developed to meet the design intent, budget, project schedule, code requirements and environmental constraints.

Prototype fabrication

The architectural façade design was modelled originally by the architect in Rhino 3D (McNeel Associates) using a Grasshopper script to alter the wavelength, amplitude and frequency of the façade 'ripples' over the curved and tilting east and west elevations of the building. Rhino models can be reliably imported into software used to guide CNC cutting tools (in this case PowerMILL by Delcam) which can be used to cut the shape of the part or its mirror image out of a block of material, thus creating a female mould directly from the architect's model (Fig. 2). Once made, this rapidly and inexpensively created mould can serve to fabricate a full-scale model of any portion of the façade.

Easily fabricated mock-ups serve as a rapid verification of material durability, process fidelity and panel weight. By early fabrication of a full-scale mock-up, such things as material cost per square foot, overall weight, repairability and strength are verified. This step improves the quality of the production cost estimate as well as the architect's and client's confidence in the material option.

Cost estimation

Although no two of the 710 façade panels were the same shape, the use of 3D computer modelling and CNC mould fabrication made cost estimating reliable. Rhino provided an accurate surface area and such key characteristics as panel centre of gravity. Knowing the materials required on a per square foot basis allowed for accurate material cost prediction. PowerMILL includes algorithms that predict milling time for each mould.

Thus, despite the highly complex and variable shapes, accurate cost estimation and scheduling was possible for the tooling phase. Through the use of digital fabrication tools and conventional material, labour and manufacturing overhead allocation methods, a reliable cost could be predicted.

Design assist

An element of contemporary construction is the ever-increasing need for collaboration between the design team and specialty contractors. Although the traditional 'design, bid, build' method is still dominant, it frequently leads to wasted time on the part of design professionals who attempt to produce a plausible 'construction document' based often on insufficient knowledge of materials or fabrication methods. Expecting an architect, or even a façade consultant, to be an expert in composite fabrication can lead to erroneous assumptions, insufficient and inaccurate documentation and faulty conclusions. At best, he or she might propose a solution that does not optimise current technology, which in turn leads to costlier and lower quality solutions.

Expensive hours are spent attempting to develop plausible construction systems in hopes of achieving a 'low' cost proposal. Too often such approaches fail to leverage the current fabrication best practices and can lead to inaccurate and higher risk results than would be the case with a negotiated contract with pre-qualified vendors based on pre-agreed budget targets. More enlightened approaches utilise design assist (Hart, 2007) services, but this method continues to suffer when the selection process is based on responses to 'conceptual' fabrication strategies, typically delivered to the vendors as 2D drawings for contract compliance purposes. When such documents attempt to describe complex shapes, this regularly leads to impossible construction details being applied to the actual 3D environment. Too often these irregularities do not reveal themselves until after the vendor selection process, leading to change orders and wasted time. Solutions to this rapidly growing problem are beyond the scope of this paper, but they must be addressed as soon as possible. These solutions must, among other things, allow a shift to the use of 3D models as construction documents. They must also insist that vendors who participate in complex architectural projects be vetted and fully conversant with mutually compatible software (Miller, 2012) – they must be fluent in the use of the latest digital tools.

Code compliance

Since 2009, the International Building Code (IBC) has recognised FRP as fibre-reinforced plastics and fibreglass-reinforced polymers in Section 2612. This section of Chapter 26 recognises FRP as a combustible material, allowing its use when the product can demonstrate the ability to meet the code requirements applied to similar architectural products. Since FRP can be formulated with a wide variety of mechanical and physical properties, formulations are available that meet most requirements. For building façades, the code allows any façade made of combustible material to be used below 40 feet provided it can pass ASTM E-84 with a Class 2 rating (or better) for flame spread and also meet the appropriate structural requirements. This is a relatively easy standard for FRP materials. Above 40 feet, the code becomes more rigorous. Although passing ASTM E-84 continues to be a requirement, any coverage over 20% of the total surface area means the material must meet the Class 1 requirements of ASTM E-84 for flame spread and smoke density and also pass, among other tests, the rigorous NFPA 285 test. For the SFMoMA addition, this was a major hurdle which had to be cleared before FRP could be seriously considered for the façade material. The specific formulation for the test panel is confidential and is in fact now patented by the panel fabricator. However, the fabricated panel did pass the test and, as a result, composite material was selected for use based on the projection of significant cost and time savings.

Code requirements for engineering of the panels to meet wind, seismic and dead load requirements, including the fixing designs, were met by following standard engineering principles and test standards for the design of similar façade products, with shop drawings stamped by engineers duly licensed to practice in the jurisdiction.

Engineering

FRP has long been the focus of reliable engineering techniques. Indeed, the development of modern FEA (finite element analysis) engineering was driven to a large degree by the need to engineer complex aircraft forms made possible by composites. Aerospace and military uses of composites started in the 1940s, followed by the large compound curved shapes found in marine applications. These applications generated a vast array of ASTM and other standard test procedures, many of which can be used for architectural composite design. The American Composites Manufacturers Association (ACMA) recently published *Guidelines and Recommended Practice for Architectural FRP*, which contains examples of relevant material properties, engineering examples and test procedures for the proper use of FRP in construction (ACMA, 2016).

Fabrication phase

As we have mentioned, one of the unique characteristics of FRP is its very high strength-to-weight ratio.

This feature led to panels whose weight was approximately three pounds per square foot (~15kg/m^2), making them light enough to be affixed to the front of the aluminium unitised panels used to form the waterproof barrier of the building (Fig. 3). This was convenient for several reasons. It eliminated the need for any penetrations of the waterproofing. It allowed the FRP to be fastened to the unitised panels offsite, which meant that the FRP rain screen was installed simultaneously with the unitised wall. This simultaneous installation eliminated the need for a back-up support system and reduced the construction time by replacing an original design that required three trips around the building by three different trades with a design that required one trip around by one contractor. Additional benefits involved less tower crane time, fewer crane moves, easier cleaning, higher quality damage tolerance, superior repairability and lower overall cost (Fig. 4). Comparative lifecycle studies done by Stanford University graduate students in a non-peer-reviewed LCA comparison (Stanford University, 2009) also suggested that the FRP had significantly less impact on the environment compared to the alternative system using GFRC or UHPC.

Another unique characteristic of FRP is that the shape and configuration can be economically customised. Conventional unitised panel systems are most economical and reliable when creating flat walls. The SFMoMA façade was anything but flat. Resolving the problem created by these two seemingly incompatible features presented a unique challenge. How do you make a flat back on an ever-varying front surface? Not only was the front wavy, but it also tilted forward and back as it rose higher and curved in plan through a wide variety of irregular radii. The solution lay in the use of digital tools to create asymmetrical return edges which were different on virtually every panel.

As the façade diverged from a conventional flat wall, the edges of the panels were moulded with edges that varied between 4 and 34in. This allowed for considerable design flexibility before running up to one of these edge dimension limits. When the curve diverged beyond these limits, a custom unitised panel was fabricated to 'twist' the flat wall into a new facet.

Again assisted by digital tools and relying on skilful craftsmanship and valuable collaboration between the FRP façade fabricator and the aluminium unitised wall manufacturer, calculation of the balance between the additional cost of these special twisted unitised panels and the cost of fabricating asymmetrical FRP panels determined the 4 to 34in edge tolerance. The contractor was able to minimise cost while retaining the original architect's shape within a tolerance of less than 1.5in throughout the entire 11-storey elevation.

5. View of SFMoMA east façade from the 5th floor sculpture garden.

6. SFMoMA east façade terrace overlooking sculpture garden.

Images: © Tom Paiva Photography.

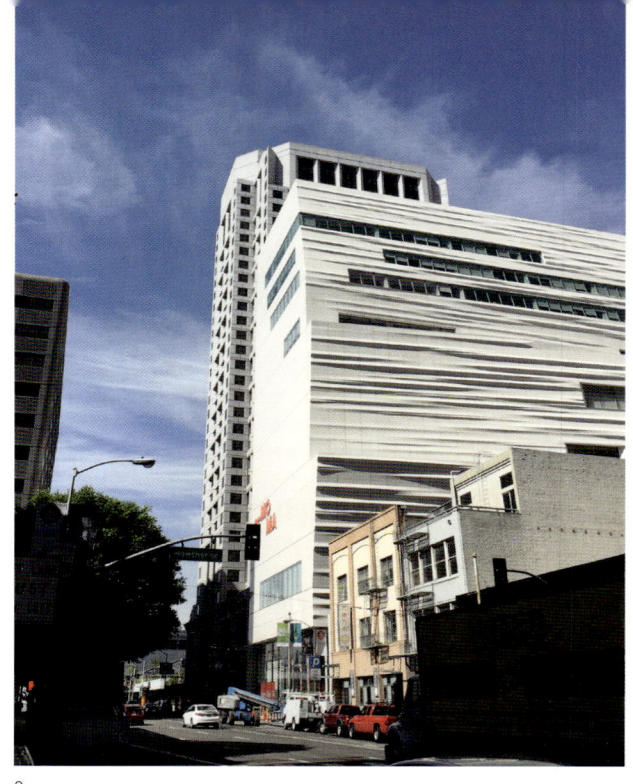

The data show that FRP can meet the IBC acceptance criteria for architectural products. Standard test methods for fire and durability can be applied. ASTM tests exist for composite materials; these tests have been in existence for many years and have proven to be reliable in assisting engineers in designing structures as well as architectural products. In addition, FRP products, in large part because of their high material efficiency, often compare favourably to conventional materials in environmental assessments such as LCA (lifecycle assessment) studies.

The future of FRP

Although somewhat new to the construction industry, FRP is a proven material with decades of successful use in demanding applications throughout the marine, aerospace and transportation industries. To date, FRP composites have been used only infrequently in construction, mainly in remote and extreme environments where the need for prefabrication, light weight and easy assembly have warranted their use.

However, since the engineering of FRP is based on internationally recognised standards, engineers have well-developed guidance to calculate and/or conduct tests. Building code obstacles to the use of FRP have been significantly reduced since the adaptation of Section 2612 of the International Building Code in 2009. Provided the fabricator can meet the requirements of the IBC for a given application, most authorities having jurisdiction will accept properly tested and labelled FRP products.

The process of qualifying FRP while maintaining the schedule and budget constraints depends on many variables. On the project discussed here, fire testing came first to verify code compliance. Passing all requisite tests took approximately five months; however, once passed, these test are valid for three years and can be used for other sufficiently similar projects. Budget constraints are more subjective, but the SFMoMA project was able to demonstrate that successful completion of testing and engineering would more than offset testing costs and would have no negative impact on the project's schedule.

Advantages to using FRP included eliminating two subcontractors and an entire steel support frame weighing over 1,000,000lbs (450,000kg), as well as the improvement of the watertight integrity of the building. Additional benefits were one pass around the building instead of three, which would have been required with the other system, and two fewer moves of lifting equipment such as the tower crane.

While offering many advantages, care must be taken to use industry standard design principles. As with any new material, the specifier of composite materials will be greeted with a wide variety of options and prices.

7. FRP being attached to unit at Mare Island. Image: © Enclos.

8. Street view of SFMoMA east façade from the corner of Hawthorne St. and Howard St. Image: © Kreysler & Associates.

Since quality is a function of fabrication, not unlike concrete, it is incumbent on the designer to exercise caution in selecting a fabricator. Conflicting information needs to be reconciled and verified. Engineers must recognise that this is a highly specialised discipline. Being an anisotropic material, there are virtually limitless options in terms of fibre orientation, fibre volume, number of layers, type of resin, resin filler options, sandwich and single-skin construction techniques and cure options. Engineers have control over a dizzying array of material properties, including even thermal expansion and contraction (CTE), which will vary from carbon fibre and its negative CTE to some resins with higher CTEs than aluminium.

Use of FRP on the SFMoMA and other façade projects in Europe and Asia demonstrates that properly executed work can result in successful outcomes. However, there are ample examples of less successful outcomes. Although FRP has been proven for decades in applications at least as demanding as building façades and often in those that are much more demanding, making decisions based solely on cost is risky and almost certain to yield poor results. With care, appropriate formulation and proper quality control, FRP can not only provide the structural properties to compete favourably with alternatives, but can also meet fire and other code requirements.

Similar to concrete, the mechanical and other critical properties are largely determined during the fabrication process. Stringent quality assurance is essential and close collaboration with a reliable and properly certified fabricator is critical. The IBC code requires that any FRP part delivered to a jobsite must have affixed to it an ICC-recognised independent test agency label certifying that it is manufactured in compliance with the code and subject to third-party inspection. Such a label is the first line of defence in the proper selection of FRP products for buildings.

Future study will need to explore structural opportunities for composites in construction. Engineering examples and ideally an LRFD model for FRP tailored to the construction industry should be developed. Durability case studies need to be assembled from the wide variety of existing examples to improve documentation. Such studies should rely on properly documented empirical evidence and science, of which there are numerous examples (Pauer, 2016).

Acknowledgements

The author wishes to acknowledge the following persons for providing information used in this paper:

John Busel, PE, American Composites Manufacturers Association.
Dr. Robert Steffen, PhD, PE, North Carolina State University School of Construction Management.
Dr. Nicholas Dembsey, PhD, FSFPE, PE, Professor, Wooster Polytechnic Institute, Wooster MA.
Jesse Beitel, PE, FPE, Jensen Hughes Consultants, Baltimore MD.
Kevin Lambrych, CE, Ashland Chemical.
Emily Guglielmo, SE, PE, Martin & Martin Consulting Engineers.
Kurt Jordan, ME, PE, Jordan Composites.

References

ACI Committee 440, 2015, *Guide for the Design and Construction of Structural Concrete Reinforced with Fiber-Reinforced Polymer Bars*, Farmington Hills, American Concrete Institute.

American Association of State Highway and Transportation Officials, 2008, *Design Guidelines for FRP Pedestrian Bridges*, D.C., American Association of State Highway and Transportation Officials.

American Association of State Highway and Transportation Officials, 2009, *AASHTO LRFD Bridge Design Guide Specifications for GFRP-Reinforced Concrete Bridge Decks and Traffic Railings*, D.C., American Association of State Highway and Transportation Officials.

American Composites Manufacturers Association, 2016, *Guidelines and Recommended Practices for Fiber-Reinforced-Polymer (FRP) Architectural Products*, Arlington, American Composites Manufacturers Association.
American Composites Manufacturing Association, 2016, *FRP Architectural Products Guidelines*, available at https://www.surveymonkey.com/r/9GMCPHD (accessed 25 May 2016).

American Concrete Institute, 2008, *Guide for the Design and Construction of Externally Bonded FRP Systems for Strengthening Concrete Structures*, Farmington Hills, American Concrete Institute.

Anon., *Delcam*, available at https://www.delcam.com/software/powermill/ (acccessed 25 May 2016).

Anon., *Robert McNeel & Associates*, available at http://www.en.na.mcneel.com/.

Hart, D., 2007, *The Basics of Design-Assist Contracting*, the AIA.
International Code Council, Inc., 2009, '2009 International Building Code' in *Country Club Hills: International Code Council, Inc.*, p.543

Lambrych, K., *300 Ft. Tall FRP*.

Miller, N., 2012, *The Proving Ground*, available at http://www.theprovingground.org/2012/09/interoperable-geometry-part-1-curves.html (accessed July 2016).

Pauer, R., 2016, *Durability*.

Stanford University, 2009, *Architectural Facades for the Heydar Aliyev Cultural Center*, Stanford.

Turner Construction, *Turner Construction Standard Contract*.

THE 2016 SERPENTINE PAVILION
A CASE STUDY IN LARGE-SCALE GFRP STRUCTURAL DESIGN AND ASSEMBLY

JAMES KINGMAN / JEG DUDLEY / RICARDO BAPTISTA
AKT II

Since 2000, the Serpentine Gallery in London has commissioned a yearly pavilion to be built and displayed during the summer months. A renowned international architect is chosen to design the installation, the only condition being that whoever is chosen has not completed a project in the UK at the time of invitation. These exciting commissions must therefore balance the opportunities for experimentation that a temporary structure affords against an extremely short timeframe: every pavilion must go from initial concept to completion onsite in less than six months.

The 2016 Serpentine Pavilion, designed by BIG (Bjarke Ingels Group) and engineered by AKT II, presents a compelling case study in the use of parametric modelling and advanced structural analysis tools in undertaking such time-constrained projects.

Concept and form

The pavilion centres on a (deceptively) simple concept: two 30m-long sinusoidal walls – one concave and one convex – undulate towards one another, before merging into a single interlocked form at their apex (Fig. 1). Each 14m-high wall is comprised of open-ended boxes and set in an inverse checkerboard pattern to its neighbour, enabling the upper reaches of both walls to overlap and interlock into one continuous cellular grid. Back at ground level, the stepping and staggering of these 40cm-tall boxes creates a 'pixelated' external landscape open to climbing and sitting, while inside BIG has taken the opportunity to sculpt a series of differently scaled spaces intended for seating, a bar and live performances.

This formal ambiguity is reinforced by the use of open-ended boxes: when viewed longitudinally, they appear solid and substantial; however, as a visitor passes through and around, they turn face-on and seem to dematerialise down to mere grids of lines and moiré interference, enabling views through and out of the pavilion to the park landscape beyond (Fig. 3).

Parametric workflow

To realise such a large and structurally complex pavilion, it was necessary to go from concept design to fully

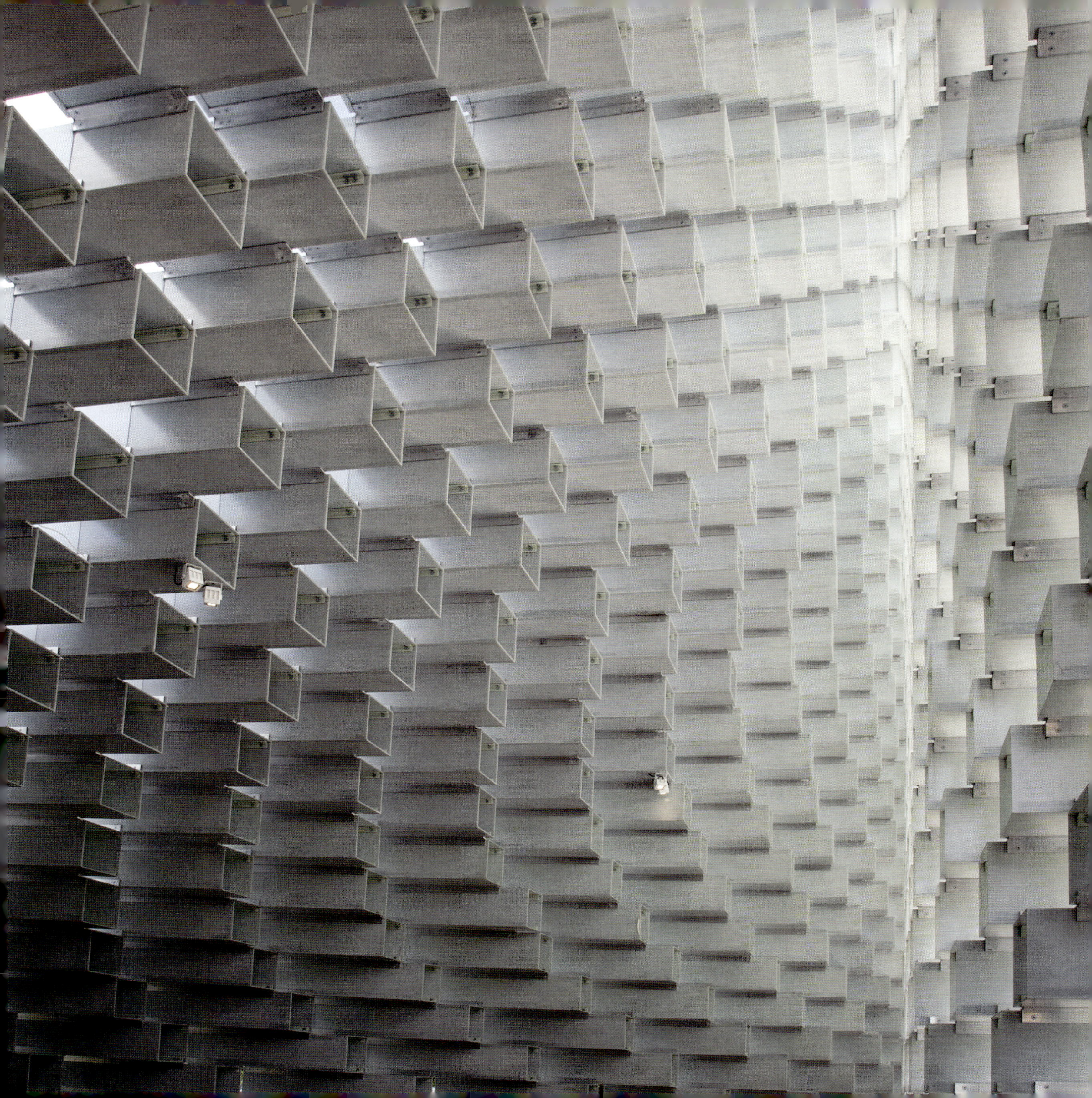

coordinated production information in less than three months. In addition to these time pressures, budgetary constraints necessitated that material topologies and quantities be optimised as far as possible without compromising the ambition of the design.

For these reasons, the BIG and AKT II design teams chose to generate the entire geometry through parametric design processes. This enabled the rapid evaluation of different options for the underlying grid early on, testing the relative merits of rectangular and square grids at different scales, as well as more complicated pin-wheel and reciprocal arrangements for the boxes. For each option, the design team could refine an array of parameters – from micro values such as the individual box height and width through to macro dimensions such as the minimum 'offset' between adjacent boxes, overall wall heights, lengths and sine wave proportions – and interrogate the resulting forms to extract quantities for material volume, number of fixings and so on. At every iteration, these metrics were passed along to fabricators to establish cost and timeframes for production and assembly.

As the initial conceptual phase moved into detailed design, these parametric models had to become more complex and take on additional structural and fabrication criteria. To aid this process, AKT II utilised their proprietary Re.AKT toolkit, which allows information to be freely exchanged and integrated between modelling, analysis and documentation software. This creates a streamlined workflow in which new information from different parties is rapidly 'folded back' into a master model, from where it can propagate outward and update other packages. Re.AKT is always configured specifically for each project based on the scale, typology and materials. In the case of the pavilion, the best Re.AKT workflow was therefore to establish connectivity between Rhino and Grasshopper (geometric modelling), Sofistik and SAP (structural analysis) and Microstation (drawing production). With this parametric workflow in place, the different design teams – spread between the US, UK and Denmark – could rapidly exchange and refine ideas.

Material development

From the earliest stages of the project, BIG emphasised that they wanted to experiment with glass fibre-reinforced plastic (GFRP) manufactured using the 'pultrusion' process. GFRP is a composite material formed of glass fibres encapsulated within a plastic resin matrix that typically has a strength comparable to that of steel, but with only around a quarter of the weight. This high specific strength has made GFRP an attractive material in instances in which weight is critical, such as aerospace and automotive applications, but the labour-intensive manufacturing process of manually placing glass fibres into custom-made moulds has historically made GFRP

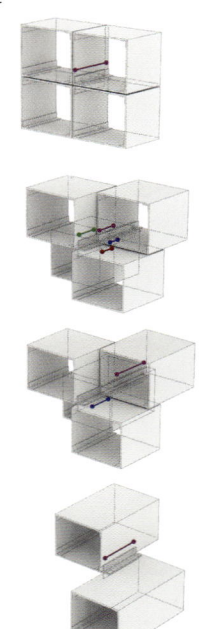

Overlap
2 conditions: true/ false

Box Length
16 conditions: 0.4 – 1.9m

Overlap Typology
5 conditions

Connection Force
3 conditions

Box Wall Thickness
3 conditions: 3, 6 and 10mm

Connector Typology
10 conditions (excluding length)

126 Unique Connector Typologies
including length

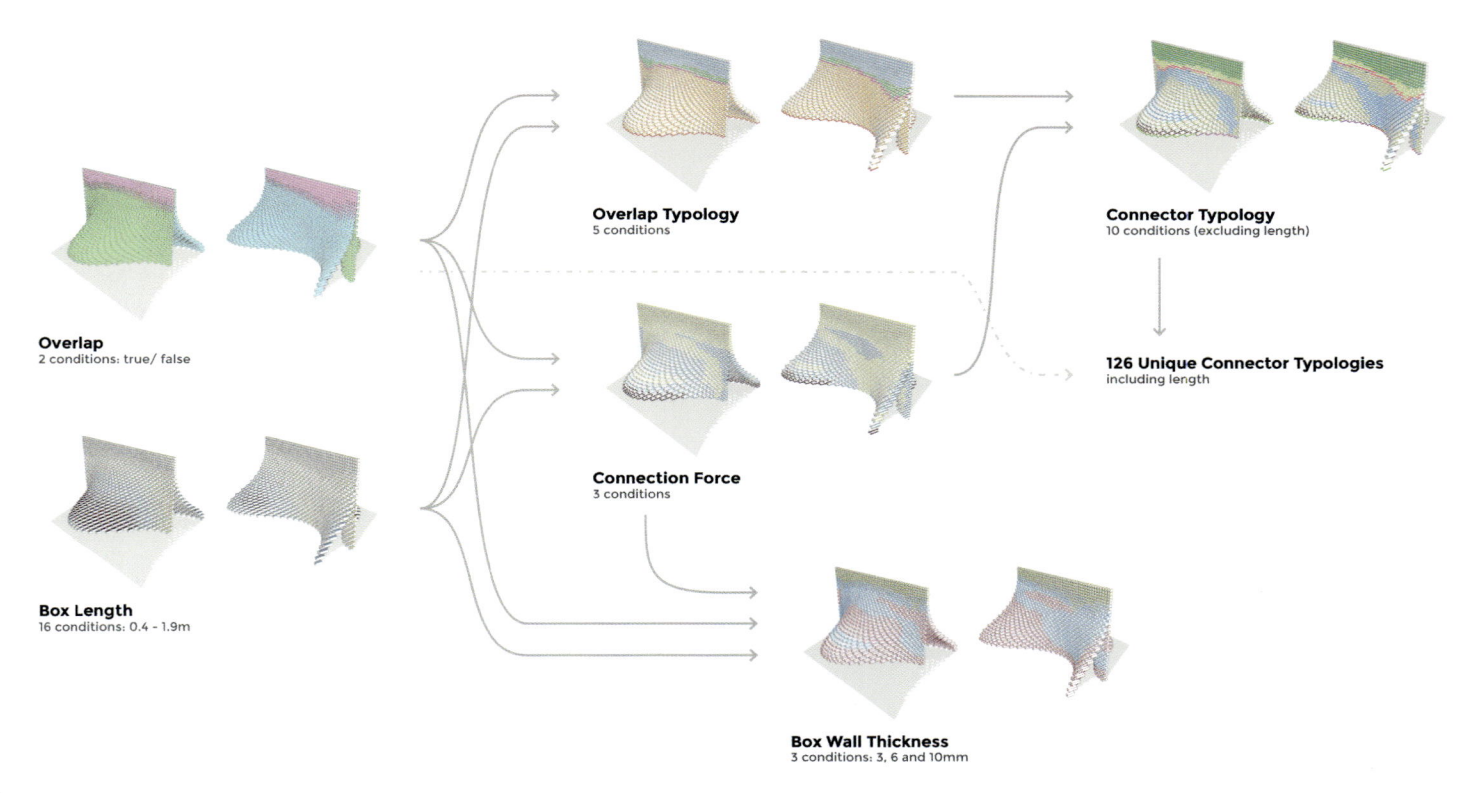

only attractive in niche areas of structural and civil engineering. Manufacturing GFRP using automated processes has been of increasing interest recently as a route to unlocking the benefits of using it at a lower cost. Pultrusion is one of these processes, involving the use of a mould through which the glass fibres are pulled and impregnated with the resin. The resulting material can be produced on a large scale, with a high degree of consistency and at a low cost.

To support our explorations with this material, BIG invited Fiberline Composites A/S to join the project. Fiberline are one of the leading suppliers of pultruded GFRP, and have developed several GFRP products with beneficial structural properties as well as unique colours and transparency levels. Initial discussions with Fiberline focused on the possibility of forming the entire pavilion from a single type of GFRP element – a bespoke extrusion designed specifically for this project that would incorporate both the open box form and corner connections. However, for speed and economy reasons, the design team instead chose a kit-of-parts solution, where each box is assembled from four GFRP plates, with GFRP angles glued in each corner to increase lateral stability and vertical load-bearing capacity. By utilising this system, the project benefited from the very fast production line Fiberline already has in place for manufacturing sheet materials, and a high-dimensional tolerance in the resulting boxes could be assured.

In parallel with this development on the GFRP boxes, the design team considered a number of different options for connecting them. They ran tests on GFRP, carbon fibre and steel connectors before settling on a 10mm-thick cruciform-shaped aluminium that provided the necessary weight-to-strength ratio.

With over 95% of the pavilion made from only these two simple elements, the expression and detailing of the fixings between them was critical. The design team worked through several different options before finally selecting one suggested by StageOne: a bespoke flat-headed bolt-and-sleeve that could be held in place asymmetrically on each box's inside face during tightening, consequently enabling a smaller offset from the neighbouring GFRP angle face. By minimising this offset, the design team could specify shorter 'arms' for all of the connector cross-sections across the

1. Internal space formed by the structural envelope.

2. Connector typologies.

3. Material lightness and translucency.

4. Parametric workflow.

Images: AKT II.

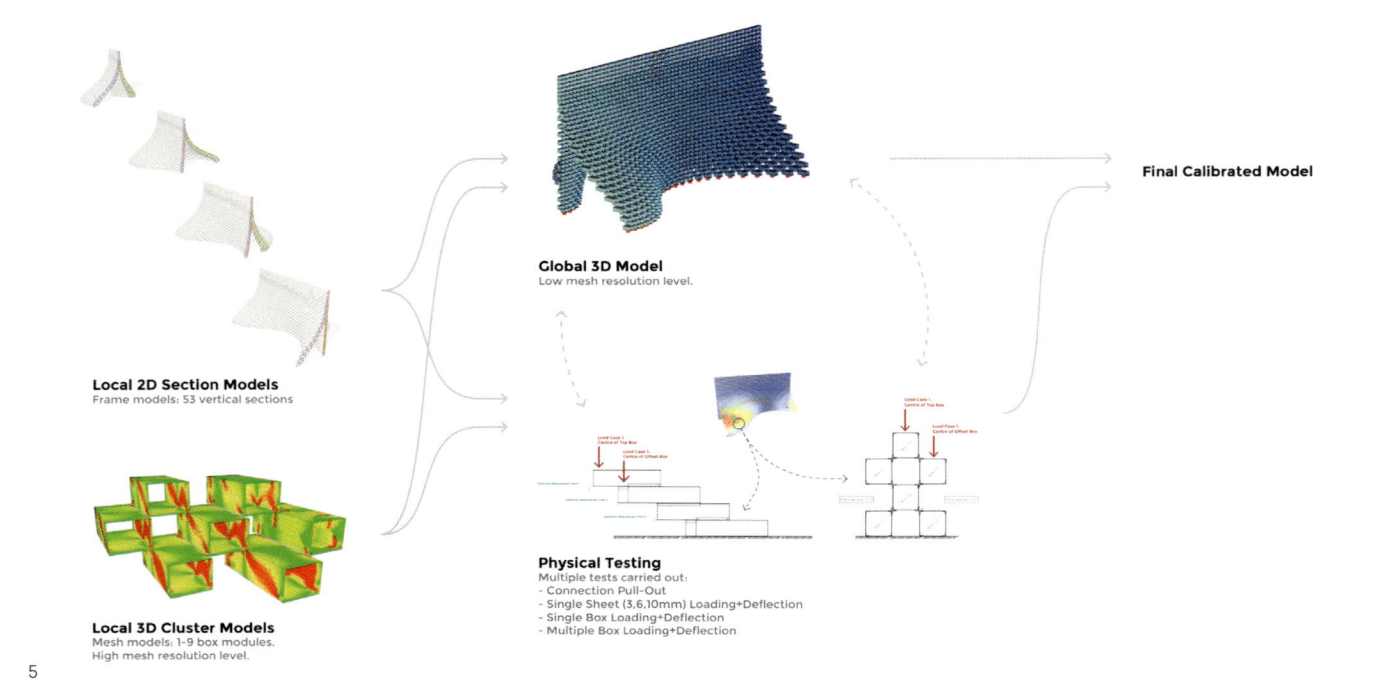

5

entire structure, resulting in faster production times, significant cost savings and reduced weight of box clusters. Advantageous cumulative effects like these were sought at every stage of the design process.

Structural design and physical calibration

Throughout the discussions with Fiberline, the previously established parametric models were used to test and provide feedback on different configurations. With Re.AKT in place, each option could be analysed simultaneously at multiple scales – both globally and locally (Fig. 5). High resolution non-linear finite element analysis (FEA) mesh models of single boxes were generated at first, and later small clusters containing up to a dozen boxes – these were then used to calibrate global 2D and 3D frame models. This process enabled the complex orthotropic behaviour of the boxes to be captured far more accurately and quickly than traditional methods, which was critical given the compressed timeframe of the project and the unusually large number of elements in the pavilion. Without this process in place, it would not have been possible to dissect the load paths and force flows within the structure so finely or to pare down the final design. It is conceivable that utilising these tools allowed the structure to be up to 20-30 percent lighter than if it were engineered in a traditional manner, with all the inherent savings this brings in material, transportation and assembly.

The existing design guidance relating to GFRP is not widely recognised in its application to primary load-bearing structures, outside of highly specific and specialised applications. To resolve this, a series of physical material tests were undertaken by Fiberline in order to provide further calibration and confirmation of the digital models. With the global models calibrated, it was clear that three thicknesses of box could be utilised: 10mm, 6mm and 3mm. This would provide the necessary stiffness where forces were concentrated, while minimising overall weight and cost and maximising the degree of translucency desired by BIG. Likewise, the varying forces present at the connection points could be transferred using either one, two or three pairs of bolts between adjacent boxes.

The final optimised design thus comprises 1,800 boxes of 16 different lengths, as well as 3,500 connectors of 126 different typologies and more than 25,000 bolts (Fig. 4). Although almost every single box and connector is unique (in its combination of length, thickness and bolthole positions)(Fig. 2), with the Re.AKT workflow in place it was straightforward to automate the production of schedules for all three elements, assigning unique codes to aid in fabrication, transportation and assembly sequencing (Fig. 6).

5. Structural calibration and optimisation.

6. Example production drawings and setting-out schedules.

Images: AKT II.

GFRP fabrication

With the major design principles in place, Fiberline began production of the first sheets in Denmark. Several hundred metres of GFRP were extruded each day, along with the matching L-profiles, and these sheets were then cut into shorter plates that matched the different box lengths. At this point, Fiberline developed a bespoke process that quickly and accurately assembled these constituent parts into a completed box. The four GFRP plates of each box were laid out flat on top of a series of ratchet straps, and the corner L-profiles were each glued to one of the plates. A chamfered block of foam was placed in the centre of one plate, and the entire assembly was folded up around this block into a rectangle and bound together with the straps. At this point, air bladders were inserted into the voids between the chamfered block and each corner of the box. By inflating these bladders with high-pressure air, it was possible to maintain a constant pressure along the entire length of the box during curing of the glue, which ensured the quality of the bond.

This process was carried out in stages, so that batches of several hundred completed boxes were regularly transported from Fiberline's facilities in Odense, Denmark, to the Serpentine's chosen contractor StageOne and their workshops in York, UK, for the next stage of production.

Pre-assembly

At StageOne, the arriving boxes were grouped by wall, and assembled into modules across several rows at a time. These modules were necessary given the significant logistical challenges that the Serpentine site poses. The Central London location immediately rules out the use of any special order vehicles and significantly constrains the time window each day during which lorries can access the site. Furthermore, the site's small footprint limits the volume of material that can be stored between deliveries. In response to these constraints and also the truncated programme of the project, the entire structure was prefabricated offsite at StageOne, and a 'just-in-time' delivery system brought small modules to the site on a daily basis. The size of these modules incorporated many factors: incoming material delivery dates, packing efficiency during transportation, reach and load capacity of the onsite mobile crane and stability of the pre-assembled modules during lifting. From this analysis, a 3 x 4 module was found to be optimal.

Even with this method established, the translation from atomised components into the final pavilion appeared to be a daunting hurdle. Setting out the 4-24 boltholes for each box and the 8-24 boltholes for each connector across the entire structure was a task inherently suited to computational working rather than human intuition. However, physically aligning and setting out these

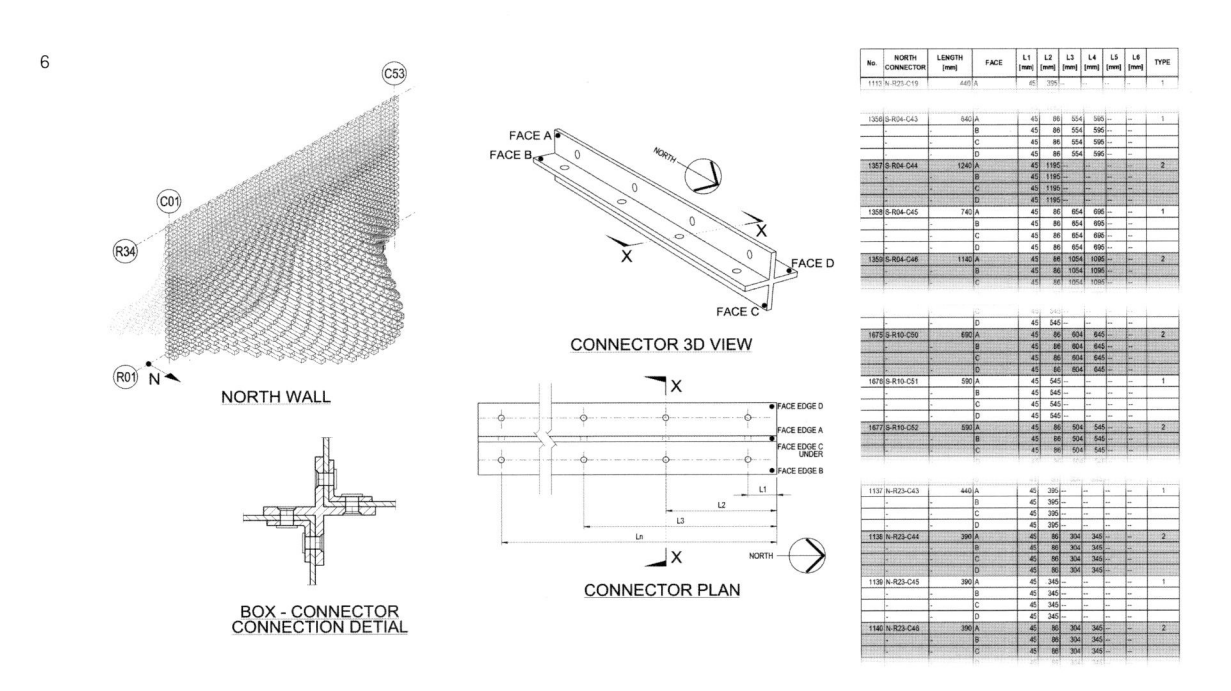

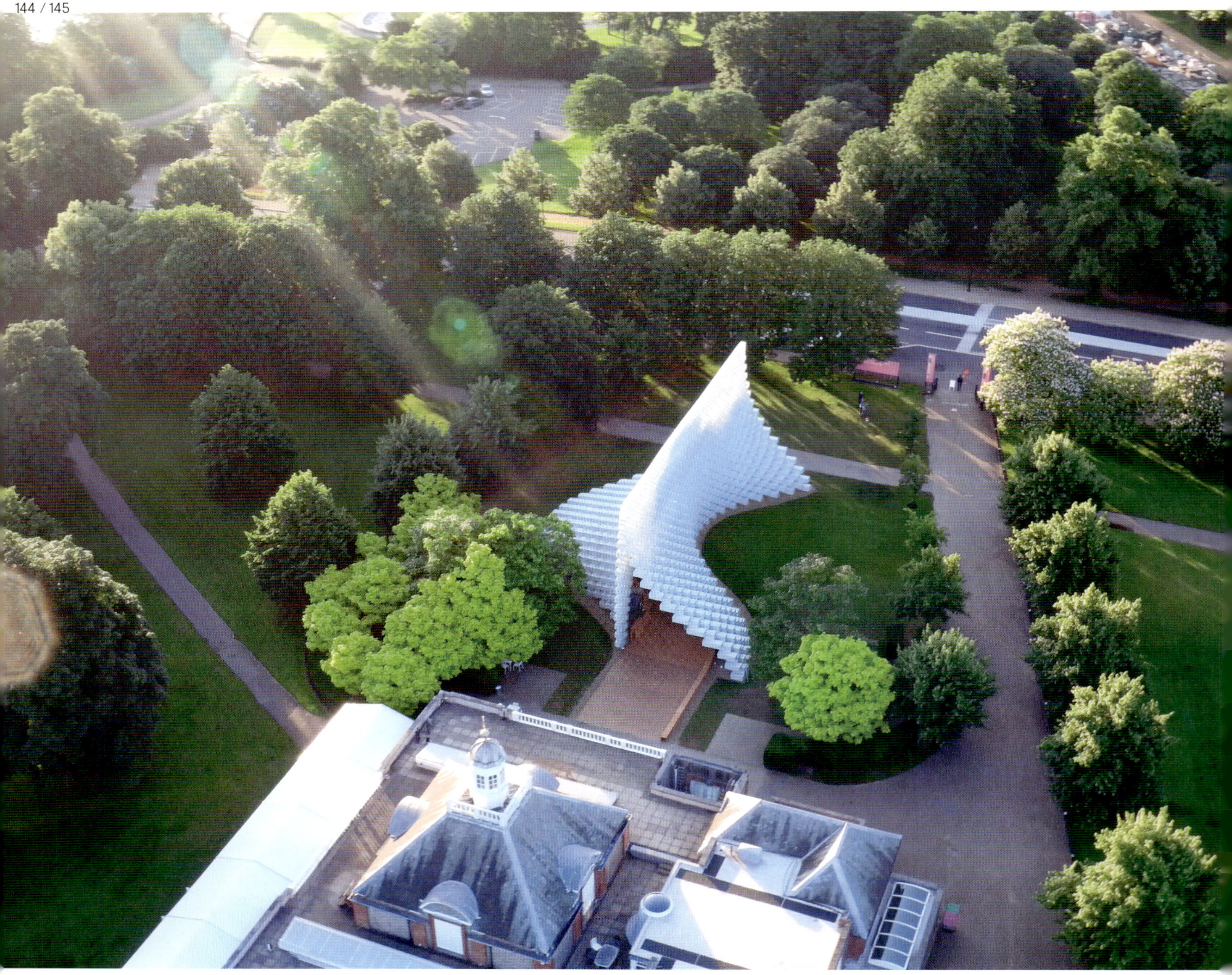

7

components into the complex, non-repetitive form of the pavilion required significantly more dexterity and flexibility than digital fabrication could provide.

This seeming paradox was overcome by fusing CNC and manual fabrication. The manageable size of each connector and more constrained bolt locations made them ideally suited to fabrication using CNC techniques. Once cut and drilled, these elements then became the template used to manually drill the more varied holes for each box. By using a single type of clip-on jig that aligned connectors against their neighbouring boxes, the setting-out was simplified by an order of magnitude.

This process had to be carried out in phases, as even StageOne's facilities could only accommodate a few rows at a time. Once a set of rows was complete, they were all shipped out to the Serpentine (except the uppermost row of boxes): the temporary bolting between modules was removed, and each one was made self-stable using ratchets and wooden props to support it during delivery to the site in London. The retained uppermost row of boxes was placed down on the ground to 'reset' the datum level – positions were checked and, using them as for setting out, the next set of rows began above.

Construction

Onsite, the lowest row of boxes for each wall was set out individually and bolted into a raft slab foundation using around 300 post-fix bolts. These connections ensured a high degree of tolerance and created a definitive datum above which the first modules could be craned into place and rapidly bolted to their neighbours.

The north and south walls rely on each other to provide stability in the form of an arching action in the final condition. While it would have been possible to design the structure for these 'cantilever' forces in the temporary condition, the increase in material thickness required was not economically or aesthetically desirable. Instead, once the pavilion reached a set height, a grid of Layher adjustable scaffolding was utilised to temporarily support specific boxes. This system enabled small adjustments to be made to the position of specific boxes and ensured a good fit where the two halves of the wall merged together.

Once the structure passed above the merge zone, the pavilion was self-stable and the scaffolding could be removed. This allowed the wooden flooring to be laid inside at the same time that the final fully merged rows were added above. Just a few hours before the opening party, the last module was craned into position and the pavilion superstructure was complete (Fig. 7). Over this phase of the project, approximately 300 modules were delivered to the site and connected together in just 25 days.

A rewarding collaboration

A holistic design approach was vital in realising this challenging concept in the time available. The collaboration that emerged between different design disciplines was of itself very rewarding, and was strengthened further by the positive critical and public reception that the 2016 Serpentine Pavilion has received since opening.

Just as significantly, it also seems that the pavilion will continue to advance conversations on material, form and structure in the future. Research is currently being undertaken on live monitoring of its GFRP elements, and the entire pavilion looks likely to tour multiple cities across the world over coming years.

Acknowledgements

We would like to acknowledge and thank the rest of the Serpentine 2016 design team for their unfaltering enthusiasm, expertise and commitment throughout the project.

Client: The Serpentine Gallery (Julia Peyton-Jones, Hans Ulrich Obrist, Julie Burnell).

Designer: BIG (Bjarke Ingels, Thomas Christoffersen, Maria Sole Bravo, Rune Hansen, Max Moriyama, Claire Thomas, Kristian Hindsberg, Maria Holst, Alice Cladet, Lorenz Krisai, Wells Barber, Tianze Li, Aaron Powers).

Superstructure engineers: AKT II (Hanif Kara, Ricardo Baptista, James Kingman, Jeg Dudley, Krzysztof Zdanowicz, Edoardo Tibuzzi, Lorenzo Greco and Stuart Sagar).

Foundation engineers: AECOM (Jon Leach, Amy Koerbel, Michael Orr, Jack Wilshaw, Frances Radford, Katja Leszczynska, Max Smith).

Material supplier: Fiberline Composites A/S (Stig Krogh Pedersen, Preben Venø Nielsen and Fritz Vinter).

Fabricator: Stage One Creative Services Ltd (Ted Featonby, Alan Doyle, James McMillan, Mick Mead).

Technical advisor: David Glover.

7. Aerial view of the 2016 Serpentine Pavilion.
Image: AKT II.

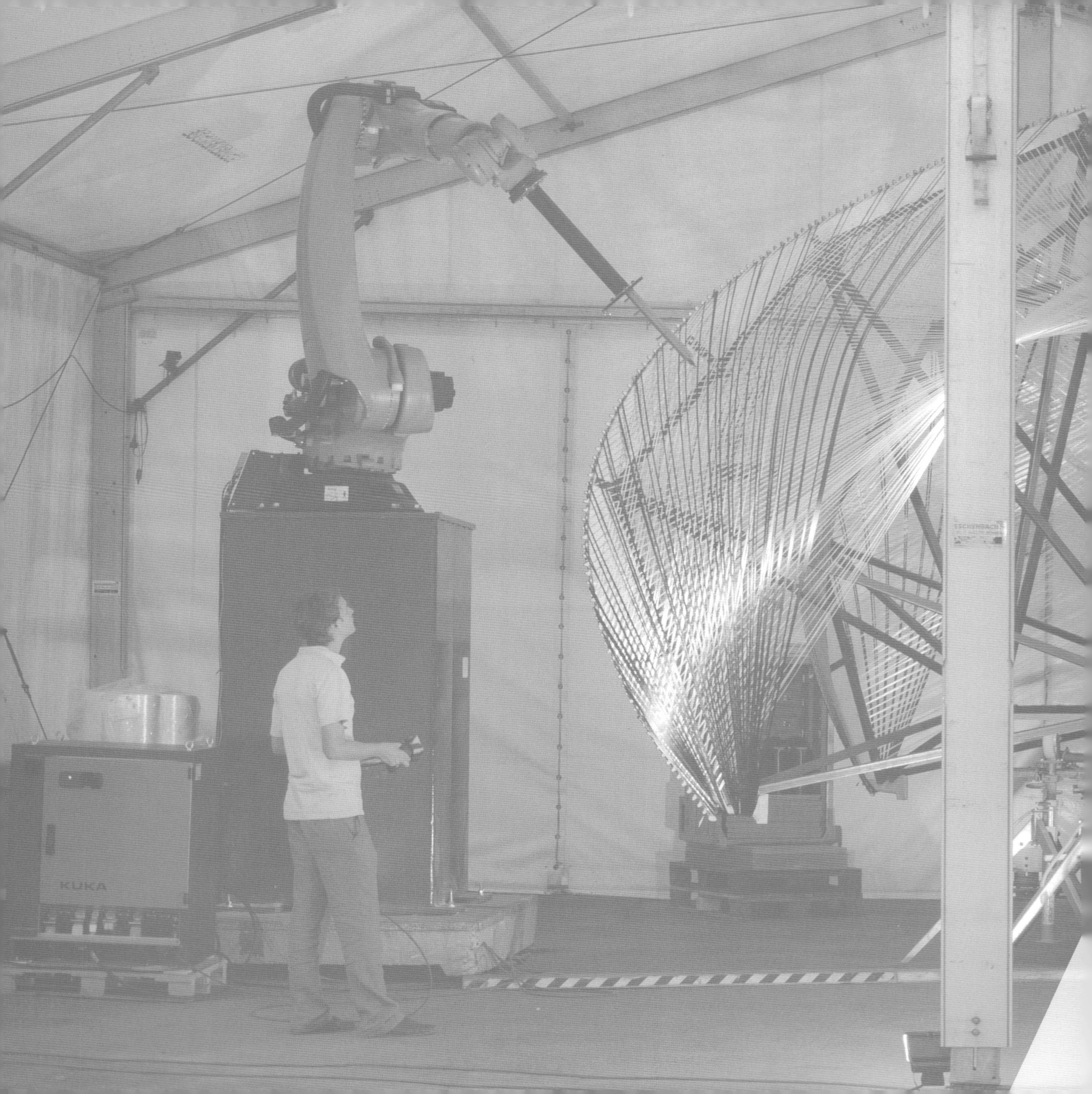
KUKA

Q&A

Q&A BIOGRAPHIES

Mario Carpo

Mario Carpo is Reyner Banham Professor of Architectural Theory and History at The Bartlett School of Architecture, University College London.

After studying architecture and history in Italy, Dr. Carpo was an Assistant Professor at the University of Geneva in Switzerland, and in 1993 he received tenure in France, where he was first assigned to the École d'Architecture de Saint-Etienne and then to the École d'Architecture de Paris-La Villette. He was the Head of the Study Centre at the Canadian Centre for Architecture in Montréal from 2002-06, and Vincent Scully Visiting Professor of Architectural History at the Yale School of Architecture from 2010-14.

Carpo's research and publications focus on the relationship between architectural theory, cultural history and the history of media and information technology. His award-winning *Architecture in the Age of Printing* (MIT Press, 2001) has been translated into several languages. His most recent books are *The Alphabet and the Algorithm* (MIT Press, 2011; also translated into other languages) and *The Digital Turn in Architecture, 1992-2012* (Wiley, 2012). His next monograph, *The Second Digital Turn: Design Beyond Intelligence*, is forthcoming from MIT Press in autumn 2017. Carpo's recent essays and articles have been published in *Log, The Journal of the Society of Architectural Historians, Grey Room, L'Architecture d'aujourd'hui, Arquitectura Viva, AD/Architectural Design, Perspecta, Harvard Design Magazine, Cornell Journal of Architecture, Abitare, Lotus International, Domus, Artforum* and *Arch+*.

Jenny Sabin

Jenny Sabin is an architectural designer whose work is at the forefront of a new direction for twenty-first century architectural practice – one that investigates the intersections of architecture and science, and applies insights and theories from biology and mathematics to the design of material structures. Sabin is the Arthur L. and Isabel B. Wiesenberger Assistant Professor in the area of Design and Emerging Technologies and the newly appointed Director of Graduate Studies in the Department of Architecture at Cornell University, where she is also establishing a new advanced research degree in Architectural Science with a concentration on Matter Design Computation. She is Principal of Jenny Sabin Studio, an experimental architectural design studio based in Ithaca, and is Director of the Sabin Design Lab at Cornell AAP, a transdisciplinary design research lab with a specialisation in computational design, data visualisation and digital fabrication.

In 2006, she co-founded the Sabin+Jones LabStudio, a hybrid research and design unit, together with Peter Lloyd Jones.

Sabin is also a founding member of the Nonlinear Systems Organization (NSO), a research group started by Cecil Balmond at PennDesign, where she was Senior Researcher and Director of Research. Sabin's collaborative research, including bio-inspired adaptive materials and 3D geometric assemblies, has been funded substantially by the National Science Foundation, with applied projects commissioned by diverse clients including Nike Inc., Autodesk, the Cooper Hewitt Smithsonian Design Museum, the American Philosophical Society Museum, the Museum of Craft and Design, the Philadelphia Redevelopment Authority and the Exploratorium.

Sabin holds degrees in ceramics and interdisciplinary visual art from the University of Washington and a Master of Architecture from the University of Pennsylvania, where she was awarded the AIA Henry Adams first prize medal and the Arthur Spayd Brooke gold medal for distinguished work in architectural design in 2005. Sabin was awarded a Pew Fellowship in the Arts 2010 and was named a USA Knight Fellow in Architecture, one of 50 artists and designers recognised nationally by US artists. She was recently awarded the prestigious Architectural League Prize for Young Architects and was named the 2015 national IVY Innovator in design.

She has exhibited nationally and internationally, including in the acclaimed 9th ArchiLab titled 'Naturalizing Architecture' at FRAC Centre, Orleans, France, and most recently as part of 'Beauty', the 5th Cooper Hewitt Design Triennial. Her work has been published extensively, including in the *New York Times, The Architectural Review, Azure, A+U, Metropolis, Mark Magazine, 306090, American Journal of Pathology, Science* and *Wired*. She co-authored *Meander, Variegating Architecture* with Ferda Kolatan in 2010. Her forthcoming book, *LabStudio: Design Research Between Architecture and Biology*, co-authored with Peter Lloyd Jones, will be published in 2017.

Ronald Rael and Virginia San Fratello

Rael and San Fratello are Professors at the University of California, Berkeley, and San Jose State University respectively. Rael has a joint appointment in the Departments of Architecture and Art Practice, and San Fratello directs the Interior Architecture programme at the Department of Design. Prior to joining the faculty at Berkeley and San Jose, they were co-Directors of Clemson University's Charles E. Daniel Center for Building Research and Urban Studies in Genova, Italy, Professors at the University of Arizona and Faculty at the Southern California Institute of Architecture (SCI-arc). Rael and San Fratello earned their Master of Architecture degrees at Columbia University in the City of New York. Rael is the author of *Earth Architecture* (Princeton Architectural Press, 2008), a history of building with earth in the modern era to exemplify new, creative uses of the oldest building material on the planet. San Fratello is the winner of the prestigious Next Generation Design Award.

Rael San Fratello, established in 2002, is an internationally recognised award-winning studio whose work lies at the intersection of architecture, art, culture and the environment. In 2014, Rael San Fratello was named an 'Emerging Voice' by The Architectural League of New York. Their work has been published in the *New York Times, MARK, Domus, Metropolis Magazine, PRAXIS, Thresholds, Log* and *Wired*, and their writing features in numerous books and journals. In the past 10 years, Rael San Fratello has won, been selected as a finalist, placed or been recognised in nine high-profile international competitions, including WPA 2.0, Sukkah City, Life at the Speed of Rail, SECCA Home/House and Descours. Research by Rael San Fratello has resulted in the start-up Emerging Objects. Emerging Objects is an independent, creatively driven MAKE-tank at the forefront of 3D printing architecture and design, where innovative materials can be printed at unprecedented sizes.

Monica Ponce de Leon

Monica Ponce de Leon is a pioneering educator and award-winning architect. Since January 2016, she has been the Dean of Princeton University's School of Architecture. Since 2008, Ponce de Leon has been Dean of the Taubman College of Architecture and Urban Planning at the University of Michigan-Ann Arbor, where she is also the Eliel Saarinen Collegiate Professor of Architecture and Urban Planning. Before her appointment at the University of Michigan, Ponce de Leon was a professor at the Harvard Graduate School of Design, where she served on the faculty for 12 years.

A recipient of the prestigious National Design Award in Architecture from the Cooper Hewitt, Smithsonian National Design Museum, Ponce de Leon co-founded Office dA in 1991, and in 2011 started her own design practice, MPdL Studio, with offices in New York, Boston and Ann Arbor.

Ponce de Leon has received the Academic Award in Architecture from the American Academy of Arts and Sciences, the USA Target Fellow in Architecture and Design from United States Artists and the Young Architects and Emerging Voices awards from the Architectural League of New York. Her work has received a dozen Progressive Architecture Awards, several awards from the American Institute of Architects (AIA) and numerous citations.

She is widely recognised as a leader in the application of robotic technology to building fabrication. Building upon her work as Director of the Digital Lab at Harvard, at the University of Michigan she developed a state of the art student-run digital fabrication lab, integrating digital fabrication into the curriculum of the school. In large part because of her pioneering work, the use of digital tools is now commonplace in architecture schools across the country.

As a practising architect who is deeply committed to architectural education, Ponce de Leon builds bridges between academia and practice, underscoring the interdisciplinary nature of architecture by encouraging experimentation and critical thinking in the curriculum. As a dean and an educator, Ponce de Leon has emphasised the connections between scholarship, research and creative practice. Under her leadership, the college's Liberty Annex has served as a think tank for faculty and student collaboration fuelled by innovative seed funding.

Ponce de Leon has also held teaching appointments at Northeastern University, the Southern California Institute of Architecture, Rhode Island School of Design and Georgia Institute of Technology. She earned a Master's degree in architecture and urban design from the Harvard Graduate School of Design and a Bachelor's degree in architecture from the University of Miami.

Carl Bass

Carl Bass is a member of the Autodesk board of directors and is presently serving as an advisor to the company. During his 24-year tenure at Autodesk, he has held a series of executive positions, including President and Chief Executive Officer, Chief Technology Officer and Chief Operations Officer. Bass co-founded Ithaca Software, which was acquired by Autodesk in 1993. Bass also serves on the boards of directors of HP Inc., Zendesk Inc. and Planet, on the boards of trustees of the Smithsonian's Cooper-Hewitt National Design Museum, Art Center College of Design and California College of the Arts and on the advisory boards of Cornell Computing and Information Science, UC Berkeley School of Information and UC Berkeley College of Engineering.

He holds a Bachelor's degree in mathematics from Cornell University. Bass spends his spare time building things – from chairs and tables to boats and, most recently, an electric go-kart.

Antoine Picon

Antoine Picon is the G. Ware Travelstead Professor of the History of Architecture and Technology and Director of Research at the GSD. He teaches courses in the history and theory of architecture and technology. Trained as an engineer, architect and historian, Picon works on the history of architectural and urban technologies from the eighteenth century to the present. His most recent books offer a comprehensive overview of the changes brought by computing and digital culture to the theory and practice of architecture, as well as to the planning and experience of the city. He has published, among others, *Digital Culture in Architecture: An Introduction for the Design Profession* (2010), *Ornament: The Politics of Architecture and Subjectivity* (2013), *Smart Cities: Théorie et Critique d'un Idéal Autoréalisateur* (2013), and *Smart Cities: A Spatialised Intelligence* (2015).

Picon has received a number of awards for his writing, including the Médaille de la Ville de Paris, the Prix du Livre d'Architecture de la Ville de Briey (twice) and the Georges Sarton Medal from the University of Gand. In 2010, he was elected a member of the French Académie des Technologies. He has been a Chevalier des Arts et Lettres since 2014. He is also Chairman of the Fondation Le Corbusier. Picon received science and engineering degrees from the École Polytechnique and from the École Nationale des Ponts et Chaussées, an architecture degree from the École d'Architecture de Paris-Villemin and a PhD in history from the École des Hautes Etudes en Sciences Sociales.

Q&A 1
JENNY SABIN
MARIO CARPO

MARIO CARPO Where is your work heading right now? What are the key ideas?

JENNY SABIN As you know, one of the driving questions and obsessions in my work is fuelled by the diminishing gap between design intent and that which is materialised – what is modelled, rendered, etc, through scripts and algorithms – and how that meets the material world via issues of fabrication and material constraints. I am really interested in that operating as a loop, both in the way I think through a design process and in the way it impacts on the tools that I produce and the projects that I generate. And at the core of that loop – which has driven an ongoing interest in, say, textiles and weaving and the origins of digital space – is, very importantly, the presence of the human and often the human hand (at least within the analogic prototyping stage).

Right now, my work is really about interventions within that loop generating feedback mechanisms. The latest paper to come out of my lab is 'Robo-sense: Context-dependent Robotic Design Protocols and Tools'. Mario, in your 2013 work for *AD*, called *The Art of Drawing*, you reference Brunelleschi's use of the turnip as a way of modelling and conveying design intent to artisans onsite. So, what is the equivalent of the turnip now? In my lab, we have been working on developing pipelines and software that allow for collaboration with machines such as Sulla, our large industrial 6-axis robot. I am interested in instilling a degree of design intuition in order for the interface to be more user-friendly and personal. This goes alongside the broader issues of fabrication and materiality, but really generates feedback and collaboration with these machines. The human is very much at the core in terms of intuition and integration within the design process. On the practice end of my work, I have been working in digital ceramics. One of the ongoing projects is called *Polybrick*, where we work with 3D printing to question how bricks are made. We have developed a way to 3D print non-standard bricks, where each brick is different and yet there is a coherence to how they assemble. This has been exciting to push forward, both in the context of how the bricks are structured and assembled and in how we conceive of the wall as an interface.

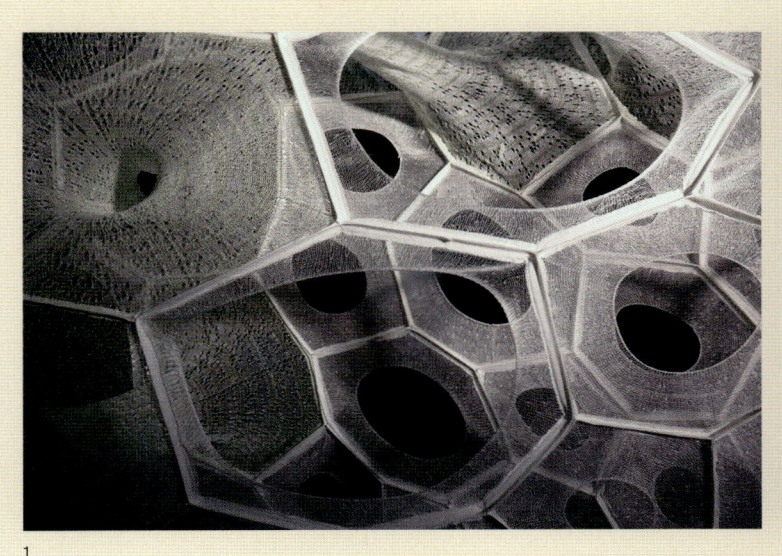

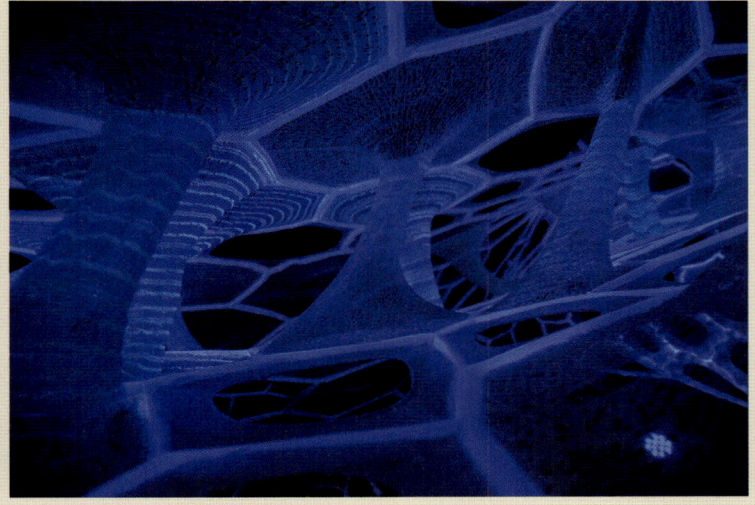

A second project I just opened recently, which continued an ongoing collaboration with Dr. Peter Lloyd Jones, is a project installed in Philadelphia called *The Beacon*. In this project, we worked with drones to dynamically weave a second exterior skin around a 20ft-tall modular steel structure over the course of 10 days. The project looks at the intersection between medicine, architecture and emerging technologies, and at the future of all three. The drones and the *Beacon* project overall served as an analogue and marker for discussion, and also as a public spectacle. It was exciting to take on something new, where you aren't restricted to the six axes of a robot but are completely freeform in space as the drones deposit threads in a generative fashion. We had some failures, but I see it as an experimental act that will be looping back into the ongoing research trajectories within my lab.

MC When we started to deal with computation for the manipulation of complex materials, i.e. of non-standard materials with non-linear behaviour, there was this idea that we could at long last engage with the indeterminacy and complexity of natural and organic matter – which is a reversal of the tradition of structural design. Architecture, even building, since the beginning of time, always tried to standardise natural materials or invented new materials in order to make them simple, isotropic and standard, so we could more easily notate them, calculate them and fabricate them. This is the story of the scientific and industrial revolution. But this trend, which started in Greece more than 2,000 years ago, came to a halt, to a sudden reversal, 15–20 years ago because with computers we could do exactly the opposite. Instead of making materials more standard, we could engage with materials just as they are, because through simulation and computers we could increasingly deal with the unpredictability, complexity, indeterminacy and randomness of materials as found – we could even design material with variable properties when needed. This was the fascination, the dream, the excitement of 15–20 years ago. Do you have the impression that we are still on the same wavelength today? Or (and this is just a suspicion I have) are the powers of computation so immense that even if we sometimes delude ourselves into thinking that we are dealing with complex materials, we are in fact only dealing with them, no matter how complex they may be, because we can reduce them, simplify them and calculate them more or less as a traditional engineer would always have done? But if that is so, then what we see as tools of vitality, indeterminacy and intuition are in fact traditional tools of notation, except that they are so powerful that we can now almost determine complexity in a sense, at least to the extent needed for some practical purpose – not reversing but fulfilling the dream of a nineteenth-century engineer. Twenty years ago, we thought we were doing the opposite. When I look at the work of some of our friends, I have an impression that the discourse they make is still a discourse of postmodernism and indeterminacy, but the practicality of what they do is the opposite – it is almost traditional engineering, amplified by the power of computational tools. Do you perceive a risk or an ambiguity in these two diverging strategies?

JS I would agree with the idea of an ambiguity. But I would also argue that 15–20 years ago there was a severe lack of any engagement with materials and materiality. I still think only a small percentage of our friends actually engage with materials and include them as part of their design protocols. I find it

1. Detail of *PolyThread* fabric structure, composed of seamless 3D digitally knitted conical forms. Image: Bill Staffeld.

2. Interior detail between the upper and lower surfaces of *PolyThread* with photoluminescent responsive threads activated. Image: Max Vanatta.

PolyThread by Jenny Sabin Studio, 2016. Commissioned by Cooper Hewitt Smithsonian Design Museum for the Beauty – Cooper Hewitt Design Triennial. Designed by Jenny E. Sabin, Jenny Sabin Studio. Design team: Martin Miller, Charles Cupples. R&D by Sabin Design Lab at Cornell University.

exciting that this is not only becoming part of their discussions, but also drives them. For example, in one of my collaborative projects, *eSkin*, we work with nano- to micro-scale features and effects, attempting to understand which material features are actually scalable. One of the primary topics is structural colour, which is wavelength-dependent colour change. There are numerous examples in nature that exhibit structural colour, so we have been extracting, synthesising and redeploying those constraints and features with the idea and hope that we can move some of it into architecture that is scalable. I found, with my team, that there was no existing software that was robust enough to render the complexity of these materials. We developed our own tools, working side by side with material scientists, to simulate and approximate the complexity of these effects, so that we could meaningfully and responsibly embed them into our design process. Having said that, I don't think we are quite there yet. Another intriguing example relates to one of my pavilion projects – a recent commission for the Cooper Hewitt, titled *PolyThread*, where I worked very closely with an engineer from Arup. I wanted to dig into the behaviour of the knitted material, from stitch to row to whole, so we did many analogue stress tests, embedding them into our simulations. And yet, despite our sophisticated testing, we were still not able to completely simulate it. So yes, I think there is ambiguity.

MC Yes, it goes both ways. We can adapt our design process to the randomness, or the animation, of the material – as if the material were a cat that behaves unpredictably, and you have to cope with it because a cat is temperamental. The other way is to use the power of simulation to diminish the area of unpredictability, and to design in a very traditional way with complex materials for which we can now model a level of granularity that a traditional engineer could never have dreamed of. And since we can also work at this hyper-granular scale in the traditional way, i.e. with a deterministic design-and-prediction methodology, we end up with two games that can co-exist. Which is more important in your work? Do you want to play with the cat? Or do you want to tame it?

JS I am partial to the unpredictable cat. I am intrigued by the unexpected, and the agency of the material that one must respond to in the design process. I think in both my core research and applied projects, there is also a process that is slow, analogue and about the integration of the human hand. This usually happens at the prototype phase and is so crucial to allowing for the emergence of the unexpected, which I then opportunistically tame, but only in pursuit of the next potential scalability. So I would say I never want to fully tame that unpredictability. It is crucial to the innovation and to beauty.

MC And this is where digital tools afford a level of interaction with the naturality of the material which until recently only an expert artisan would provide. An expert artisan can deal with whatever irregularity is found in a chunk of timber because that is his skill, his intuition – he doesn't need to make an x-ray of a log. If the log has a hole inside, he can just feel it or, by just tapping on it, can hear the reverberation of the sound. Likewise, if a particular log has some irregularity, he can work around it. Machines traditionally couldn't work this way, so we invented plywood or we converted timber into an industrial material which is the same all over because mechanical machines cannot deal with anything else. But sensors and computers can now increasingly interact with irregular materials almost as well as a skilled artisan could. But the point is that there are not many skilled artisans that can still do that, whereas every computer can with the right programme. So that is our advantage today. And it is a reversal of the traditional science of materials. Until recently, the rule was to make materials standard; and now it is to take them as they are,

3. Detail of responsive *eSkin* prototype featuring structural colour change. The goal of *eSkin* is to explore materiality from nano- to micro-scales based upon an understanding of the dynamics of human cell behaviours.

4. *eSkin* prototype featuring dynamic switching of structural colour and transparency change over time.

eSkin, 2010-2014 / Jenny E. Sabin and Andrew Lucia (architecture), Cornell University; Shu Yang (materials science), Jan Van der Spiegel and Nader Engheta (electrical and systems engineering), Kaori Ihida-Stansbury, Peter Lloyd Jones (cell biology), University of Pennsylvania. This project is funded by the National Science Foundation Emerging Frontiers in Research and Innovation, Science in Energy and Environmental Design, and is jointly housed at the University of Pennsylvania and Cornell University. This prototype was originally on view as part of the 9th ArchiLab, FRAC Centre, Orléans, France.

Images: Jenny E. Sabin.

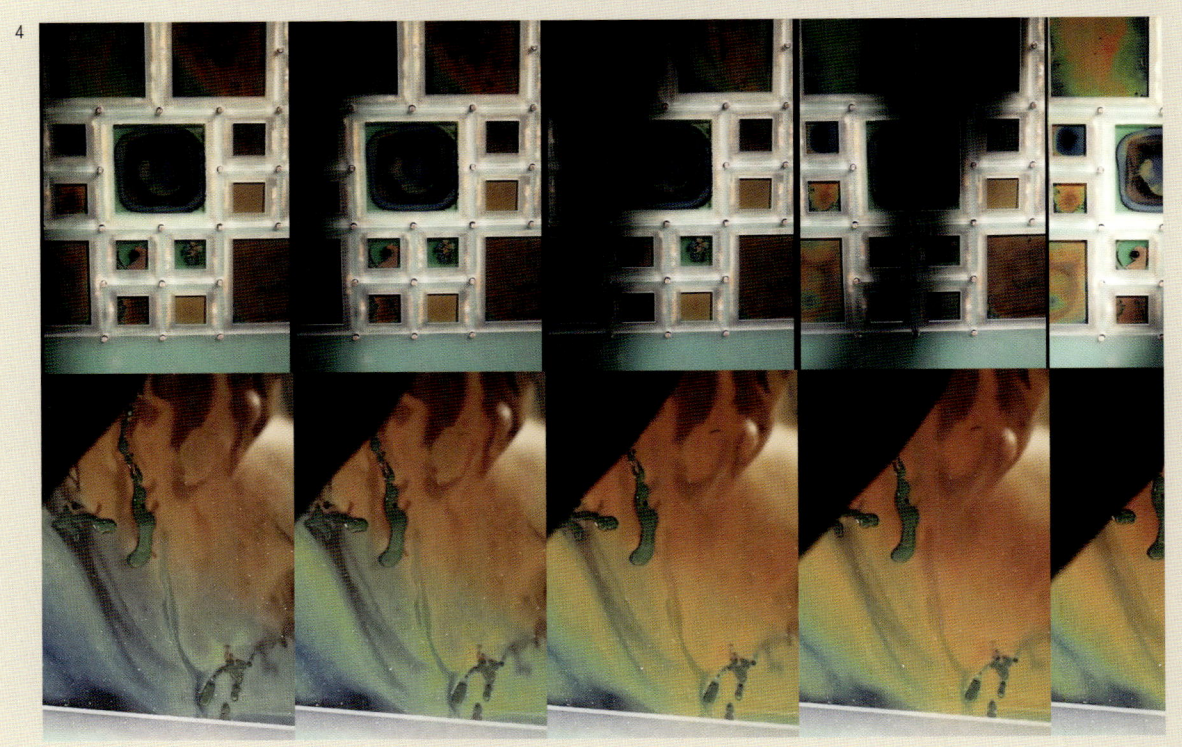

154 / 155

5

6

5. *PolyBrick 2.0* is based on the rules and relationships governing human bone formation. Detail of a highly porous 3D printed brick.

A project by Sabin Design Lab, Cornell University, 2016. Principal investigator: Jenny E. Sabin. Design research team: Jingyang Liu Leo and David Rosenwasser. Currently on display at the Cooper Hewitt Design Museum as part of Beauty – Cooper Hewitt Design Triennial.

6. *PolyBrick 1.0* makes use of algorithmic design techniques for the digital fabrication and production of nonstandard ceramic brick components for the mortarless assembly of 3D printed and fired ceramic bricks.

A project by Sabin Design Lab, Cornell University, 2015. Principal investigator: Jenny E. Sabin. Design research team: Martin Miller and Nicholas Cassab. Currently on display at the Cooper Hewitt Design Museum as part of Beauty – Cooper Hewitt Design Triennial.

Images: Courtesy Cooper Hewitt Design Museum.

because if you scan them and push the resolution of the scan as far as needed, there is a level where the material becomes predictable again and you can design at that scale. But again, at that point intuition is replaced by calculation, so the magic of the game may be lost in favour of the predictability of the result. So it is a difficult game, which again can go both ways. But what worries me is that we are often still within the frame of mind of postmodernism; we still interpret non-linearity in the way that Manuel DeLanda taught us all so well. But at the same time the tools we use achieve levels of design predictability that are no longer those of traditional structural engineers, but rather those of a surgeon or a dentist. This is where I think there may be a divergence between our frame of mind and the tools we use: we see them through the lens of an ideology which no longer applies to the way these tools actually work. In their thinking, Cecil Balmond and Hanif Kara celebrate the magic of the material in a way that a traditional engineer would find delirious. But the teams of engineers that work at Arup do not really see their work that way.

JS Cecil Balmond is one of my most important mentors. I was a student of his, then I taught a seminar with him called Form and Algorithm, and then we taught a research studio together for several years at PennDesign before I came to Cornell. I agree with your worry, and I think, for me, the focus is on the integration of intuition and the technologies, and also that references my ongoing interest in biology, starting with the foundation I formed around a decade ago with Dr. Peter Lloyd Jones, who is a matrix biologist by training. For me, it has very much been a process; it is only recently that I have been able to articulate this for myself, in terms of positioning the use of matrix biology in design. For me, it is about thinking and the impact it has had upon my design process. For example, in matrix biology the big idea is that half the secret to life resides outside of the human cell. So its active morphologies and form are specified by protein events within the dynamic extracellular matrix. So very early on, I was presented with a series of incredibly powerful ecological models for us to consider. And so, it is less about bio-inspiration as formal expression, although there is a residue of that in the work, and more about a way of thinking that engages feedback loops, where events – within a material, or inspired by intuition, or through a collaboration – together create a dynamic choreography that produces the form. The tools I am developing now are pushing this forward in a really sophisticated way. But I agree that we are in an ambiguous moment – and we get caught sometimes.

MC Yes, and it is inevitable, because we are in fact to a large extent dealing with the unknowns of artificial intelligence. Furthermore, the non-standard paradigm that the architectural community has been nurturing since the 1990s is now going mainstream. So there are a lot of technological applications for these ideas that 20 years ago were seen as wacky, impractical and impossible to exploit. Now, with the ubiquity of digital tools, they are the rule.

JS I see the most impact not necessarily in industry and the built environment, but in other fields. For example, in my collaboration with Peter, he is looking at how this thinking can impact medicine. But also, given that architecture has been familiar with these ideas for over 20 years, what are your ideas on the role of the architect in pushing forward a discussion that is now in every field?

MC Paradoxically, I think the traditional role of the architect, as it has been known since the Renaissance – even if the tools and traditional models are gone – is to have a bigger picture, to be in charge simultaneously of notation/geometry, calculation/structure and fabrication/technology without being a specialist in any of these fields. This is still a role for which some architects are uniquely prepared, but it is a rare position. Of 100 of our students, 95 will become specialists, and they will sell specialised skills in a specialised marketplace to earn their living. The remaining 5% will be those who will have this general holistic view of how we make things. We train them knowing that most of them will end up being specialists and a few will end up being architects. And this is good, because we need the generalist and the specialist.

JS So what you are stating is that we need both?

MC Well, 90-95% of our students will only be as good as they need to be to become specialists, but it was always that way. Is 5% pessimistic? You train designers – I only teach history, so I don't know.

JS I would say the number is higher in terms of those who become architects – whether they are acting within an office or leading projects, or whether they go off to start their own practices. I hesitate to put a number out there, but I definitely think it is higher. At Cornell, there is a long history in the art of making and drawing. Students graduating from Cornell Architecture have a comprehensive understanding of the art of building, making and drawing. Many go on to become successful and impactful architects.

MC To have a holistic view, to be the master builder who can determine notation, calculation and fabrication, you need to use ideas, not just notations or numbers or script. It is true that, in the last 15-20 years, I have seen the best and brightest of students make brilliant careers in different fields because they were the precursors, the early experimenters of many digital technologies.

JS I would like to go back to the turnip. I can't remember the last time I initiated a design process with plan, section and elevation. Sure, I still use these techniques and notations because they are necessary in the way we communicate, but that is not how I work at all. Effectively, Brunelleschi turned the turnip into a piece of technology that allowed him to communicate information about form.

MC Yes, notating a three-dimensional model was difficult at a time when parallel projections did not exist. But then he had another problem: he wanted these instructions to builders to remain secret, which is why after showing the turnip he ate it! And the builders would need him on the scaffolding, onsite every day – he was a modern designer, but he was designing as a medieval artisan and not like a Renaissance architect. That would come one generation later with Alberti. Alberti came up with the idea of making as many drawings as needed, then putting a name and date on them, so the builders would just follow the drawings. And when the building really looks like the drawing, then the designer can claim, "It is 'my' building; not because I made it, but because I made the drawing." This is the act of foundation of the modern architectural profession. Brunelleschi was not yet there, because he wanted to have the building built according to his ideas, but he didn't want to make drawings – he wanted to keep his ideas as secret as possible. He still had the mentality of a medieval craftsman, which he was – he was a goldsmith by training. And by the way, we still don't know how he built the dome – it was a secret, still is!

JS I think many of us engaged in this type of work have a strong interest in the return to the site – I think that's also why in so many projects the work exists currently within either a gallery or a museum as an installation, because we are still at the nascent stage of how this can move into the built environment as architecture. There are still so many constraints.

7

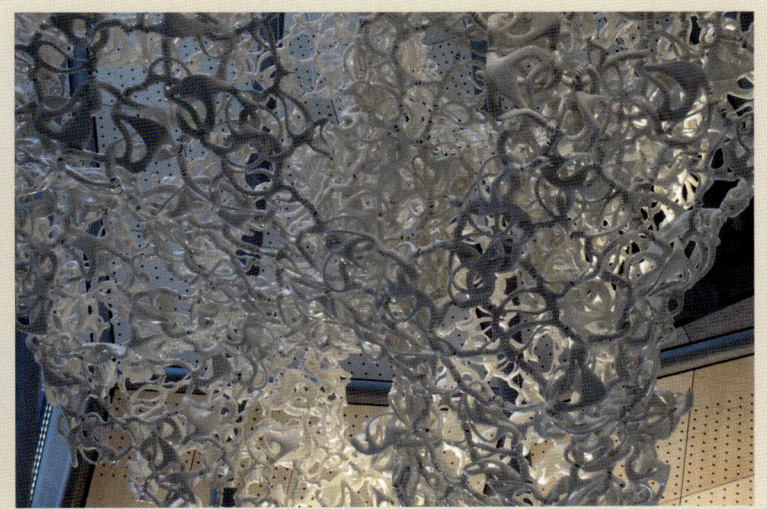

8

7. *PolyMorph*, a large suspended spatial structure, interrogates the physical interface between networking behaviour and fabricated material assemblies in order to address novel applications of non-standard ceramic components towards the production of 3D textured prototypes and systems.

8. Detail of *PolyMorph* installation. The large spatial structure is composed of 1,400 digitally produced and hand-cast ceramic components held in compression with a continuous interior network of tensioned steel cable.

Architectural designer and artist: Jenny E. Sabin, 2013. Design and production team: Martin Miller, Jillian Blackwell, Jin Tack Lim, Liangjie Wu, Lynda Brody. On view as part of the 9th ArchiLab, FRAC Centre, Orléans, France. Now part of the FRAC permanent collection.

Images: Jenny E. Sabin.

MC Yes, because now, in many ways, the distance between the designers who make the notation but don't materialise it and the builders who materialise the notation but don't invent it is being eliminated by the technical logic of digital tools. With digital design and fabrication, this distance has already collapsed. And so we go back to the medieval and pre-notational way of thinking and making at the same time – this is what we call digital craft, which is why I think we are much closer to the way we made physical things 500 or 600 years ago. We are reviving a pre-Renaissance, pre-Albertian, pre-Brunelleschian way of making. I think we are closer to the model of a medieval city where master builders were members of guilds, who had to conceive and make at the same time. The separation between the thinker and the maker – this great invention of modernity – was not yet there. We are now reverting to this intellectual model, and I wonder if we are also reviving the social and political model which went with it. That would be an interesting parallel, because in a sense the first phase of the digital turn reversed the industrial revolution, eliminating the need for mass production, standardisation and economies of scale. Artificial intelligence now suggests an almost pre-scientific intuitive approach to making. We really don't know much about this magic power of digital intuition. From a distance, it is clear that its closest parallel is not the nineteenth-century engineer, it is not the twentieth-century designer, it is not the modern scientist – it is the medieval master builder, the artisan who can manipulate materials and can conceive and make without designing. Paradoxically, we are returning to this.

JS Some of the tools I developed with Peter and with my ongoing collaborators, material scientist Dr. Shu Yang and bio-engineer Dan Luo, allow us to generate models to simulate certain natural systems and material phenomena and then to visualise them. This involves working with abundant dynamic data. I still think that these developments have the most relevance in terms of both navigating this return that you describe – the medieval master builder, where making and designing is a collapsed condition – and, at the same time, navigating the future. I do think that – in designing with tools and simulations that work with these abundant data, that in turn allow us to develop intuitions in a design space which references these loops – there is real, albeit ambiguous, potential there.

MC I agree, there is something big we are all trying to define: a new kind of science. The science that entrenched the power of the West – the science we studied at school – mostly does not apply any more. We do not use this science and there are probably large swathes of it that we don't need any more. Often today, when students make a structural model, they can test it in simulation right away and get immediate feedback, so they can tweak their first model tentatively, by trial and error, as many times as needed. When I was a student, each one of those trials would have required three months' work from an engineer. So the traditional feedback from engineers, an indispensable part of the design process when I was a student, is no longer needed. This is good in a sense, because we can almost do away with the science of engineering, but then almost inevitably the next step might be that the end user can do away with us! Today, cab drivers complain that Uber is making them obsolete, and Uber drivers say they are making cab drivers obsolete. But in a few years, driverless cars will make Uber drivers obsolete. Gloat in the glory, but you will be the next in line.

Q&A 2
MONICA PONCE DE LEON
VIRGINIA SAN FRATELLO
RONALD RAEL

MARILENA SKAVARA Welcome to each of you and thank you for making time for this conversation. Our wish is that you develop a conversation between you. As convener, I will only prompt you here and there if necessary. To get started, may I ask a straightforward opening question: where is your work heading right now, what are the key ideas and questions driving it?

RONALD RAEL We ask questions about material – particularly in terms of material provenance – in other words, where materials come from, where they are going and how they are filtered through various kinds of media. We are on a continual journey of exploration as we think about how particular materials – such as salt, recycled grape skins, recycled car tires – could be used in 3D printing to make building materials. To demonstrate this, we are doing several proof-of-concept pavilions, as well as integrating them more and more into architecture.

VIRGINIA SAN FRATELLO We are trying to find practical applications for 3D printing in the near future, and we also want to start putting these materials together in the same building. With so much 3D printing, it seems that only one material is used at a time. We are developing a small house that uses several different 3D printed materials, including 3D printed clay and 3D printed cement, and we are even thinking about how to mix materials within one print, which is not something that has happened much yet beyond metals.

MONICA PONCE DE LEON I am interested in the fit between technology and construction, and how everyday construction is affected by both the realities of technology and the myth of technology. We tend to think that technology provides a certain level of precision, but in reality both it and the way we deploy it are imprecise.

VSF You are widely recognised as a pioneer in applying robotic technology to fabrication within architectural education. How do you imagine buildings will be made in the next hundred years? And what are the changes that need to happen for these predications to come true?

MPdL I cannot predict the future at all. If you had asked me to guess where the discipline would be today 20 years ago, my guess would probably have been far away from where we are now. One of the exciting things in architecture is that there is no formula, and when you think everything has been exhausted, people come along and provide new twists and turns. So I don't have predictions. I am interested in how, through building projects in the nitty-gritty of the everyday, that puts pressure on the kind of research that we do in academia. Ron and Virginia are very interested in real projects, whether they are test cases for a particular exhibition or whether they are for clients with budgets. Often, clients do not want the kind of material that one is researching, and you have to convince them that your research is culturally relevant for them and for the discipline at large.

RR I think about past projects of yours such as Casa la Roca – that was an amazing project because it almost predicted things that people are attempting to do today with technology, but in traditional ways. Can you tell us about that?

MPdL We were very much thinking parametrically – the kind of rule geometry that generated the figures of those walls and the layout of the bricks of those walls is part of the zeitgeist of what came afterwards. It is interesting to see those same explorations executed through robotic fabrication, as opposed to by hand, and to see the research coming out of ETH Zurich, where robotic fabrication was used to do the geometry we had imagined with a different technology. This goes to show that there is again a loose fit between technology, execution and the culture of contingency that was intended for the house in Venezuela, which used vernacular materials – with terracotta tiles and terracotta bricks everywhere.

RR What does that say about the development of architecture? Are we moving forward but yet not moving forward at the same time? That project in some ways is much more advanced in terms of thinking about parametric brick stacking, and yet the way parametric brick stacking has now entered into the profession is very banal – we'd much rather have a robot that stacks straight bricks in straight courses – and is almost retrogressive. How does technology help us move forward? How does it prevent us from moving forward?

MPdL Your hacking of technology poses an alternative to the status quo. I think there is a difference between accepting tools as they are and misusing the tool – and in misuse, the tools create a new way to think about materials, the relationships between them and their culture and context. This is one of the things I love about your work – you are always misusing the materials, misusing the technology, defamiliarising materials and methods of fabrications. The history is still embedded, but we're asked to think about it differently. I think this offers a way to bring technology into question, and to destabilise the way we think about building in contemporary culture.

VSF I'm reminded of the Helios House and the Tectonic Argument at MoMa, and how those projects referenced fashion design – and I'm thinking about overlaps between architects and other design disciplines including computer science, perhaps as a way of destabilising building or misusing material. On your website, you say you work with

1

product designers and fine artists – I was wondering how interdisciplinarity affects your work and research?

MPdL If we are going to misuse material and technology, I think it's helpful to look outside the discipline for techniques that can be appropriated and reinvented within architecture and construction. I tend to work opportunistically – if I see something in a different field that looks like it might work, I try to adapt it. Like tailoring, for example, that's something I pursue. There are also different relationships between how one draws a project and how one builds a project. I think drawing is a way of bringing techniques from different disciplines into architecture. Through model-making you can bring analogous techniques from other disciplines as a way to explore cultural concerns. I think you do this extremely well – the way you have reinvented the vernacular by applying techniques that do not necessarily belong to the history of a material. You mention 3D printing and how, by hacking the equipment, you use materials that would not normally be 3D printed – this is another version of using techniques outside of a particular mode of construction.

RR We have also looked closely at building traditions – one of the things we did prior to 3D printing was travelling around the world to look at traditional vernacular buildings and learn from them. I think one thing that certain technologies have allowed us to do is figure out how to collapse many of the systems within vernacular constructions into new systems; we came up with a brick that can absorb water and passively cool a space by having ventilation in it, but where that comes from is a much more complex and beautiful demonstration of many different techniques – the creation of wooden screens, ceramic vessels, traditions of collecting water, massive constructing rooms. In many ways, much is lost through these translations and much is gained. We always struggle when thinking about how old traditions are lost and new traditions emerge. What are the new traditions that will emerge in the technological era? I don't believe in the idea that giant 3D printers will replace all the building traditions that exist. I think there needs to be an integration of older and newer traditions. In that hybrid moment, beautiful things emerge.

MS What do you think are the most valid terms of reference to think about design? Is it performance, narrative or scarcity – or is it something else?

MPdL One of the challenges, if you think about architecture only in terms of the immediate present, is that you end up with a series of buzzwords which can be very transitory. I can only imagine in the long run that scarcity, performance, etc, are not actually going to matter. What I always care about is whether a piece of work is culturally relevant and can be understood as part of a wider context, and for that it has to engage with a long understanding of a place, reflecting on the past, present and future. Architecture both constructs a particular idea of culture and reflects upon it. So in that sense, I think categories can sometimes get in the way.

RR I agree; I was thinking about two particular categories that are creating a split in architecture culture. For example, there is a split between the social project and the parametric project – I understand those kinds of projects are divided, but why do they never attempt to cross over?

VSF That's a good point – designing for performance using a particular Grasshopper script and merging that with social concerns about community or beauty, for example, might allow for new culturally relevant works to emerge.

RR I think there are cultural tendencies toward technology that suggest that its output must do something or perform

1. Helios House Station, Los Angeles, CA. Image: Eric Stuadenmaier.

2

something – it should have feedback the way an iPhone does. I think this limits architecture to having a very singular role. Things can be more multi-levelled and expansive than doing something in a particular way. The machine or technology might be very complex, but architecture is even more complex in its cultural associations, and we often overlook that – we overlook how complex the making of a building can be, and all the references that both inform and appear in its materialisation.

VSF How do you teach students this at your school?

MPdL I think always asking "why?" or "why not?" is important. Nothing makes me more impatient than when someone says something is impossible – well, why? Or when a student takes a particular direction – why? For me, a key component of architectural education is to demythologise the process of design, fabrication and construction so that your student is really focused on imagining alternative scenarios, speculating hypotheticals. This is particularly true outside city centres – the state of building today and the state of the landscapes and the sites around us is deplorable, so if we don't teach our students different ways of operating within these conditions, and if we don't push ourselves to imagine alternatives, then change won't take place. For me, it's not about becoming proficient with technology or methodologies, but actually demythologising all aspects of architecture so it opens up the imagination.

RR The demythologising surmounts the impossible. For some students, architecture is a myth – it seems impossible to achieve or attain because of its complexity, but this is also part of the 'demystification' – asking how we can achieve the impossible. Another aspect of this might be the emergence of female leaders in the field (such as Jenny Sabin, Neri Oxman, Liz Diller). I was wondering if you saw this notion of demythologising, or changing societal forces, or even the paradigm of fabrication technology itself, as contributing to their strong and increasing presence?

MPdL Women have been around forever; we have been doing stuff forever. I was the Director of the Digital Fabrication Lab at Harvard in 2003, over 13 years ago – and I was working with fabrication prior to that. I became a Dean at Michigan eight years ago. The presence of women in the profession has a very long history – perhaps the media is highlighting it more today, so it seems as if we are more present. But I think women have been interested in technology from the very beginning, just as men have. Perhaps there is more of an effort now to make sure they are equally represented in the media, which may make it seem as if they are only now emerging in the field.

VSF For me, the paradigm of fabrication has been very significant, it's allowed me to be a craftsman and use materials that otherwise I had never worked with – I wonder if other women feel the same.

MPdL I think we are all individuals. I worked in a mill workshop before studying architecture, so I have the opposite experience. I ended up pursuing digital fabrication because soon after I graduated I realised that the same mill workshops were no longer doing things the way I had done them myself, but yet this was not a conversation we had in academia. So I became interested in digital fabrication precisely because I saw it as an emerging context for the building industry that was being ignored by the academy. For me, it was not a way of enabling me to do things that I otherwise couldn't. I think that your earlier question about Casa la Roca is very relevant, though. We were drawing by hand and then it became easy to draw with Grasshopper. But it is really a question of how long it takes – we are still drawing by hand, it just takes five times as

2. G-Code clay vessels.

3. G-Code clay vessels and seed stitch tiles.

Images: Courtesy of Emerging Objects.

long. But I think that applies to everyone, men and women equally and all generations equally. One of the things I am very excited about in your work is that combination between material invention and advanced technologies. I think one of the challenges for me, as a designer and educator, is that there has been a divide between material interest and advanced fabrication (it seems as if there are those only interested in advanced fabrication and those only interested in advanced materials). What I love about your work is that you are unapologetic about bringing together – and allowing the history of – sourced materials to be understood as part of a continuum with the more recent generation of tools. That opens the door to a future which I think is very exciting; we no longer have to compartmentalise what is high tech and what isn't. So there is a conflation of ways of looking at materials that wasn't part of the discipline before!

RR *Smithsonian* magazine came out a couple of years ago with a list of the top 40 things you must know about the future, and number one was that advanced buildings would be made out of earth. This is not an anachronistic material – it is a technological material that has undergone 10,000 years of human development. If we look at every moment in history when there was some sort of global crisis, the scarcity of materials often asked humankind to review materials that they could already use – I think that we are now in that cultural moment. We are looking at materials that we are good at, and that's why there is a tremendous interest in the relationship between ceramics and technology. We are talking about larger cultural connections and ecologies of material. This is one of those moments where we can step aside and say, sure, it might be easy to put clay in a 3D printer or in a robot, but there are reasons why we are doing it culturally and historically – the availability and the plasticity of the material, but also our ability

4

5

as humans to engage a material which has evolved with us over the course of human civilisation.

VSF Perhaps the same can be said of salt. For example, in South America there are towns, hundreds of years old, built entirely out of salt. Salt is an ancient material that has the potential to be used as an advanced building material in the future in places around the world where salt is harvested. In the Bay area, it's local, it's renewable and there are both historical and ecological reasons for building with salt. Instead of shipping sand and cement all around the world to make concrete buildings, architects have new opportunities to revisit old materials and new manufacturing techniques for thinking about the evolution of material and building.

MPdL I am curious about your attitude towards precision. I have always been fascinated with the fact that architecture seems to rely on the concept of precision for its own disciplinary existence. You have the notion of tolerance and use certain details as a way of hiding the lack of precision – base boards are used to hide the gap between the wall and the floor, ground mouldings to hide the gap between the wall and the ceiling, and so on. We operate with tolerances, and of course in digital fabrication each tool has its own level of imprecision and you are actually fabricating with a certain level of tolerance. In your use of vernacular materials, through the use of fabrication tools, I am wondering what your approach is toward precision?

RR One thing we realised is that there is a fundamental difference between machine precision and material precision. The machine wants to do one thing, and the material wants to do another. So many of our experiments are wrestling with or negotiating between these two conditions – and this means we can have a certain degree of prediction about what the machine will want to do, and yet no degree of understanding of what the material wants to do, because we are hacking these materials. I think the most recent experiments in clay finally gave some resolution to that negotiation, in that we are accepting errors – and errors fundamentally become the vehicle with which we explore the making of forms and the materialisation of objects. They become glitches – we accept the fact that we will never make the perfect object we conceive of onscreen, so why don't we just hack that notion of precision itself? And that becomes – in our early cases – a kind of aesthetic agenda. More and more, we discover that there are some structural logics to these imprecisions, and so we find a series of ceramic objects that can be crush-tested because we realise that the way we lay down ceramic material increases or decreases its strength, and this is all through a series of controlled errors.

VSF He is talking about vessels – we call them the 'G-code vessels', which are mostly cylindrical in shape, and instead of modelling in form we use the G-code itself to design. And we have no idea how big these loops are going to be, or if the printer can do it, or if they're going to break, so we keep pushing it until we see at what point it will fail. And at the same time, when we 3D print with cement, it is fairly accurate. We are currently working with engineers who are helping to develop a strong cement material which has more water in it, but then the cement prints come out bigger, puffier. We have had to keep working back and forth between the digital and the physical to figure out what the limitations are and we build those into our material specifications.

RR What are your thoughts on precision?

MPdL I think my work has focused more on precision of ideas – understanding that there is a loose fit between how tight an

4 & 5. Star Lounge.
Images: Courtesy of Matthew Millman Photography.

idea is and its execution, so peeling away, for example, is something I am interested in, because you can peel at different moments and you don't have to precisely peel in a particular spot, which is a way of hiding the lack of precision of the equipment. I have also been interested in certain geometries in which the pattern can also hide the fact that the equipment will never be completely precise. I used to give a lecture called Zero Tolerances where I used the relationship between the semi-precision of a piece of equipment and the imprecision of the site where the equipment will be installed to rethink designs, the structure of the design process and the actual architectural object. So to me, imprecision is something that makes us very human and also makes us very beautiful, and we can really gauge the relationship between building a space and those who occupy it and experience it.

RR So do you see this as a larger critique of some of the agendas of fabrication that explore precision? Do you see it as a humanist moment in fabrication when many are attempting to achieve what is on the screen – hyper-smoothness, for example, or seamlessness – how do you see that?

MPdL One of the first exercises I give my students when we are working with a robotic arm is to ask them to design a stool with very few cuts and very little waste, and it is interesting to see how they struggle with the fact that the robotic arm is not actually as precise as they had assumed, and that it actually affects the set of the pieces. The more they had assumed that a piece of equipment was precise, the harder it is to deploy it, and the harder it is to come up with a cultural object, like a stool. So the critique is also embedded within the equipment itself – which means you have to pretend that the piece of equipment is precise to even talk about precision. When one accepts the implications of the piece of equipment, you will end up with a more interesting conversation.

RR Given that we accept that there are not categories – not of curation, design or education – how would you define yourself as an individual in our field?

MPdL Architects are always curating and educating – whenever we choose one material or form over another, we are curating. We have an arsenal of history, we have an arsenal of what is available to us today – and in that we are choosing from these things, we are constantly curating. This is our primary disciplinary trajectory. At the same time, in terms of educating, I think buildings educate the public. When I imagine someone looking at your hay house, the public will look at that and learn what it means to build today within a larger frame of reference. So I have always seen all of these things as one and the same – I see them as all related and somehow all architecture.

VSF That's beautiful. I love the way that it turns out all the same for you – it's all about culture and imprecision and imagination!

Q&A 3
CARL BASS
BOB SHEIL
ACHIM MENGES

ACHIM MENGES I think what makes the FABRICATE conference unique is that it brings together people from both academia and industry. It revolves not only around research findings, but also around projects. It's not just about submitting a paper, but also about presenting your research or practice through the project, which is quite a unique format in our world. Being in Stuttgart, we tried to give it a special focus, presenting it as one of the heartlands of manufacturing and fabrication, with strong connections to industry. So we have also established special industry talks by cutting-edge companies – for example, the robot manufacturers KUKA. So it is really about bringing together leading practitioners, academics and industries, and exposing their conversation to a broader audience who come equally from practice and academia.

BOB SHEIL Looking back on the work you presented in 2014 and where your work is now, what do you feel has been the major stepping stone in the last three years? And where are you headed next?

AM What we have seen is that computation is becoming closely related to materialisation, with the physical world rapidly emerging from within the digital domain. This is a very interesting situation. We have realised we cannot focus on computational design exclusively, but instead that it is inseparable from construction. This is the reason why we have changed our name to the Institute for Computational Design and Construction. We are working towards a higher level of convergence between design and making, with ramifications on how we conceptualise designs, how we work with designers and where the industry is going. For us, this is decisive. We like to borrow from robotics (where it has also been recognised that you cannot separate your design method from your hardware or software, but in fact that you can co-design these three things in unison). This is what we have increasingly realised: we are co-designers of design and manufacturing processes.

Carl, I am incredibly pleased that you are doing the keynote for FABRICATE, particularly because FABRICATE is about the future of making, which on the one hand is one of the key concerns and the key ambitions of Autodesk. This way, I think, we can share the vision that design and making are undergoing fundamental, radical changes, and that we need to update our knowledge and our tools. On the other hand, we agree that this challenge is also an enormous opportunity for both industry and education. What I find interesting is a basic but profound question: how can we negotiate the fact that the future of making software is so directly linked to the future of making things?

CARL BASS To a degree, design has gone from documenting design to informing design, and it is now moving to a place where the act of designing is going to be one of co-creation between people and machines. Only five years ago did we start asking: if you had all the computing power in the world, how would you design things differently? Until then, we had treated computers as a scarce and precious resource, as opposed to an abundant one. As the price of computation plummeted and computational power increased exponentially, you started asking how you would design and engineer things differently. We are getting to a place where various practitioners all along this spectrum, having been barely informed by computation before, are now moving to designs that are completely co-created with a computer. Another profound change was the introduction of the micro-processor, so that you could now make high-quality, low-cost objects or products in small batches. As unique as a single building or a single structure, these objects would be low volume or custom-run. This inverted the basis of the industrial revolution, where you could make a huge quantity of high-quality, low-cost things if they were all the same – all of a sudden, by making the process digital and controlled by micro-processors, you are able to do it in a completely different way that allows for low volume and high quality at a reasonable cost. The third thing tying these together is the availability and soon-to-be ubiquity of sensors. Something performed in the real world can be measured to see if there is fidelity between the digital and the real, as well as to gather information that can be fed back into the next iteration of the design cycle. So if you look at computation for design and engineering, micro-processors evolved in the process of manufacturing, and sensors closed the loop between design and fabrication – this is the kind of landscape we are in.

AM When we look at technological developments, we note that there is an initial phase where new technologies are used to mimic old processes and products. This is true for almost all technologies; for example, material technologies – there are composite materials initially employed to mimic old processes and products – but it also applies to software technologies where, in the first generation of commercial CAD applications, the screen mimics the drawing boards and the mouse mimics the pencil. It is also true for production technologies. CAM was primarily used to automate and better control fabrication processes that existed before. One can argue that we are currently transitioning from this first phase of using digital technologies for designing and making things that are essentially pre-digital products to a second phase where we are beginning to explore processes and related products that are genuinely computational – things that we could not have made or even conceived of in pre-digital days. This enables us to tap into the potential of computation as it becomes increasingly ubiquitous, and to come up with

168 / 169

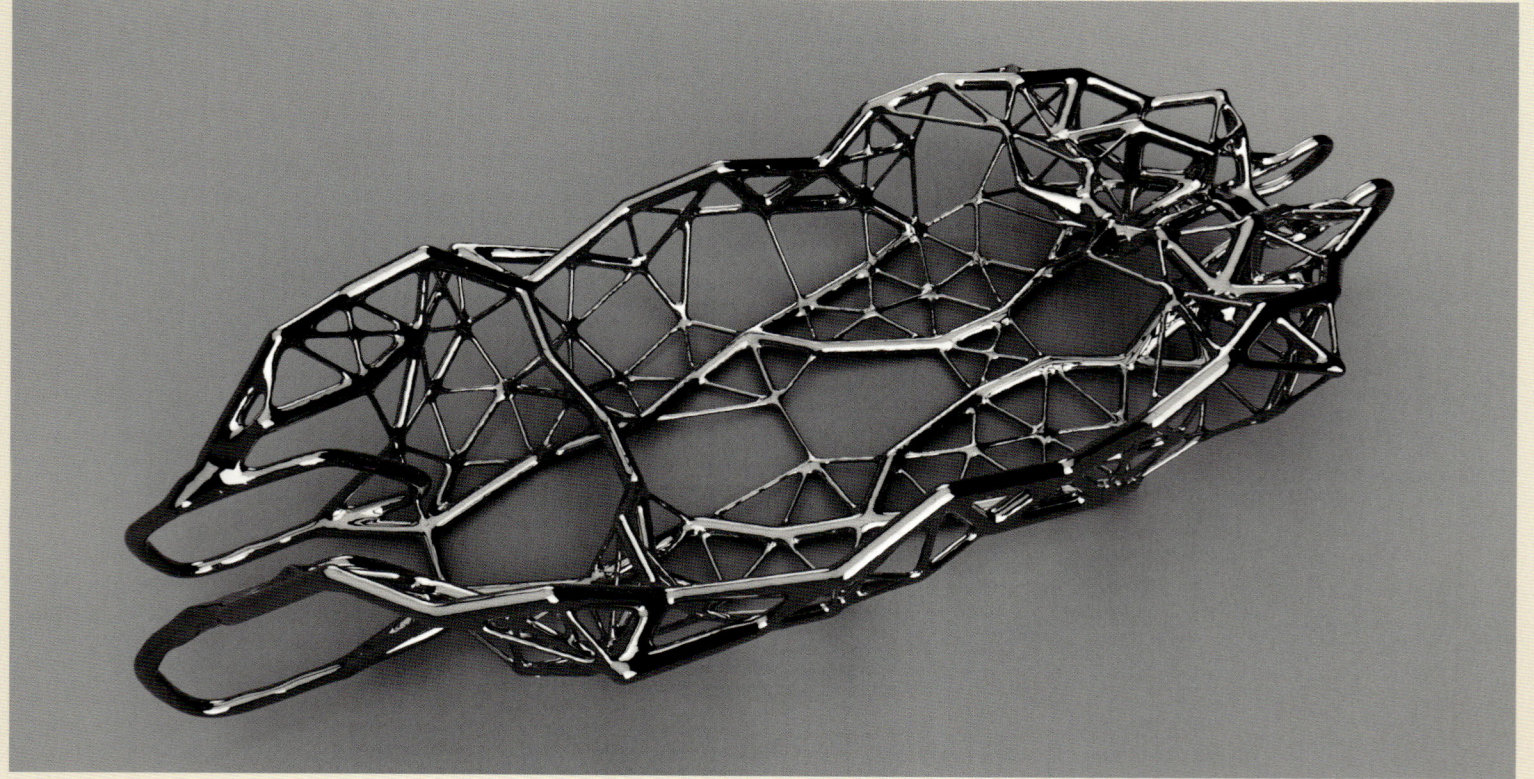

3

radically new ways of designing and making. This has great promise, but also offers quite a challenge because it means that we, as designers, have to adjust our design thinking. It is not just about updating the design tools and techniques; design thinking also needs to be fundamentally updated.

BS Do either of you envision a point in the future where we stop prototyping? In a sense, computation, simulation and the design process have become so complete that manufacturing is only about the delivery of the final piece. Or will prototyping remain as the middle ground between design and making?

CB I don't think there's an absolute. As people become more fluent and proficient with their digital facsimiles, they will be able to go without prototyping for things of greater complexity. At a certain point, this will break down and you will want to see, feel, smell or experience the thing you are building. If you look at CAD software, just like every other technology it tries to mimic the technology that came before it found a life of its own – CAD technology started out mimicking the drafting table. Now, the goal of most CAD software is to build a digital model, a replica of the thing you are going to build. We are only partially there, but if you think about a building in CAD software, we now have a fairly good understanding of what it will look like, what that structure is, how the air will move in the building, how it will sound, how you feel when you move in the space or how it will react to environmental conditions. But there is no reason to presume that, over the next 10 to 20 years, we won't be able to get very good approximations of the things we build. In essence, in manufacturing, I think prototypes that are small and manageable will continue to get built because it is easy. But many buildings, specifically any one-off building, are prototypes in themselves and we can only prototype parts of them. I think that is where we are headed.

BS Looking at your work, Achim, I enjoy what you say about adding the word 'construction' to the lab. Your recent work is becoming increasingly performative, in that the spectacle of making is a wonderful thing to watch. It shows that the performance of making is a part of design. This performance opens up the imagination for other things that we can make. Do you have a conscious view on performance as being part of the act?

AM Yes! I see it in two ways: 1) the performance of the process itself, and 2) the performance of the object or structure. Especially interesting is the way in which digital fabrication processes become more open-ended, flexible or, in other words, designable. When we talk about a prototype, we like to prototype not only the actual product, but the processes, too. Today, designers actively engage in developing new fabrication processes as part of the design process, instead of just using existing products and technologies. That leads to new modes of what one may call the co-design of processes and products, which is a different way of going about design. For me, this is one of the essential aspects of robotic fabrication – it extends your possibility as a designer beyond the product, beyond the building, to the processes in which the buildings and the products come about. We have a lot of collaborations here with production engineers – people that come from manufacturing – and it is interesting to see how designers bring a different agenda to the table as opposed to someone who is trained traditionally in this field. It really broadens the spectrum of what we refer to as 'making', in the broadest sense, to a kind of design thinking. The other aspect that is of interest to me is how we can conceptualise this convergence of design and making. In recent years, one of the most radical changes is that the line between what we call 'making' and what we call 'design' is beginning to blur. This relates to the prototype, because the prototype is what

1. A Hack Rod test drive with sensor data points. Image: © Autodesk.

2. A generatively designed chassis of the Hack Rod, after the sensor points were fed into the programme. Image: © Autodesk.

3. A Hack Rod test drive in the desert.

we see as a step between design and making. With the arrival of what we here in Germany call 'cyber-physical systems', we see that design and making can happen in a kind of feedback loop, where they co-exist and co-evolve. This is also part of what Carl mentioned as the possibility to equip our fabrication environment with an abundance of sensors, which means that all of a sudden what you actually make becomes the model for what you want to make. This means that a machine is no longer just executing a control code taken from previously established models, but actually has a far more active interaction with the process of making, to the point where it can begin to make its own decisions so that the designer designs conditions and performances that need to be fulfilled, and a certain level of decision-making can happen on the level of the machine (taking into consideration the fact that these machines are increasingly capable of learning sophisticated ways to operate in the physical world). So I think we can overcome this idea that design comes to an end, and then we prototype, and then we make – instead, these things start to co-exist in the same space. This begins to challenge some profound aspects of architecture, as well as our conceptualisation of what a designer is and does.

CB There's an example from some work we did recently which shows the way in which design and fabrication become more of an inter-related cycle. We wanted to build a new kind of vehicle chassis, so we built the frame of this car in a very traditional way and then hired a bunch of drivers from Hollywood to take it into the desert and drive it, aggressively, for 10 days. The vehicle was monitored for the duration. When we were finished, we had enough knowledge about the forces that acted during extreme stress testing. We took that information and put it back into an algorithm that generated an ideal structure for that vehicle, and then we added in three different fabrication techniques and said: given this idealised form, how would you realise it through different fabrication techniques? One was an improved version of something that was made out of tubing, and the others are these two wild-looking designs that were intended to be done with additive manufacturing, one out of chromoly and the other out of titanium. What's interesting is that you have this form that you want to get to, and then you have three different kinds of material and processes to actually realise the design.

AM One example that I like to mention from our work is one of our recent research pavilions, where we inflated a membrane to look like a big balloon which we then reinforced by gluing carbon fibre inside it, and therefore turned this flimsy membrane into a building envelope that is actually supported by the fibres. The interesting thing is that during the fabrication process, the structure changes shape constantly, so you no longer have a finite design. In this case, the robot has the capacity to sense the stress in the membrane and actually see where the membrane is in space, adjusting its carbon fibre layout path accordingly. So there is direct feedback between the environment in which the robot operates, the structures it builds and the way it is controlled. This is something you can't fully predict. It's also something you can't predefine in a sort of representational geometric model; it's really about forces, structures and predictive simulation and also about real-time sensing and the constant exchange of all that data. In that case, it is really interesting, because sometimes the robot makes semi-autonomous motions which obviously leave traces of carbon fibre, which become part of the design. It is difficult to determine where the design ends here and where fabrication starts – it is a kind of coalescing of the two.

BS How can we look for a gear shift in the construction industry at a more general level, and how can a designer's playfulness and inventiveness have an impact on a much

4. ICD/ITKE Research Pavilion 2012, Stuttgart, Germany, 2012.
Image: © ICD/ITKE, University of Stuttgart.

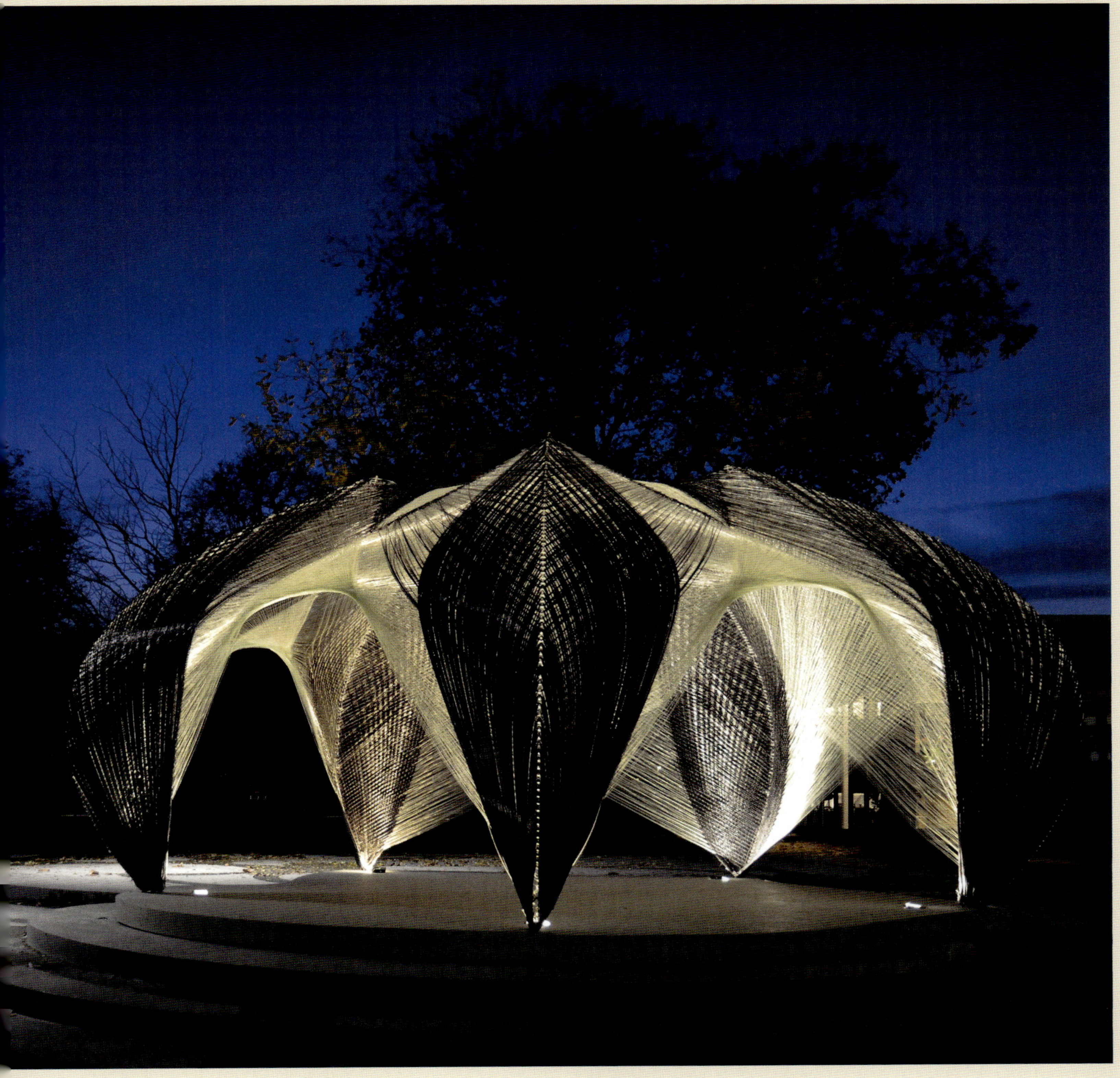

5

broader scale? I wonder if you could talk about your readings of industry and your projections and forecasts for design experimentation at a much greater scale?

AM We get a lot of approaches from industry asking us about the possibility of updating their fabrication and construction processes by digital means. Usually the conversation revolves around benchmarking digital processes versus established processes. This is fundamentally a problematic approach which holds the construction industry back. I think what we really need to ask is: what can we produce that we couldn't before? Where are the real benefits that only come from these technologies? Obviously this is not something you can resolve now. This is something that needs a more profound rooting in research and will take more time. However, ultimately I think this will be rewarding, because these are truly disruptive technologies and you are short-served if you consider them just as a kind of digitalisation of what you have done in the past. This is what we have to get industry to realise. In Germany, this is not so easy, because there is a very strong construction industry and not a lot of incentive to change established business models. Computational technologies have incredibly disruptive potential, and inevitably the construction industry will have to reinvent itself. We need to lay the foundation for that change and, at least from my academically biased perspective, make sure we can tap into its true potential – not just its design potential, but also the ecological and environmental perspectives, which very much need addressing in the construction industry very soon. We will not be able to go on as we do, because we consume more than half of the resources and energy on the planet. So we need that long-term vision and the persistence to see it through. It is the burden and the privilege of academic research that we can engage with these longer-term problems which are difficult to engage with if you have to run a construction company.

CB I don't have the same prejudice as Achim, because I am not in Germany. I am slightly more optimistic, due to the timeline. The construction industry is bottom-line driven, so whenever we can build better things more cheaply, construction will pick it up. Yet construction companies are actually very resistant to change. If you look at job sites today, they look nothing like jobs sites thirty years ago – the skills, the people, the tools they use, the processes, the materials they are working with – they move with the times quite effectively. There is a kind of capitalistic tech approach that serves as a counterpoint to what happens in architectural artistic practices. Just as I can go to my workshop and dream up and build any crazy thing I want – and it does not have to make economic sense – I think many firms can build that way, and I love this experimentation and it should continue. On the other hand, the construction industry offers a check and balance on this, saying what makes sense and what is sustainable. In that sense, I am pretty optimistic about construction companies moving forward with digital fabrication, because the right incentives are underlying their choices.

BS Do either of you see a future in which the construction industry gets challenged by lots of micro design and maker industries, similar to the way in which artisan beer makers are prepared to take risks based on notions of quality and distinction as opposed to mass production and profit?

CB My initial response is no. What I would say is that spending six dollars for a beer instead of three and a half is a decision that millions of people can make every day. The stakes involved in the cost of a building are so much higher, and what we see when things become more expensive and discretionary is that the number of owners who are willing to incur that extra expense are few and far between. Obviously there are all kinds

5 & 6. ICD/ITKE Research Pavilion 2014–15, Stuttgart, Germany, 2015.
Image: © ICD/ITKE, University of Stuttgart.

of wildly innovative projects being built, but I don't think it could ever become a mass market thing. However, once we get to a point where we can build more unique designs for the same kinds of prices, then all bets are off.

BS Is this prospect on the horizon?

AM Well, I also have quite an optimistic outlook on how the construction industry might change. Very often, we have the hand-laid brick wall and the robot-laid brick wall, which is something you have mentioned. It boils down to which is cheaper to produce. I think the real question to ask is: does the robot really want to build a brick wall? And the answer is: probably not. I think the construction industry tends to benchmark digital processes on pre-digital construction systems. But as we are just making the transition from the stage where we employ computational technologies to mimic traditional processes to the stage where we start to uncover radically different solutions, we need to challenge norms and established ways of doing things. How do we want to build when we have computational construction, cyber-physical systems and man-robot collaboration? Obviously, the goal cannot just be the automation of the building site and the automation of existing offsite processes. Accordingly, how do we move it towards following the logics and economies of the digital age? This is where we really need to get the construction industry to. This is a challenge, but also a new opportunity – we might perhaps be able to democratise what the ordinary or extraordinary is.

Q&A 4
ANTOINE PICON
BOB SHEIL

BOB SHEIL We are delighted that you have agreed to be a keynote at FABRICATE 2017, extending the tradition, which began with Mario Carpo at FABRICATE 2014, of having a historian speak. Where is your work heading right now – what are the key ideas and questions driving it?

ANTOINE PICON Between 2010 and 2015, I devoted three books to looking at how the rise of digital culture links to transformations within urban architecture: *Digital Culture in Architecture* (2010), *Ornament* (2013) and *Smart Cities* (2015). These books identified a series of theoretical issues that I would like to concentrate more specifically on in the years to come, such as the question of materiality and the links between the evolution of architecture and subjectivity in the digital era.

Alongside these lines of investigation, I plan to focus on techniques themselves – on software in particular and its influence on the design process. If the first line of inquiry is akin to a philosophical investigation, the second would be closer to an anthropology of techniques.

BS How have your work/interests evolved over the past decade?

AP I have gradually shifted more towards urban and societal issues. For instance, the need to reconcile the quest for sustainability with digital advances appears to me to be a major challenge. More generally, I am perhaps less interested in architecture as such and more in broader issues of space, technology and society.

BS Looking ahead to the context of FABRICATE and your forthcoming keynote, do you believe we are witnessing a new era in computation/design/making?

AP The rise of digital fabrication represents a major turning point, even if there are still a lot of ideological discourses that obscure the path it is taking. Not everyone will become a 'maker'. I also think that the notion that thanks to digital fabrication the designer will become a kind of postmodern craftsman is also ideological. Comparing oneself to Ruskin seems to me to be profoundly dubious.

BS So what lies ahead?

AP Another crucial evolution will stem from the urgent need to reconnect the digital with the quest for sustainability. Also, what does it mean to design in a true context of augmented reality – at the level of the articulation of atoms and bits?

BS What are the most valid terms of reference for new ways to think about design?

AP As I have argued repeatedly, one of the main consequences of the digital revolution is to make design appear more strategic, commensurate with a form of action, rather than being about the revelation of some pre-existing formal idea. Design becomes synonymous with event-making and with the production of scenarios.

Making and speculating tend to become more and more intimately linked, but not in a 'craftsman' way. They are linked more by a common inquiry into the foundations of materiality.

Nostalgia is inevitable, since the digital has separated information from matter while pretending that it does the contrary. Material computation is actually permeated with nostalgia. For me, this is part of what makes it interesting beyond its claim to a new objectivity.

3

RETHINKING ADDITIVE STRATEGIES

DISCRETE COMPUTATION FOR ADDITIVE MANUFACTURING

GILLES RETSIN / MANUEL JIMÉNEZ GARCÍA / VICENTE SOLER
The Bartlett School of Architecture, UCL

Large-scale discrete fabrication

The research presented in this paper, based on two projects, investigates design methods for discrete computation and fabrication in additive manufacturing. The first project, *CurVoxels* (Hyunchul Kwon, Amreen Kaleel and Xiaolin Li) introduces a discrete design method to generate complex, non-repetitive toolpaths for spatial 3D printing with industrial robots. The second project, *INT* (Claudia Tanskanen, Zoe Hwee Tan, Xiaolin Yi and Qianyi Li) proposes to make this discrete approach physical, suggesting a fabrication method based on robotic discrete assembly. This discrete design and fabrication framework aligns itself with research into so-called digital materials – material organisations that are physically digital (Gershenfeld et al., 2015). The suggested methods aim to establish highly complex and performative architectural forms without compromising on speed and cost. Both projects propose design and fabrication methods that are non-representational and do not require any form of post-rationalisation to be fabricated. The research argues that, compared to 3D printing, robotic discrete fabrication offers more opportunities in terms of speed, multi-materiality and reversibility. The proposed design methods demonstrate how discrete strategies can create complex, adaptive and structurally intelligent forms. Moreover, by moving computation to physical space, discrete fabrication is able to bridge the representational gap between simulation and fabrication. This representational gap is a result of a two-step process usually associated with computational design strategies, where a design is first developed digitally and then passed on to be fabricated.

Analogue and digital fabrication

The projects described in this paper are produced in a research-through-teaching context within The Bartlett Architectural Design Programme (AD) – Research Cluster 4 (RC4). RC4 is a part of BPro, an umbrella of postgraduate programmes in architectural design at The Bartlett School of Architecture, UCL. The research can be situated in the context of robotic manufacturing and the automation of construction processes. The two projects presented are based on the use of industrial robots, but these are assumed as abstract, notional machines. The projects could potentially be more efficiently implemented with other types of custom-made robots, but the research in question here is first and foremost focused on design methods. Both projects should effectively be understood as research into design methods, rather than as research into robotics and manufacturing itself. In terms of fabrication, both projects are additive fabrication processes: *CurVoxels* (Fig. 2) is a 3D printing process, and *INT* (Fig. 3) is an additive assembly workflow.

There have been significant research efforts into robotics and automated construction, especially in the context of additive processes. Gramazio Kohler has developed additive projects such as *The Programmed Wall* (2006), *Complex Timber Structures* (2013) and *Mesh Mould* (2014). However, these attempts to automate construction have had little impact and are caught in a conflict between complexity and speed (Gershenfeld et al., 2015). Neil Gerschenfeld argues the need for digitising not just the design but also the materials (Gershenfeld et al., 2015). In this context, The Centre for Bits and Atoms has developed the notion of digital materials – parts that have a discrete set of relative positions and orientations

2

(Gershenfeld et al., 2015). These materials are able to be assembled quickly into complex and structurally efficient forms. These digital materials establish material organisations that are digital rather than analogue. Following Gershenfeld's distinction between analogue and digital materials, most of Gramazio Kohler's robotic fabrication projects are to be considered analogue. Despite the use of discrete elements for assembly, these elements tend to use analogue connections, which are continuously differentiated. Unlike digital materials, every element has a unique connection possibility, increasing the degrees of freedom and possibilities for error. Continuing from Mario Carpo's distinction between continuous and discrete design processes, the notion that structures can be physically digital, rather than analogue, becomes an important driver for the work presented in this paper (Carpo, 2014). However, as a design method, these digital materials present some fundamental problems. In order to be considered digital, the elements necessarily need to be serialised. As a result, the digital materials proposed by The Centre for Bits and Atoms are highly repetitive and homogenous. From an architectural design point of view, these structures are efficient, but not complex in terms of formal possibilities. A possible solution could come from combinatorial design strategies.

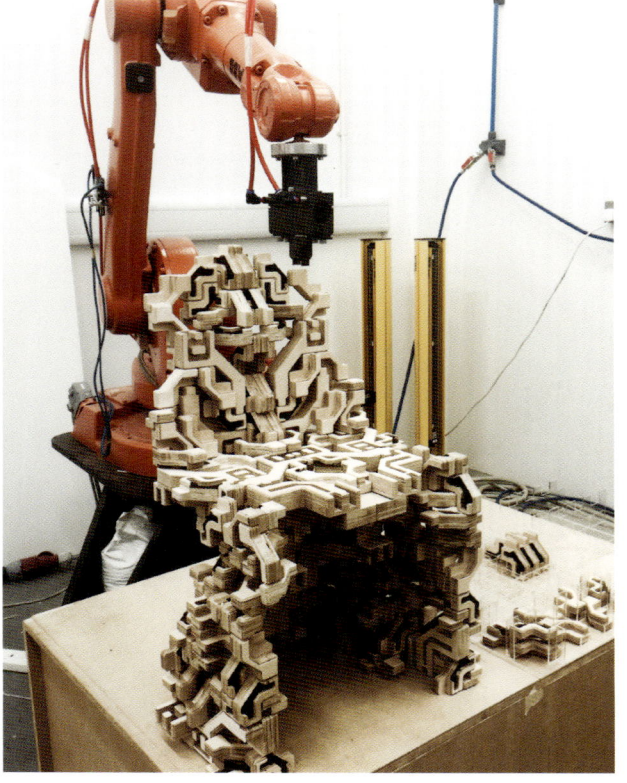

3
Jose Sanchez demonstrates how standardised, serially repeated elements can result in differentiated and complex wholes (Sanchez, 2014).

Towards discreteness

In the first instance, this research is driven by the question of how the notion of discreteness can make the automation of construction processes more efficient while also allowing for more complexity and differentiation. It attempts to combine the efficiencies of digital materials with combinatorial design methods. Secondly, as a broader question, the projects presented develop design methods that remove representation, resulting in structures that are digital both in the design process and as a physical product. The research first introduces discreteness as a design process in the *CurVoxels* project, and subsequently as a fabrication process in *INT*. Both projects can be considered additive manufacturing processes.

More specifically, the *CurVoxels* project questions how discrete computational processes can make spatial printing with robots more effective, while also opening up more formal possibilities. It demonstrates how the serialisation of toolpath segments allows for efficient

1. Digital prototypes of vertical elements, using different arrangements of discrete pieces.

2. *Curvoxels* (RC4 2014-15). Half 3D printed chair v3.0. Image: Curvoxels.

3. *INT* (RC4 2015-16). Robotically assembled chair v2.0. Image: Manuel Jiménez García.

4. *Curvoxels* (RC4 2014-15). Robotic extrusion of ABS filament. Image: Curvoxels.

5. *INT* (RC4 2015-16). Human-robot collaboration for the assembly of chair prototype. Image: INT.

error mitigation and prototyping. The project presents a case for combinatorial design as a method to create structures with a differentiated material distribution, and complex formal articulation that bypasses the repetitive and homogenous grids usually associated with spatial printing (Fig. 4).

The second project, *INT*, explores the implications of physical discreteness and discrete assembly as an alternative, non-continuous method of additive manufacturing. The *INT* project questions the benefits of discrete assembly compared to 3D printing. *INT* sets out a framework for discrete fabrication with combinatorial building blocks, investigating both the design of the units and their assembly procedures. The work experiments with human-robot interaction and questions the consequences of moving computation to physical space. The fabrication procedure proposed by *INT* aims to resolve some of the problems associated with continuous additive manufacturing, such as the lack of speed and mono-materiality (Fig. 5).

CurVoxels and INT

CurVoxels is a team of students in RC4 who developed the project *Spacial Curves* (2014-15). The project is a continuation of research into spatial printing, which started with a previous team of students called *Filamentrics* (2013-14). Spatial printing is now a popular method for robotic printing, but first appeared in the *Mesh Mould* project (GHack, Lauer, Gramazio, Kohler, 2014). The printing process is based on a tool head, mounted on a robot arm, extruding hot plastic along a spatial vector. This method saves a lot of time in comparison to layered methods. Preferably, the robot does not stop during the process, but continuously extrudes material. Robotic spatial printing has a number of limiting constraints, the most important being that the robot can never intersect with previously deposited material. There are also structural constraints: material can only be extruded in the air for a limited range – at some point, support structures are needed. Therefore most spatial printing projects make use of a highly repetitive toolpath organisation, based on parallel contours connected with a triangular toolpath. The formal possibilities are limited, and the toolpath organisation is not very complex.

CurVoxels developed a design method which is aimed at controlling the toolpath constraints and developing new freedoms within these constraints. *CurVoxels*' computational approach is based on discretisation: a voxel space is developed, where every voxel contains a toolpath fragment. It was decided to use a Bézier curve as a unit to compose the toolpath. The team then developed a process that cycles through the voxel space in a layered and linear fashion, simulating the trajectory of the robot. Every time a voxel is accessed, the Bézier curve inside the voxel is rotated to connect to the line in the previous voxel. In principle, there are 24 rotations possible but a number of these are not printable, as the extrusion tool would intersect with the curve. The logic of combining separate toolpath fragments is essentially combinatoric: there is a discrete set of options for how curves can connect without losing continuity. The printing process can be prototyped on a few voxels, rather than having to compute and prototype an entire toolpath. The error space is not continuously differentiated, but discrete and limited. After the toolpath is tested for a single curve-voxel in 24 different rotations, it can be used to assemble thousands of toolpath fragments together into one continuous, kilometres-long, printable line. The size of the voxel itself also introduces a structural parameter: if the voxel is smaller, more material is deposited and the structure becomes denser. If it is bigger, the structure is more porous and less strong. This observation was translated into an OcTree subdivision for the voxels, linked to structural data. In areas that carry more load, voxels are subdivided and more material is deposited. The design method was tested on the generic shape of a panton chair. The shape of the chair itself is not questioned and has to be understood as a generic placeholder, similar to the Stanford bunny or the Utah teapot. In total, three chairs were printed. The last chair, which made use of the OcTree subdivision logic, was strong enough to withstand up to 80kg of load.

The next project, *INT*, combines discrete design with discrete fabrication. Similar to *CurVoxels*, a combinatorial unit is developed, but this time as a physical building block that can be aggregated and assembled. This unit is able to combine with itself in different ways and can be robotically assembled. Similar to Neil Gershenfeld's digital materials, the unit is serialised and has a discrete set of connection possibilities. The 'digital' building block, or tile, has a geometry which can be inscribed in a voxel space: one L-shaped unit is comprised of three voxels. The tile is further defined by a series of subtractions so that it can be picked up by a gripper tool in different orientations. It is also marked with multiple reflectors that help a camera system to track the elements in physical space. The project is based on multiple scales of CNC-milled timber blocks. The smallest can be gripped at the outside boundary, and the largest from specifically designed gripping spots. A combinatorial logic was developed to combine tiles into structurally stable forms. Different combinations of blocks are structurally evaluated in terms of surface area. In areas of the design that require more strength, combinations with a larger area of shared surface are privileged. The robot is given a specific boundary and total amount of tiles to fabricate a structure. One tile is placed as a start and then, for every robotic action, the position of the next tile is calculated (Fig. 6). Users can intervene in the process by placing tiles themselves. These are tracked by the camera system and evaluated in respect to the other tiles. The robot can then subsequently add new tiles to complete the structure. In case the new tile would, for example, break the boundary of the design, the robot can remove the tile again. The robot is able to address imperfections in the assembly process – for example, if a tile falls off the structure, it can re-evaluate its position and add a new tile. The design process was tested on two different chairs. The first chair is purely a product of automated decisions, without human intervention. The second chair is an authored product, where the students decided to favour symmetry. The fabrication process is significantly faster than for the 3D printed chair, both of *INT*'s chairs being completed in under 45 minutes (Fig. 7).

From continuous to discrete

Both projects question established design methods based on discretisation. They capitalise on the efficiencies emerging from that process, and can be considered less representational.

The *CurVoxels* project enables the efficient generation and evaluation of complex toolpaths. After optimising one toolpath fragment in one voxel, an entire structure can be generated without further problems. This serialised approach reduces the amount of unique problems to solve. Toolpaths with continuous formal differentiation, on the other hand, also have to deal with continuously differentiated problems, which all require unique solutions. The proposed combinatorial method, in combination with the OcTree logic, allows for complex differentiation and adaptability to structural criteria. Through embedding a combinatorial logic, the repetitive character commonly associated with digital materials is avoided. Through always combining the initial toolpath unit into different patterns, complex structures with differentiated formal qualities and structural behaviour can be designed. This introduces a fundamental shift away from the paradigm most usually associated with digital design: from mass customisation and continuous differentiation of parts to discrete, serially repeated elements.

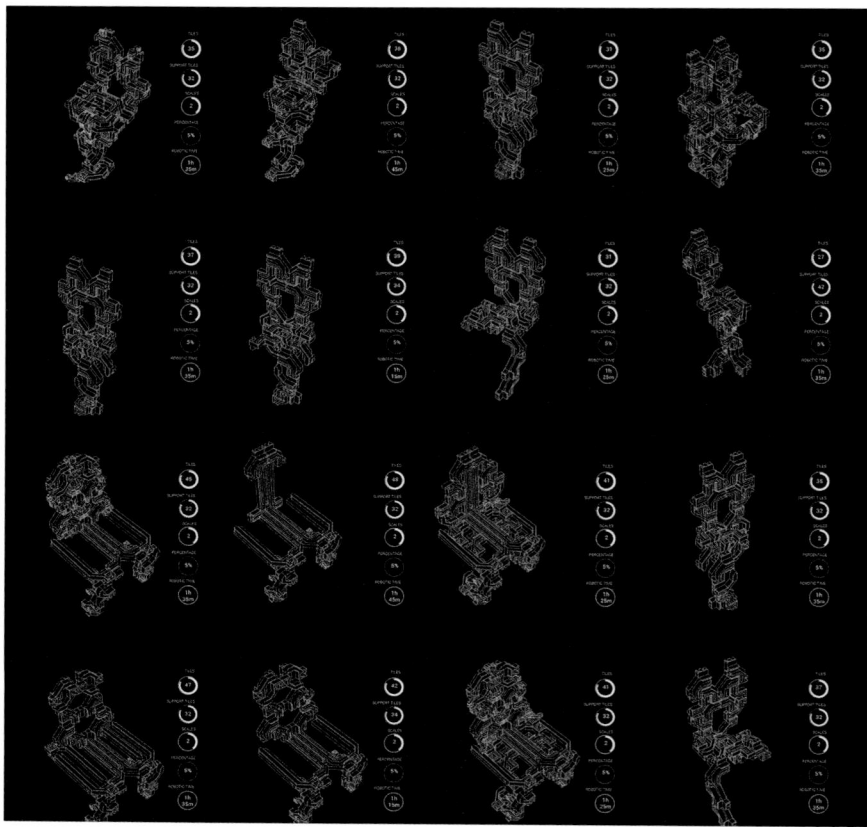

The resulting objects are not a result of a post-rationalisation, where a shape would be first designed and then sliced into layered toolpaths for printing. The design method operates directly on the toolpath

6. *INT* (RC4 2015–16). Catalogue of emergent objects using custom-made applet developed in processing.

7. *INT* (RC4 2015–16). The Bartlett B-Pro Show 2016, physical prototypes. Image: Manuel Jiménez García.

itself. However, the project does not manage to bridge the gap between design and fabrication. Essentially, the design first has to be generated and then sent off for fabrication. As a consequence, if there is a problem in the fabrication, the entire object has to be printed again. There are significant logistical constraints to the project: multi-materiality is hard to achieve, and the printing time is slow.

On the other hand, the *INT* project allows for the efficient assembly of objects which could potentially be multi-material, while also maintaining a high degree of complexity and heterogeneity. The research establishes structures that are physically digital. It introduces interesting new questions about the potential interaction of robots and humans in the design process. The project could, however, benefit further from more advanced feedback loops between the simulation and the physical assembly. The use of heavy, compression-based material introduces an added difficulty to the assembly process, presenting a whole range of structural problems. More significantly, the project makes use of joints, but in the end relies on a significant amount of glue in order to be assembled. The use of glue prevents the reassembly and reconfigurability promised by the project. The problem with the joint is one of the main limitations of discrete fabrication: the smaller the elements, the more joints are created. Potential solutions could attempt to make the element itself interlocking, but this would inevitably increase the complexity of the robotic assembly process and again severely limit the formal possibilities.

Discrete fabrication

The design methods developed in the *CurVoxels* and *INT* projects have significant implications for additive manufacturing, the automation of construction and architecture. The proposed discrete design methods establish a series of efficiencies while also enabling complex material organisations. The shift from continuous fabrication to discrete fabrication moreover introduces a series of advantages, such as multi-materiality, structural performance, speed and reversibility. The proposed combinatorial method allows for complex differentiation compared to the repetitive character commonly associated with digital materials. Formal differentiation no longer relies on the mass customisation of thousands of different parts, but can be achieved by the recombination of cheap, serialised units (Fig. 1). The use of cheap, prefabricated building blocks, in combination with increased assembly speed, reduced error space and vast formal possibilities, provides a firm ground for additive manufacturing techniques to scale up. From a design point of view, in moving computation to physical space, discrete fabrication is able to bridge the representational gap between simulation and fabrication. Digital data and physical data are aligned. Computation and fabrication can happen in parallel, and design decisions can be made during the fabrication process. This versatility makes the process more robust and adaptable to demanding scenarios such as onsite fabrication.

The potential for reversibility has implications reaching far beyond automated construction. Architectural building elements that are recombinable could significantly change the lifecycle of buildings. The combinatorial aspects can help to introduce complexity and adaptability in prefabricated building systems, without losing the benefits of seriality and standardisation.

References

Dillenburger, B., Hansmeyer, M., 2014, 'Printing Architecture: Castles Made of Sand' in *Fabricate 2014*, Zurich, ETH, p.92-97.

Carpo, M., 2014, 'Breaking the Curve. Big Data and Digital Design' in *Artforum* 52, 6, p.168-173.

Hack, N., Lauer, W., Gramazio, F., Kohler, M., 2014, 'Mesh-Mould' In *AD 229, 84, Made by Robots: Challenging Architecture at a Larger Scale*, p.44-53.

Ward, J., 2010, 'Additive Assembly of Digital Materials', PhD thesis, Massachusetts Institute of Technology.

Gershenfeld, N., Carney, M., Jenett, B., Calisch, S. and Wilson, S., 2015, 'Macrofabrication with Digital Materials: Robotic Assembly' in *Architectural Design*, 85(5), p.122-127, doi: 10.1002/ad.1964.

Sanchez, J., 2014, 'Polyomino – Reconsidering Serial Repetition in Combinatorics' in *ACADIA 14: Design Agency*, 23-25 October 2014, Los Angeles, ACADIA/Riverside Architectural Press, p.91-100.

CILLLIA
METHOD OF 3D PRINTING MICRO-PILLAR STRUCTURES ON SURFACES

JIFEI OU / GERSHON DUBLON / CHIN-YI CHENG / HIROSHI ISHII
MIT Media Lab
KARL WILLIS
Addimation Inc.

Throughout nature, hair-like structures can be found on animals and plants on many different scales. Beyond ornamentation, hair provides warmth and aids in the sense of touch. Hair is also a natural responsive material that interfaces between the living organism and its environment by creating functionalities like adhesion, locomotion and sensing. Inspired by how hair achieves those properties with its unique high-aspect ratio structure, this project explores ways of digitally designing and fabricating hair structures on the surfaces of manmade objects. Material science and mechanical engineers have long been investigating various methods of fabricating hair-like structures[1,2]. In this paper, we present Cilllia, a digital fabrication method to create hair-like structures using stereolithographic (SLA) 3D printing.

The ability to 3D print hair-like structures would open up new possibilities for personal fabrication and interaction. We can quickly prototype objects with highly customised fine surface textures that have mechanical adhesion properties, or brushes with controllable stiffness and texture. A 3D printed figure can translate vibration into a controlled motion based on the hair geometry, and printed objects can now sense human touch direction and velocity. In this paper, we will focus on introducing the fabrication pipeline and the emerging mechanical adhesion property of the printed hair surface.

The 3D printing revolution

3D printing is rapidly expanding the possibilities for how physical objects are fabricated[3]. Its layer-by-layer fabrication process has tremendous potential to enable the fabrication of physical objects not previously possible. High resolution 3D printers have become increasingly affordable and widely available, enabling the fabrication of micron-scale structures. Cilllia is a bottom-up printing pipeline intended to fully utilise the capability of current high resolution photopolymer 3D printers to generate large amounts of fine hair on the surfaces of 3D objects. We introduce methods, algorithms and design tools for the fabrication of Cilllia and explore its capabilities for mechanical adhesion.

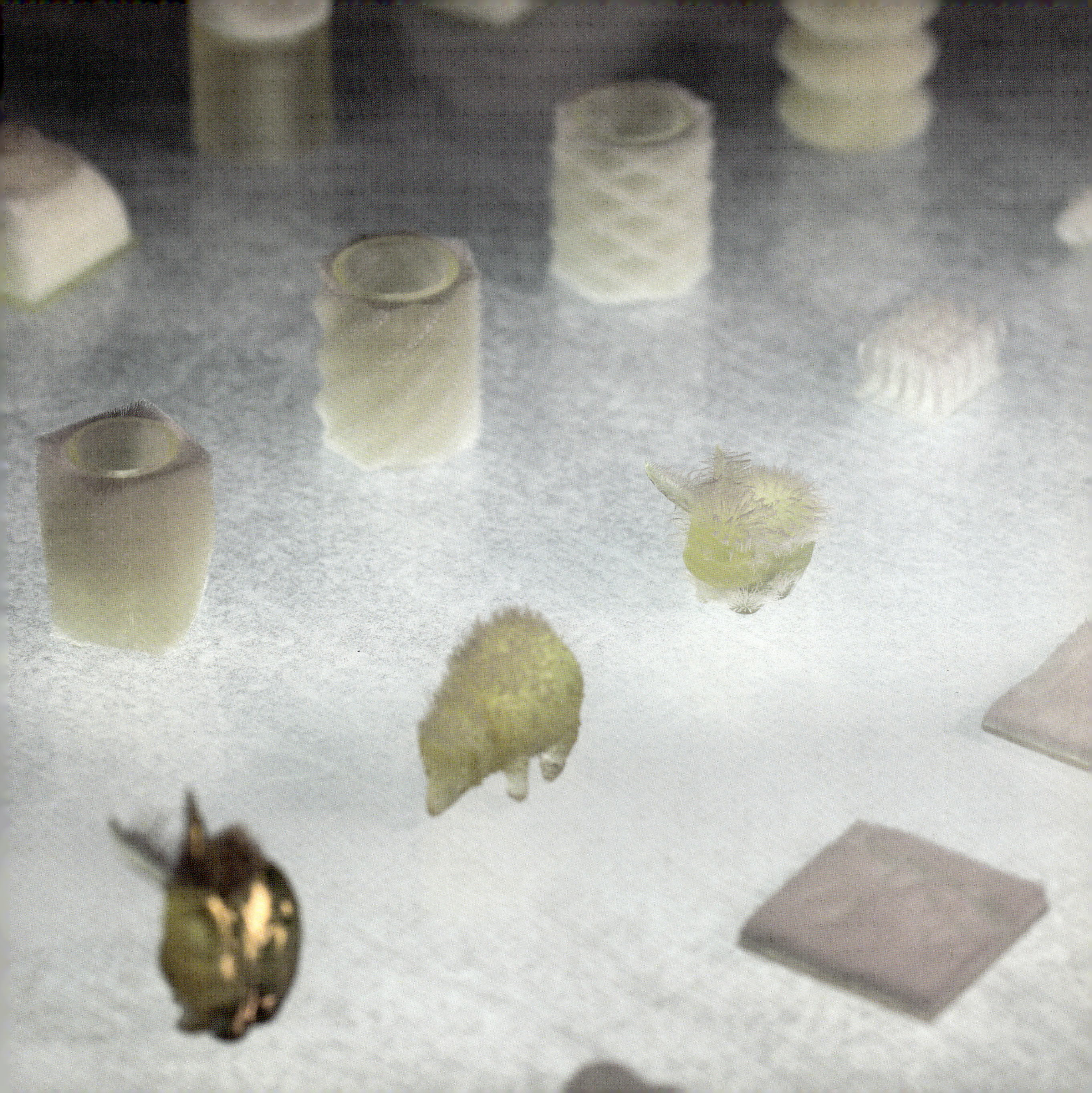

In this paper, the following contributions are presented:

1. A bottom-up approach for generating 3D printable micro-pillar structures.
2. A simple graphical interface that allows users to easily design hair structure.
3. Examples of encoding mechanical adhesion property into hair structures.

As high resolution 3D printers become increasingly available and affordable, we envision a future where the properties of physical materials, whether optical or mechanical, electrical or biological, can be encoded and decoded directly by users. This allows us to customise and fabricate interactive objects as needed.

The challenges of Cillla

Although the resolution of recent 3D printers has been improving, it is still considered impractical to directly print fine hair arrays on object surfaces. This is due to the lack of an efficient digital representation of CAD models with a fine surface texture[4]. Most of the current commercially available 3D printers use a layer-by-layer method to deposit/solidify materials into shapes that are designed in CAD. The process follows a top-down pipeline, in which users create digital 3D models, and then a programme slices the models into layers for the printer to print. In the field of computer graphics, the standard way to represent surface texture is through lofting bitmaps on the CAD model to create an optical illusion. These representations do not actually capture the three-dimensional structure. It is difficult and impractical to create many thousands of small hairs with real geometry using conventional CAD systems.

The data for describing the total geometry become extremely large and rendering such complex structures can also be computationally expensive.

To overcome these challenges, the goal of the project is to bypass the modelling and slicing process of the 3D printing, and instead to directly generate machine-readable files that reconstruct hair-like structures.

3D printing hair-like structures

We introduce a bottom-up approach to 3D printing hair-like structures on both flat and curved surfaces. Our approach allows users to control the geometry of individual hairs, including aspects such as height, thickness and angle, as well as properties of the hair array, such as density and location. We then present three example applications to demonstrate the capabilities of our approach.

All the tests and examples shown in this paper, unless stated differently, are printed on a commercially available digital light processing (DLP) 3D printer (Autodesk Ember Printer). The DLP printer takes stacks of bitmap images from the CAD models and directly projects the image onto the liquid resin layer by layer. The printer has a feature resolution of 50μm on the X and Y axes, and 25μm on the Z axis. The build volume is 64 x 40 x 150mm. The print material is near UV light photopolymer.

Printing hair-like structures
The bottom-up 3D printing approach presented here allows one to design and fabricate hair-like structures without first making a 3D CAD model. The user directly generates printing layers that contain hair structure

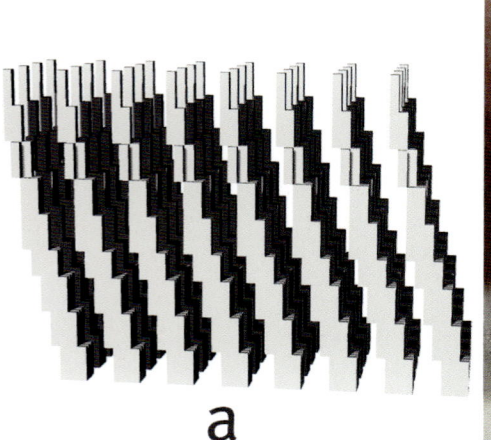

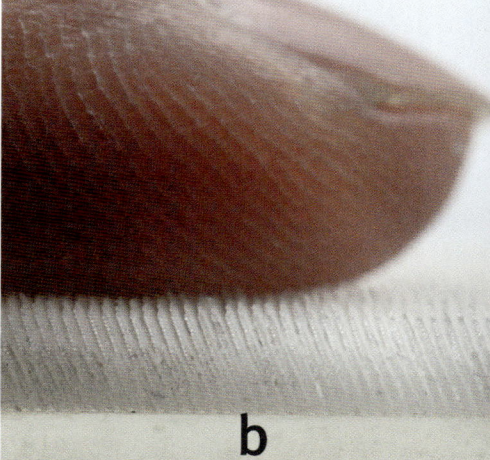

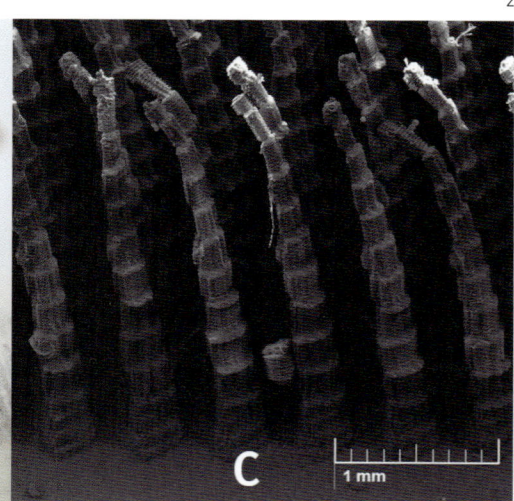

information for the 3D printer. The method can be viewed in three layers:

1. A single hair's geometry (1D): height, thickness, angle and profile.
2. Hair array on flat surfaces (2D): varying single hair geometry across the array on a 2D surface.
3. Hair array on curved surfaces (3D): generating hair array on arbitrary curved surfaces.

Single hair geometry

Compared to other surfaces textures, such as the wrinkle, hair is simple to describe mathematically. It usually comprises a high-aspect ratio cone that is vertical/angled to the surface, although the height, thickness and profile might vary. As we know, the diameter of a cone continuously decreases from the base to the tip. However, the smallest unit in the DLP printer is a pixel. Therefore we need to find a way to construct a model that could approximate the geometry of a cone. We set the base of a pillar to be a matrix of array (e.g. 3 x 3 pixels). As the layer increases, the pixels linearly reduce in a spiral stairs manner, leaving the top layer with just one pixel. This method gives us the highest resolution control of the printed cone shape. We can also add acceleration to the base pixel, reducing velocity to create hair with a different profile.

For tilting the hair to a certain angle, we can offset the pixel group in the X or Y direction every few layers. As the printer has the double resolution on the Z axis compared to the X and Y axis (25μm vs. 50μm), the relationship of tilted angle and layer is:

$$\tan\theta = (L/2) \times P$$

where L is the number of layers and P is the number of offsetting pixels. We successfully printed a series of sample surfaces with oriented hair. Fig. 2 shows that our printed geometry matches the computer visualisation.

Users can easily change the parameters of the hair geometry through a graphical user interface that we designed. It visualises the hair structures as well as generating bitmaps for printing.

We can also generate curved hair by offsetting the pixel group in a spiral layer by layer.

Hair array on flat surfaces

The ability to individually control hair geometry can be applied to thousands of hairs across a flat surface. In order to do this quickly, we use a colour mapping method to make an RGB bitmap in Photoshop, then turn it into a hair array. The values of the R, G and B of each pixel correspond to one parameter of hair geometry. The algorithm checks the bitmap every few pixels to create a new hair based on the pixel's colour. One can therefore easily vary the density of the hair by changing how frequently the bitmap is checked.

Based on our experience, height and angle are the most common parameters that need to be varied frequently. We therefore map the R-value to the angle of the X axis, the G-value to the angle of the Y axis and the B-value to the height of the hair. We use this method to create the conveyor panels in the later section. In the future, we plan to develop a more general approach to encode hair geometry information into one bitmap image, where other parameters such as profile and thickness can be included as well.

Hair array on curved surfaces

In order to apply the presented techniques to a variety of models, it is desirable to print hair on an arbitrary curved surface. To do that, we developed a hybrid method, where users create the curved surface in CAD software, then generate bitmaps that contains pixels of hair array.

To do this, we first import the STL file and position it in the correct printing position. We then find the centroid location of each triangle on the mesh and shoot a ray along the direction of the triangle's normal. A plane moves along the Z axis to intersect with the mesh to create bitmaps of the CAD model, and to intersect with the rays to draw pixels for the hair. In this way, we created bitmaps that contain both CAD model and hair array information. This method allows us to apply the control of hair geometry while slicing as well. However, the generated hair array is highly dependent

1. A collection of 3D printed hair-like structures on flat and curved surfaces. The voxel-based printing approach allows one to define each hair's geometry. Image: Jifei Ou, MIT Media Lab.

2. (a) Computer visualisation of printed hair; (b) close view of actual printed hair; (c) SEM photo of (b).

3. Successful printed hair arrays on curved surfaces.

on the distribution and amount of the triangles. For the examples in this paper, we try to use meshes that have dense and evenly distributed triangles. One can use publicly available online tools (e.g. Meshlab) to create more uniform models. We should also notice that the 3D printer allows only 60° of overhang, so rays beyond that range are ignored. There might also be parts of the hair that penetrate the nearby surface if the surface is curved inwards. We can eliminate this by reducing the hair length correspondingly (Fig. 3).

There are three advantages when directly generating bitmaps of hair structures:

1. By manipulating a single pixel, we can control aspects of a single hair's geometry, such as height, thickness and angle, with a precision of 50μm.
2. Without a CAD model of the hair and slicing process, it becomes possible to print a high density hair array. In our test, we successfully printed 20,000 strands of hair on a 30 x 60mm flat surface.
3. Hair array can 'grow' on any arbitrary CAD model while the model is being sliced.

Printing with laser beam-based SLA

We also experimented with the layer-by-layer method on a laser beam-based SLA printer (Form1+). In the experiment, we directly manipulate the exposure time and the moving path of the laser beam to create an array of laser 'dots' for polymerising the liquid resin. We move the laser beam to the spot where we would like to have hair structure and turn on the laser for two milliseconds, then move to the next spot and turn it on for another two milliseconds. Based on our experiment, two milliseconds is the minimum exposure time one needs to fully polymerise the resin. It produces a dot with a 100μm diameter. To increase the size, one can increase the exposure time. However, we discovered that as one increases the exposure time, the polymerised dot forms into a long oval instead of a circle shape. This is due to the shape distortion of the laser beam. Although the Form1+ has a larger build platform and potentially can be useful for more applications, we decided to use the Ember printer, as it produces more uniform results.

Applications for designers

To show the capability of our printing method, we created three types of possible application for designers.

Objects with fine surface textures
As we can generate hair on curved surfaces, we can now 3D print animal figures with such features. We can also vary the thickness of the hair to create jewelry pieces with controllable stiffness (Fig. 5).

Customised brushes
We can also directly 3D print brushes with customised textures and different densities. With the colour mapping method, one could create a more complex shape of brush for increased and varied artistic expression. In our example, all brushes are 30mm in diameter. The length and density vary based on the input bitmap.

4. A DLP printer is used for hair printing. Image: Jifei Ou, MIT Media Lab.

5. Printed figures with fine surface texture.

Mechanical adhesion

One interesting phenomenon we found during our exploration is that two panels with dense hair can tightly stick to each other when their hair is pressed together. This is due to the large amount of contact surface on the hair that creates friction. To demonstrate this, we printed several hair panels (40 x 40mm) and glued them into boxes. These boxes can be easily attached to each other. In order to keep the hairs on two panels fully in touch with each other, the gaps between the hairs must have the same size as the diameter of the hair base. In our example, the hair base and the gap are both four pixels (200μm).

We tested the strength of the adhesion in relation to the tilting angle of the printed hair. In our experiment, a pair of hair panels (30 x 30mm) were glued onto a solid truncated pyramid (30 x 30 x 30mm). We pushed the hair surfaces against each other and measured the force that was needed to pull them apart. Our test shows that as the tilting angle of the hair increases, the adhesion force rises as well.

Successful fabrication of customised hair-like structures

To summarise, we present a method of 3D printing hair-like structures on both flat and curved surfaces. This allows a user to design and fabricate hair geometry at the resolution of 50μm. We built a software platform to let one quickly define a hair's angle, thickness, density and height. The ability to fabricate customised hair-like structures not only expands the library of 3D printable shapes, but also enables us to design surfaces with mechanical adhesion properties.

While we demonstrated methods and a possible design space for 3D printed micro-pillar structures, we are aware that the technique is very much limited by the physical constraints of current SLA 3D printers. For example, if we had to create an arbitrarily shaped object fully covered by hair, we would have to split the object so that the curvature of the surface could still be printed without a supporting structure. The printable materials are also limited in terms of colour and stiffness. Our current algorithm for generating hair on curved surfaces is also highly dependent on the amount and distribution of the triangles of the CAD model. This means that to print high quality hair requires either a clean mesh model or a preprocessing step for the model. In the future, we will add re-mesh functions to our software platform to control hair distribution. It would also be very interesting to test if tilted hair is mechanically weaker than straight hair, as there is less contact area for each layer of the voxel.

Notes

1. Amato, L., Keller, S.S., Heiskanen, A., Dimaki, M., Emnéus, J., Boisen, A. and Tenje, M., 2012, 'Fabrication of High-Aspect Ratio SU-8 Micropillar Arrays', *Microelectronic Engineering*, Vol. 98, ISSN 0167-9317, p.483-487.

2. Paek, J. and Kim, J., 2014, 'Microsphere-Assisted Fabrication of High-Aspect Ratio Elastomeric Micropillars and Waveguides' in *Nature Communication*, 5.

3. Willis, K., Brockmeyer, E., Hudson, S. and Poupyrev, I., 2012, 'Printed Optics: 3D Printing of Embedded Optical Elements for Interactive Devices' in *Proceedings of the 25th Annual ACM Symposium on User Interface Software and Technology* (UIST '12), ACM, New York, USA, p.589-598.

4. Hopkinson, N., Hague, R.J.M. and Dickens, P.M., 2006, *Rapid Manufacturing: An Industrial Revolution For The Digital Age*, Chichester, Wiley, p.43-45.

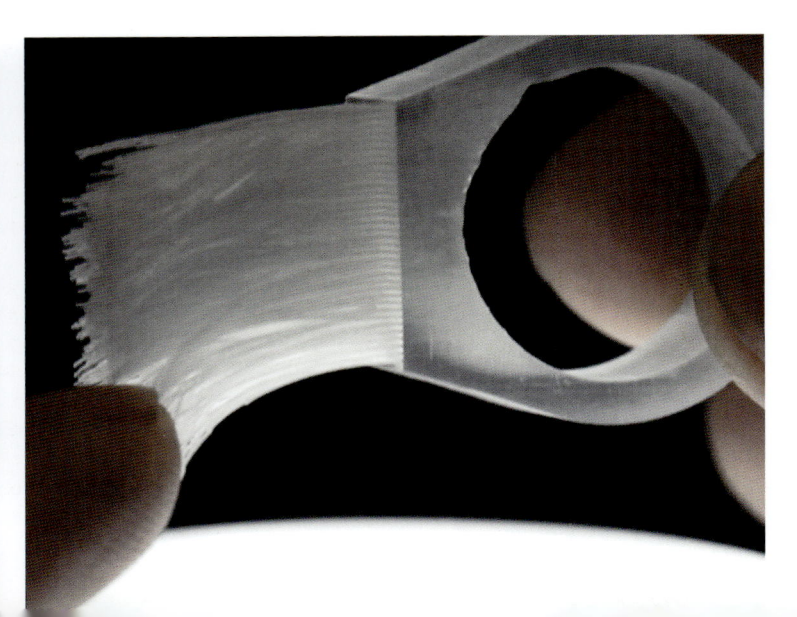

FUSED FILAMENT FABRICATION FOR MULTI-KINEMATIC-STATE CLIMATE-RESPONSIVE APERTURE

DAVID CORREA / ACHIM MENGES
Institute for Computational Design, University of Stuttgart

The aim of this project paper is to present new findings in the implementation of a design for fused filament fabrication (FFF) and custom material development methods, alongside a stimulus-responsive 3D printed prototype (Fig. 1). The project demonstrates control in the design and manipulation of new composite materials to create architectures capable of complex kinematic deformations in response to environmental stimulus. The project highlights hygroscopic, doubly-curved, shape change apertures capable of autonomous climate-adaptive kinematic response.

While recent research into stimulus-responsive materials (SRM), such as timber composites (Wood et al., 2016), bimetals (Sung, 2008) and multi-material composites (Tibbits, 2013), have had to rely primarily on multiple fabrication steps in order to assemble structures capable of double curvature shape deformations, the presented research can build custom directional deformation within a single fabrication process. Moreover, unlike Poly-Jet matrix approaches to SRM, the presented FFF method, using fibrous fillers, enables anisotropic properties through the deposition and the make-up of the material itself. The project showcases a methodology that couples a design-oriented and computationally enabled method that integrates 3D printing via FFF, custom SRM composite polymers and bio-inspired material syntax strategies. The project highlights precise kinematic geometric deformation with multi-directional curvature made possible via a precise understanding and negotiation of the material properties and behaviours inherent to the FFF process. Variations on material properties through the development of a custom polymer composite provide an outlook into the capacity to programme differentiated kinematic response time and material performance for bespoke applications. The research builds upon over seven years of previous work by the authors into hygroscopic actuators using wood composites and bio-inspired 3D printed architectures, furthering this research by transferring and expanding the functional principles and material intelligence of these mechanisms.

Double curvature as a functional principle

The primary research question for this project concerned the integration of double curvature as a functional

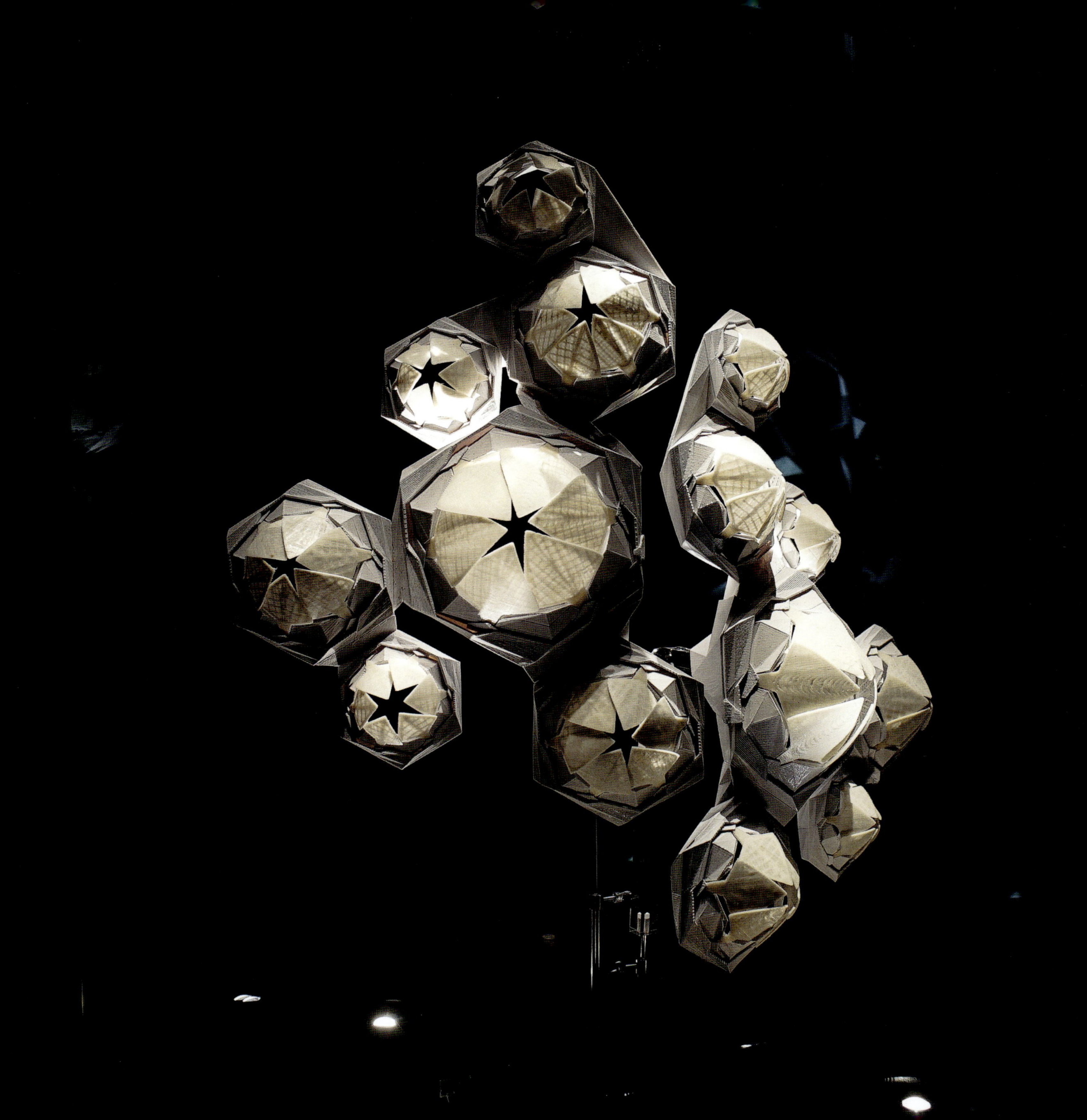

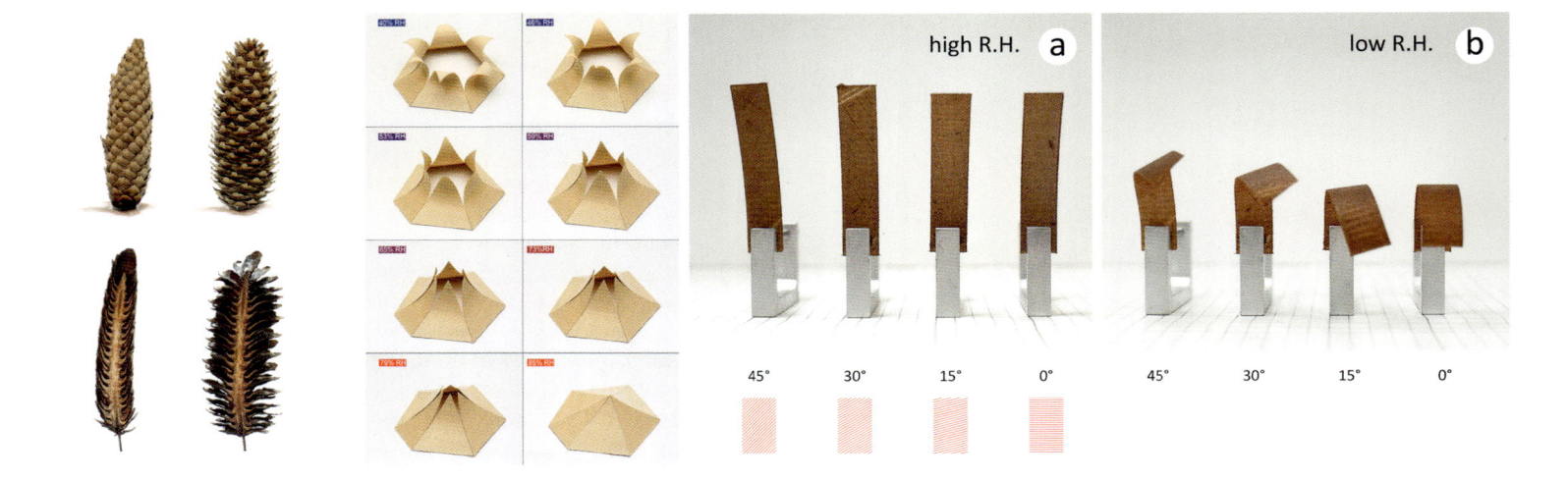

principle into the development of a multi-hierarchical system architecture. Previous research by the authors for 3D printed SRM presented reliable shape change actuation with single curvature deformations using wood fibre composites (Correa et al., 2015). Early tests highlighted that shape change curvature direction could be further directed into controlled twisting angles through the differences in the dominant angle of deposition orientation of each layer (Fig. 2) (Correa et al., 2015). Similar to previously developed veneer composite bilayers (Reichert, Menges & Correa, 2015) or hygrogels (Erb et al., 2013), varying the angle of deposition can enable twisting of the sample through global manipulation of the composite architecture. That is, the changes in material orientation apply homogeneously across the whole sample. The key principle that allows SRM composites to shape change is the ability to direct small expansion forces from the SRM material over a non-SRM substrate. Therefore single curvature shape change deformations are most effective when all expansion forces are directed along a single axis. To achieve double curvature, it is therefore necessary to further expand the understanding of these principles by investigating material organisation methods and composite architectures that can negotiate the interaction of expansion forces in multiple directions. For the presented project, the development of an architectural aperture capable of double curvature shape change was selected as the medium to investigate multi-component interaction and mechanism scale.

New insights through collaboration

Using a customised additive manufacturing process, the double curvature in the hygroscopic-responsive SRM flap is achieved in a single step. Fabrication of the complete multi-aperture assembly involves two steps: first, the printing of the SRM flaps with the fastening support attachment; and second, the printing of the non-responsive understructure that positions the flaps into apertures.

Research into two-stage, doubly-curved pine scale actuation, in collaboration with the Plant Biomechanics Group at the University of Freiburg, provided novel insight into the kinematics and functional material differentiation that allows double-curved shape change in pine scales (Poppinga et al., 2015). This functional principle of double curvature actuation was abstracted into a double-curved flap component with two integrated curling axes. Consequently, for each individual flap, the performance goal is to have a dominant and a secondary curling axis. The primary axis is responsible for the opening of the aperture, while the secondary axis facilitates the lateral expansion and the resulting double-curved shape. As the flaps are configured concentrically within the aperture, their lateral interaction is facilitated by this secondary double curvature deformation. In the closed state, the double curvature deformation allows for the aperture to form a segmented dome geometry, while in the opening state the flaps push each other further into the open position, enabling a wider aperture diameter. Moreover, the flap actuation within the aperture is further supported by a secondary functional region located at the base/stem. This region is both responsible for the fastening of the flap to the aperture understructure and also designed to have a single curvature shape change along the dominant curling axis.

1. Completed piece in controlled climate chamber exhibit at the '+ultra. gestaltung creates knowledge' exhibition at the Martin Gropius Bau in Berlin. Closed apertures indicate a high R.H.% state.

2. Left, pine scale actuation and veneer bilayer system (Reichert et al., 2015). Right, 3DP shape change curling direction in relation to material deposition angles (Correa et al., 2015).

3. Multi-kinematic-state climate-responsive aperture time lapse shape change from high R.H.% environment (left) to low R.H. environment (right).

For the hygroscopic shape change actuation, the composite architecture of the flap is constituted of a hygroscopic SRM material and a secondary constraint material, which has negligible hygroscopic expansion characteristics. While several FF plastics with a limited moisture expansion coefficient can be used, acrylonitrile butadiene styrene (ABS) filament was selected in order to facilitate fuse bonding of the flaps to the aperture understructure. For the hygroscopic SRM material, two materials were tested: commercially available wood composite polymer (WCP) filament and a custom-developed cellulose composite polymer (CCP). Both materials rely on the fibrous cellulose or wood fillers for hygroscopic expansion, and the presented SRM composites make use of the shear-induced alignment of the fibres in the 3D printed beads to define the anisotropic properties of the composite architecture. While single curvature deformation only requires a primary angle of material deposition perpendicular to the shape change axis, the presented double curved flap mechanisms must negotiate different directions of expansion and corresponding constraint. Early tests indicated that positioning two mirrored angles of deposition at 10° from the main curling axes was effective at enabling double curvature, but it achieved a limited curling angle along the primary axis. It was speculated that the central region, along the central axis, did not provide enough material alignment perpendicular to the primary axis. In this initial test, the stem region was designed to be longer in order to compensate for this limited curling angle along the primary axis (Fig. 3).

While this is an effective strategy for the aperture opening, a second approach was developed that could adaptively change the deposition angle to meet both axis requirements. In this second approach, a paraboloid curve was implemented in the toolpath that allowed for the material to be deposited at 90° from the main axis, at the apex; the deposition angle then changed to meet the 10° angle for the lateral sections. This approach functionally distributed the material in relation to the desired curvature and reduced internal stresses resulting from different areas of expansion meeting at a narrow angle along the central axis. As a result, the flaps were successful at achieving a narrower curling shape change angle along the central axis, without compromising curvature changes on the secondary curling direction (Figs. 4 and 5). After achieving the target performance for the main functional region of the flap, the stem region was reduced in order to be more seamlessly integrated within the substructure; it therefore plays a more limited role in the overall angle of opening of the aperture. For the non-hygroscopic constraint components, the material organisation along both the primary and secondary axis followed the same corresponding functional distribution. The main constraint material beads are deposited in line (0°) with the primary curling axis, while the secondary constraint follows a 90° angle. Additionally, in order to better integrate the stem and the main flap region, a boundary edge is implemented.

While the WCP provided the desired hygroscopic shape change performance for the doubly-curved mechanisms, a second custom SRM material was developed and tested. A cellulose composite polymer (CCP) filament was developed, using isolated wood cellulose fibres embedded in a proprietary co-polyester polymer. It was of interest to test the effect that the isolated cellulose fibres had in relation to the SRM composite response time and shape change characteristics. Furthermore, using a second composite polymer with hygroscopic fibres allowed us to test the effectiveness of the previously developed shape change methods.

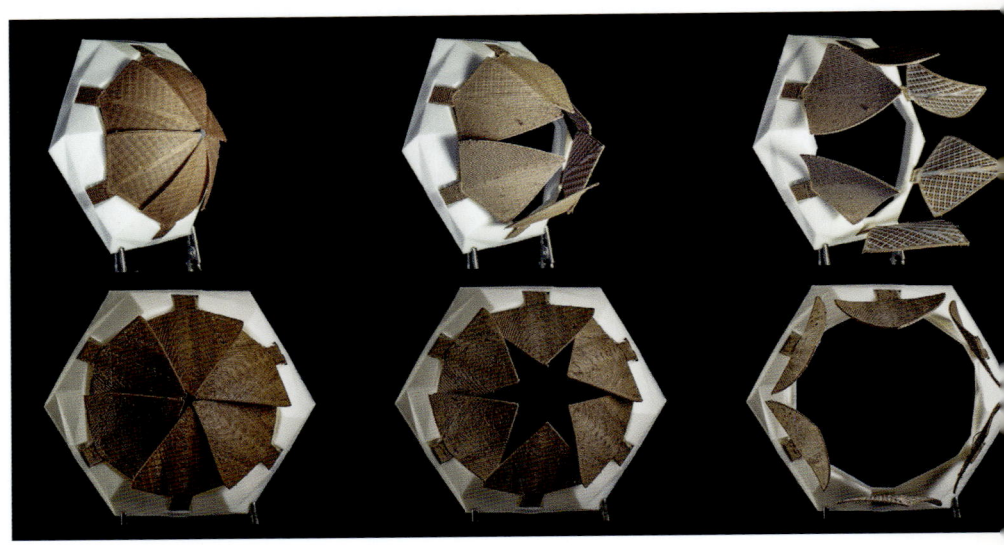

The design of a bespoke composite filament allowed for the verification that no additional colourings or foreign fillers were added that could directly or indirectly affect the performance of the SRM composite. While technical characterisation of the material falls outside of the scope of this project paper, empirical tests, with both isolated single-curved samples and with the presented double curvature aperture, indicate a small increase in moisture absorption and desorption in the samples, resulting in a faster stimulus response time. However, the colour change resulting from the removal of the lignin appears to have an impact in tests using radiation from light sources. As opposed to the darker WCP samples, the whiter CCP samples can reflect most wavelengths of light, resulting in reduced temperature increase. Due to the complex interaction that relative humidity, radiation and localised surface evaporation can have in moisture desorption, the samples have a uniquely different performance profile. When subjected to moisture desorption tests, the WCP samples can have a faster response time under exposure to light radiation due to their colour, while the CCP samples can be faster in low light environments with equivalent low relative humidity.

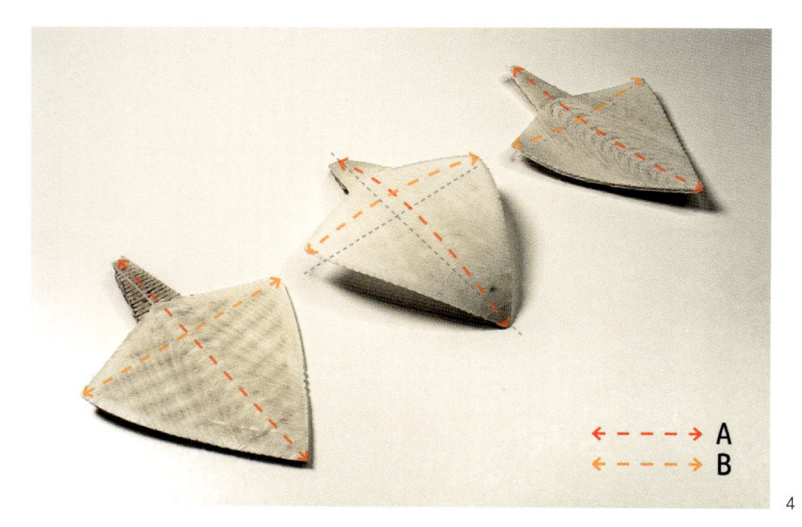

The ABS 3D printed substructure was designed to provide a support structure that can accommodate three scales of flap mechanisms ranging from 38 to 72mm (measured along their primary axis). The piece is composed of two halves, containing a total of 14 apertures. Small changes in angle direction allow the piece to generate a sense of enclosure while exposing each aperture to slightly different light angles.

Developing a new 'smart' material

Wood composite 3D printed filament enabled the application of a found material 'wood' into a new fabrication process, using a thermoplastic polymer to bond the particles and enable the deposition of the material in a directed and controlled matter. In other words, the method hijacks the precise deposition of the FFF 3D printer and the hygroscopic properties inherent in the material to enable the development of a new designed meta-material/'smart' material. By isolating cellulose, the active hygroscopic component of wood, the new custom cellulose composite highlights the possibility of selectively choosing desired performance properties. In collaboration with material science experts and industry partners, additional aesthetic or functional performance characteristics can be further integrated to meet desired applications. The integration of 'smart' functional material performance into a multi-hierarchical architectural system enables closer insight into the conception of truly smart and adaptive buildings, whereby the function, material and form are intrinsically programmable to respond to and anticipate user performance needs.

In this content, it is evident that in addition to furthering research into SRM shape change architectures, more research and testing is needed for the adequate characterisation of both composite filaments and the resulting meta-material composites. The potential of FFF for form generation continues to be widely investigated, but the physical and material intricacies resulting from the material interactions of FF layered deposition remain

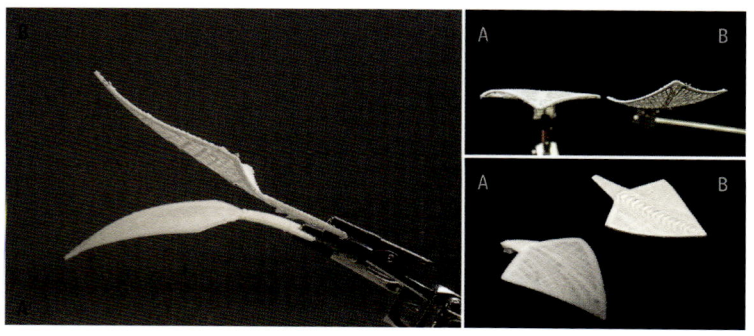

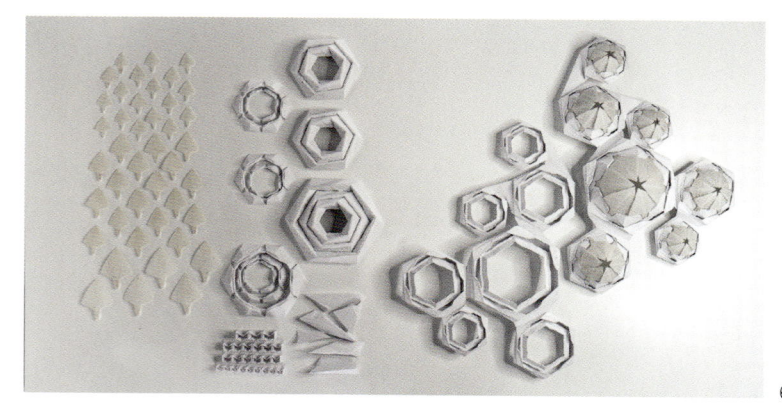

4. Multi-kinematic-state climate-responsive flap with doubly-curved deformation along primary axis (A) and secondary axis (B). Three states are presented: left, flat state after FFF under controlled environment; centre, convex double curvature under high R.H.%; right, concave double curvature under low R.H.%.

5. Three viewing perspectives of multi-kinematic-state climate- responsive flap under high R.H.% (A) and low R.H.% (B).

6. From left to right, three sizes of responsive flaps: support understructure components, complete understructure and completed multi-aperture assembly piece.

poorly understood. While there is substantial research into the base constituent polymers used in FF plastics, little is known about their final physical performance once they are chemically modified/optimised for FFF criteria. Long-term testing for flexural strength, material fatigue or UV decay will be required in order to be able to consider possible technical applications. Moreover, meta-material composites with SRM multi-material architectures offer additional layers of complexity and opportunity, requiring a wider scope of investigation in a multidisciplinary context (Le Duigou et al., 2016).

New directions for material intelligence

This novel approach to generating shape-changing architectures using FFF provides new opportunities for architectural design that can further access material intelligence through programmable and adaptive responsive systems. The competence of architects lies in the conception of material organisation strategies that are functionally integrated through geometric and material interdependence. Reciprocities in form, structure and material differentiation require a sound conceptual understanding as a formal and material assembly, in order to implement effective and adaptive multi-hierarchical functional structures capable of performance-driven local differentiation. Nevertheless, the challenge for architects and designers is that while material science forms a critical component of material-oriented architectural research, it does fall outside its core field of expertise and professional scope. It is only through a truly interdisciplinary research approach that both competencies and professional expertise can yield innovative approaches and applications.

The outlook of this research presents the possibility of applying and expanding the presented FFF methods into other movement mechanisms with additional curvature direction or the integration of synclastic and anticlastic curvature changes within a single piece. Additionally, material development of the composite filament offers great potential to include bespoke performance characteristics that can enable further control of the actuation response. Development of testing methodologies to evaluate feasible applications into architectural applications can foster better understanding of desired technical performance and limitations. Moreover, considerations of the lifecycle of the material systems is of particular concern; further studies into the incorporation of bio-based polymers and additives that can also be biodegradable is of critical importance for this research.

Acknowledgements

D.C. doctoral research was supported by the Natural Sciences and Engineering Research Council of Canada, PGS D Award and the JONAS Research Network. We would also like to thank Dr. Simon Poppinga and Professor Thomas Speck from the Plant Biomechanics Group Uni-Freiburg for their invaluable insight, Kai Parthy for support and expertise on FFF hardware and material sample development and Abel Groenewolt and Derik Gokstorp for their support with Python development of custom G-code script.

References

Wood, D.M., Correa, D., Krieg, O.D. and Menges, A., 2016, 'Material Computation – 4D Timber Construction: Towards Building-Scale Hygroscopic Actuated, Self-Constructing Timber Surfaces' in *International Journal of Architectural Computing*, 14(1), p.49-62, doi: 10.1177/1478077115625522.

Tibbits, S., 2013, Self-Assembly Lab. Available at http://www.selfassemblylab.net/ProgrammableMaterials.php (accessed 9 October 2016).

Correa, D., Papadopoulou, A., Guberan, C., Jhaveri, N., Reichert, S., Menges, A. and Tibbits, S., 2015, '3D Printed Wood: Programming Hygroscopic Material Transformations' in *3D Printing and Additive Manufacturing*, 2(3), p.106-116, doi: 10.1089/3dp.2015.0022.

Reichert, S., Menges, A. and Correa, D., 2015, 'Meteorosensitive Architecture: Biomimetic Building Skins Based On Materially Embedded and Hygroscopically Enabled Responsiveness' in *Computer-Aided Design*, 60 (ISSN 0010-4485), p.50-69, doi: 10.1016/j.cad.2014.02.010.

Erb, R.M., Sander, J.S., Grisch, R. and Studart, A.R., 2013, 'Self-Shaping Composites with Programmable Bio-inspired Microstructures' in *Nature Communications*, 4, p.1712, doi: 10.1038/ncomms2666.

Poppinga, S. and Speck, T., 2015, 'New Insights into the Passive Nastic Motions of Pine Cone Scales and False Indusia in Ferns' in *Proceedings of the 8th Plant Biomechanics International Conference*, 30.11.-04.12.2015.

Le Duigou, A., Castro, M., Bevan, R. and Martin, N., 2016, 3D Printing of Wood Fibre Biocomposites: From Mechanical to Actuation Functionality' in *Materials & Design*, 96, p.106-114, doi: 10.1016/j.matdes.2016.02.018.

3D METAL PRINTING AS STRUCTURE FOR ARCHITECTURAL AND SCULPTURAL PROJECTS

PAUL KASSABIAN / GRAHAM CRANSTON / JUHUN LEE
Simpson Gumpertz & Heger Inc.
RALPH HELMICK / SARAH RODRIGO
Helmick Sculpture

3D metal printing is developing rapidly into a viable structural technology for architectural and sculptural projects. The benefits of 3D printing can be married with the strength of metal to achieve high performance levels. Successful implementation requires the development of new computational design and analysis tools that become integrated into the design process, as well as an understanding of metallurgical processes for controlling material properties. We have pushed forward with this on two projects: one sculptural and the other architectural. In both examples, the standard approach would have been cost-prohibitive or physically impossible. This paper covers the technical background and the detailed design process for using 3D metal printing as a structure.

There is an ongoing convergence of low-end and high-end 3D printing technology that has a far-reaching impact on the structural performance of sculptural and architectural projects. At the low end (defined here as 'low' structural strength and ductility), we have seen the rapid rise in the creative design space of toys, jewelry and fashion (wearables, clothes, etc)[1]. These approaches have excelled at rapid generation and production of form and resulting accessible cost levels, so the market continues to expand in both volume and quality. At the high end, we have seen specific uses of medical and aerospace designs[2] with an accompanying high cost of design development and production for the unique performance market; so far, this market has been limited by cost.

This is a prime time to take advantage of this current convergence, and we have developed computational approaches, designs and prints for real projects as proven examples. Below, we outline the current state of 3D printing for the AEC industry and provide details of two case studies.

A vision for 3D metal printing in the AEC industry

The upfront benefits of 3D printing (mass customisation, complexity at low/no cost, reduction in assembly effort, production of forms previously not feasible, etc) are valuable to the AEC industry, in which most projects are essentially one-off unique buildings, pavilions, sculptures, etc. Uptake in the industry has been slow, as the material and design technologies have had to develop to match need.

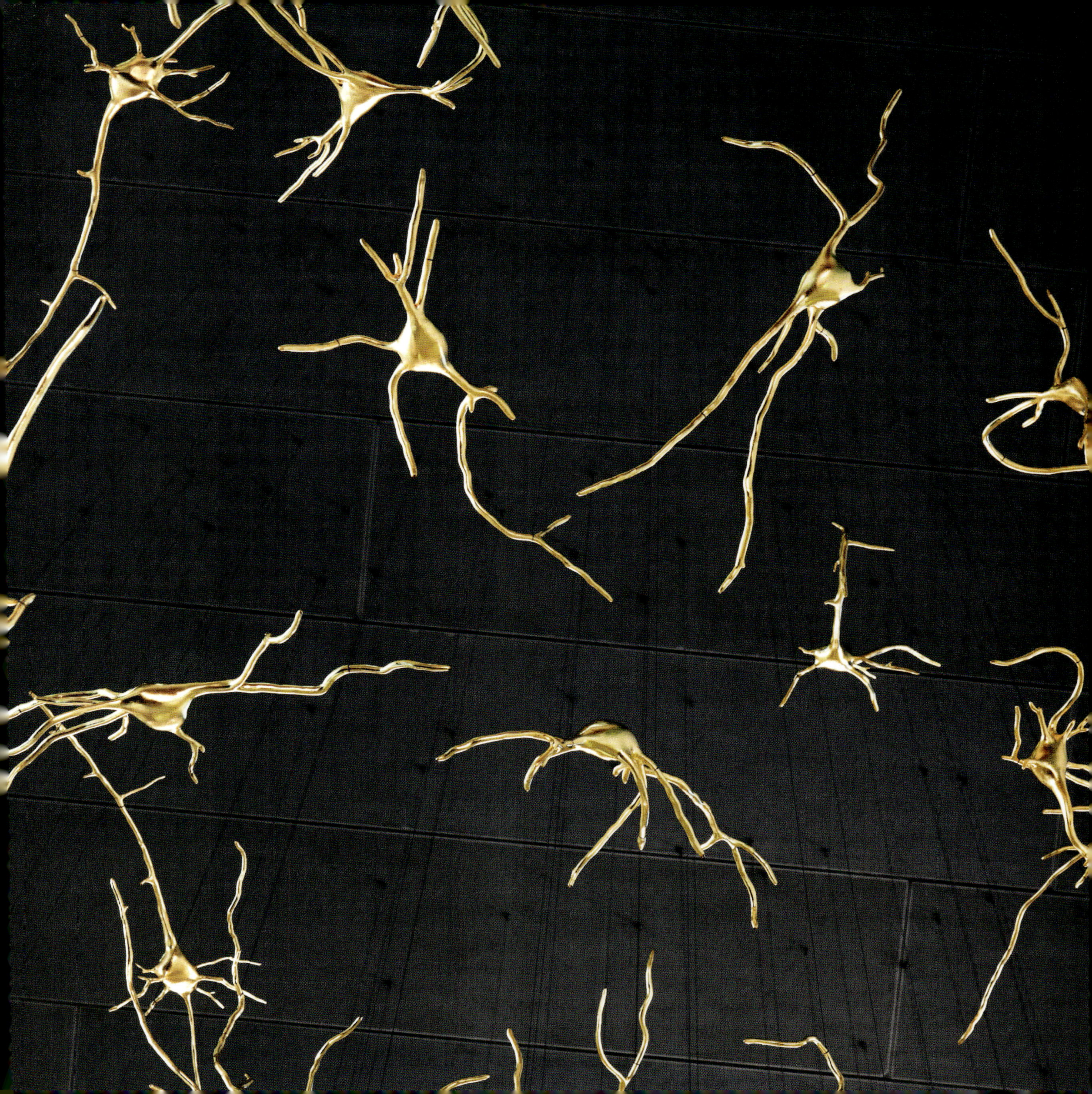

Tools and materials currently available

The material world is broad. For structural performance, the AEC industry typically works with a limited range of materials dominated by steel, reinforced concrete and wood. We believe 3D printing will open up opportunities for other materials where the technology will provide greater performance possibilities (examples include structural plastics[3], which have existed and been used for decades on industrial structural applications, fibre-reinforced composites where placement of the fibres can be optimised, etc). Contour crafting[4] developed at USC, and other similar approaches, have developed printed concrete technology and there will be much more innovation going forward. For this paper and the immediate next step in development, we focus on the wider range of metals.

Metal 3D printing can be divided into powder-bed fusion and deposition-based approaches. In the former, the metal powder is sintered layer by layer via high-energy methods such as electron beam melting (EBM) and direct metal laser sintering (DMLS). For the latter, metal is deposited in a manner similar to continuous welding in air.

Powder-bed fusion approaches allow for high-quality control and production of the resulting metal, with alloy powders of particle sizes between 45 and 100 microns. This process allows for fine detail, albeit within a current limited build envelope of approximately $0.05m^3$. Deposition-based approaches produce a more coarse build, but the build envelope is essentially limitless.

Certain metals, such as stainless steel, titanium and aluminium, are more conducive to this approach than typical softer metals such as bronze or copper. As the cost of the EBM and DMLS processes are driven more by energy requirements than by bulk material cost, we are seeing lower costs in the market for 3D printing of titanium than for stainless steel – a different paradigm than normal for the architectural/sculptural design market in AEC.

While printed metal can reach equivalent tensile strength to standard metal product, there is ongoing research work for all these metals aimed at achieving sufficient ductility. The sintering process forms a granular structure that is not directly equivalent to typical structural metals, resulting in less ductility. To date, this has been addressed either by limiting sustained deformations of the printed item to the elastic regime or by annealing or treatment by hot isostatic pressing (HIP). Both remedies have cost implications.

How can these benefits be made real on AEC projects?

As stated above, there are real benefits to 3D printed structural components that are valuable to the AEC industry, including mass customisation, complexity at low/no cost, reduction in assembly effort, production of forms previously not feasible and others. We see two areas of technological development that can make the benefits of 3D metal printing real:

1. Development of integrated design tools: most structural software is focused on analysis only and is inflexible to the changing design process. For 3D printing, the generation of a form and its iterations are intrinsic to the process, so new methods and approaches are required.

2. Understanding of the metallurgical process: currently, the resulting printed metal is not identical to metal product and design standards, as they do not yet exist for 3D printed metal. Hence, to provide an equivalent performance, the metallurgical process and its effect on the design must be understood and integrated accordingly.

Overall, there is no better learning than doing. We have worked on two recent projects, one sculptural and one structural, to address these issues head-on and drive the technology forward.

Sculptural project:
Schwerpunkt at MIT, Cambridge, Massachusetts, USA

Helmick Sculpture was commissioned to create a 3D anamorphic sculpture for MIT's McGovern Institute for Brain Research in Cambridge, MA. The sculpture is comprised of a hundred individual neuron sculptures ranging in size from 305mm to 915mm in the longest dimension, and suspended in a three-storey atrium with viewpoints throughout the space. The individual neuron sculptures must 'read' from every direction, with primary views from below and the exterior entry plaza on a main pedestrian and vehicular artery of the area known as 'Technology Square'. The composition not only had to work in the round but create a culminating moment when one reaches the third-floor entry to the Institute, where all the neurons visually coalesce into a 'drawing' of the human brain.

Design approach

One of the primary guiding principles in Helmick's approach to sculpture is the human eye's remarkable ability to collate disparate data points into a recognisable

1. Schwerpunkt: view from beneath the sculpture.

2. View from the side: neurons distributed in space.

3. The sculpture seen from the anamorphic 'sweet spot' viewpoint.

image – essentially connecting the dots to quickly recognise patterns.

Creating a 2D image in space consisting of 3D lines is a challenging undertaking, with the added complexity of creating a compelling 3D sculpture. The initial models for the sculpture were simple and concerned only with 'proof of concept' rather than sculptural complexity. Once we began 3D modelling each of the neurons in Rhino[5], we realised that preserving the dynamism of the sculpture from all angles would require increasing the palette of forms. Any repeat forms, notwithstanding changes in size and orientation, became visually apparent. Over time, the creation of the culminating brain image required specific shapes from various dendrites to create a unique collection of forms from various viewpoints.

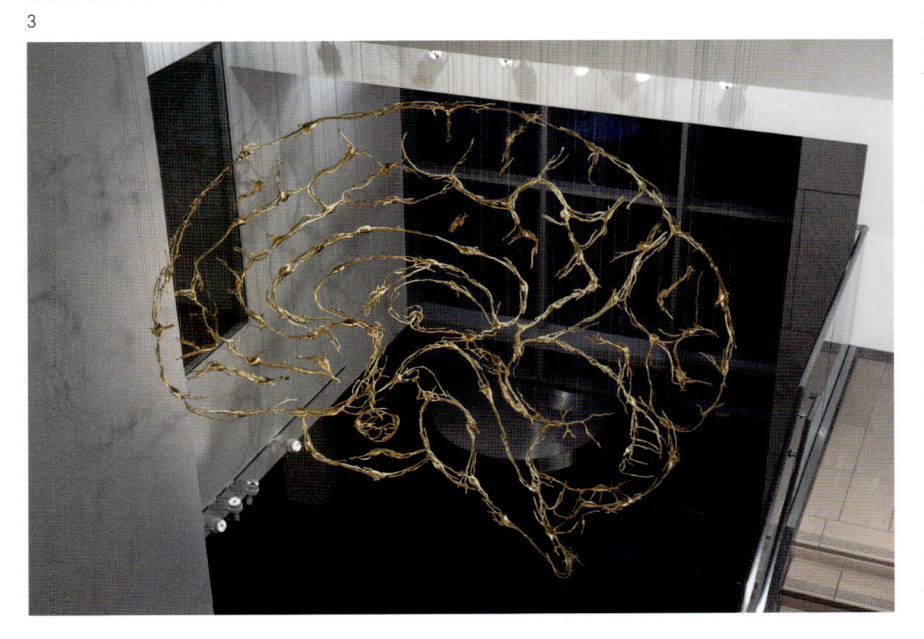

From a formal perspective, unique neuron forms, and specifically unique dendrites, would be the most efficient and successful approach to the sculpture design, but the fabrication of a hundred unique organic forms using traditional methods was daunting. Previous Helmick sculptures with similar forms were made using bronze casting, although this does not allow for specialisation and one-of-a-kind pieces without prohibitive pricing. After creating a physical suspended ¼ scale 3D sketch of the entire sculpture in the studio, we also explored other methods such as a 'kit of parts' approach, but we found the number of components needed in order to achieve the appearance of individuality made the kits impractical in scale. Is it really a kit of parts if you produce three each of three hundred parts? It may be, but it's certainly not an efficient or cost-effective kit.

Finally, we turned to the possibility of direct 3D printing.

3D metal printing and design
Helmick Studio has long combined digital tools – 3D laser scanning, rapid prototyping, CAD modelling – with traditional sculpting methods, but this approach created an entirely new fabrication sequence for us. Instead of starting with a hand-sculpted object and digitising it, we would start with a digital object and hand-finish it. Helmick Studio modelled each neuron individually in Rhino, and Simpson Gumpertz & Heger (SGH) expanded this same model to include both the existing structure above and real-time structural analysis within the Rhino model itself. Through an iterative approach, the Helmick-SGH team simultaneously evaluated varying parameters, including the three-dimensional location of the neurons combined with strength and deflection of the suspended ceiling (which was a typical lightweight suspended ceiling not typically rated for the loads of the sculpture). SGH also performed physical material and system testing at their in-house materials lab of sample framing and panels to quantify their strength and stiffness. The testing provided accurate information for the design iterations and also confirmed the need for an additional design element: adhering thin plexiglass to the ceiling panels where required for added stiffness.

Helmick Studio developed the Rhino model further by breaking each neuron into a number of pieces (five to ten) based on bed sizes and material parameters, and had all the pieces direct printed in a bronze/stainless steel alloy. This approach had several advantages: we maintained weight tolerances for each neuron by hollowing larger pieces and making smaller neurons solid; we made the cell body to dendrite connections structurally robust in a quantifiable way; we labelled each individual component in CAD for easy assembly and tracking; and we were able to fabricate a hundred unique organic sculptures at a significantly lower cost than creating the same pieces using traditional sculpture fabrication methods.

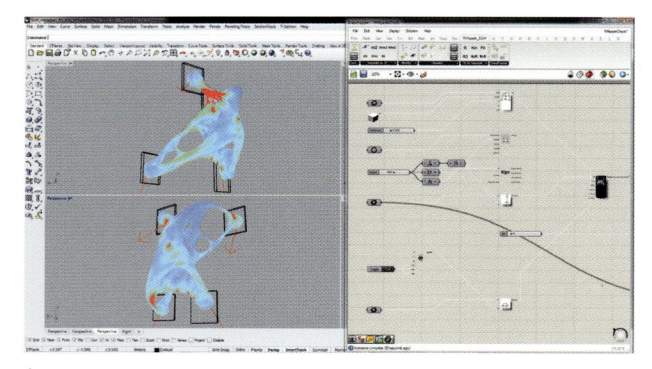

Once printed, the neurons were hand-finished, assembled, primed and finally goldleafed, again by hand. The final result is a sculpture that would not have been possible to create on this budget even a few years prior.

Structural project:
Entrance Building, Northern Massachusetts, USA

This ongoing project is a new two-floor entrance pavilion building to an existing office headquarters in northern Massachusetts. The client is confidential, but is associated with design and the AEC industry. Thus the entrance pavilion building is to be a statement of their current and future standing. The architect is NADAAA of Boston, MA.

The building façade consists of articulated glass panels that relate to both the internal structure and an architecturally defined skin. The combination of façade articulation and varying plan shape results in each glass panel being at a different distance from the slab edge. At the first-floor slab edge, where four glass panel corners meet, each glass panel corner is at a unique distance and orientation from the normal vector to the local slab edge geometry. As a result, each connector must be unique.

Initial design iterations focused on modulating the slab edge geometry and geometric options for the glass panels. Although these options reduced the amount of geometric variation, they did not result in any practical benefits to cost or fabrication complexity. The connectors would still be a variety of welded plate or individual unique castings. Thus 3D metal printing was a clear candidate for the design and fabrication of the façade connections.

The SGH team focused on developing design and analytical approaches that could provide valuable insight and information on the connectors without limiting design creativity and iteration, and where the resulting designs could still be analysed to an accuracy required for refined shape-forming and printing. The team recognised that the integrated and unique nature of architectural and structural requirements for these components would need addressing during the design process rather than being left to a delegated design procurement method. Although we noted above the potential cost benefit of using titanium, we chose stainless steel for the material design to provide a more direct comparison as well as a level of comfort to other parties involved. For future projects, we expect to be designing with 3D printed titanium and developing forms and performance not feasible with other standard production processes.

Design approach
SGH developed two workflows that were used to take the project from concept through to the final print process.

Workflow 1: the 'quick and dirty' method. This approach allows, as a starting point, for rapid generation of the range of structural topologies based on geometry and wind loads. SGH developed custom C# scripts within Grasshopper[6]/Millipede[7], using the building façade geometry as input to automatically generate a finite

4. Wind forces and stresses on optimised model.

5. Refined topological optimisation and finite element analysis.

6 & 7. Images of final 3D stainless steel print connector.

number of element models to calculate the effect of the forces at the connection points for the range of wind load cases. In essence, this provided the envelope of force vectors (magnitude and direction) at each of the unique connection points. Starting with a similar structural 'block' or design space at each connection point, we used 3D topology optimisation to remove material of low stress and computationally iterated on the analysis as an integrated geometric and structural process until the family of connections were produced.

The result of this stage was a series of 3D-related forms that satisfied overall structural requirements but would need more design and analytical refinement.

Workflow 2: the 'detailed and accurate' method. This stage followed from the above family of 3D-related forms and design discussions with the team, and allowed for refinement of the design, analytical accuracy and geometric refinement for the printing process. We exported the generated geometry to ANSYS[8] and performed multi-objective topology optimisation and a detailed finite element analysis. The resultant mesh geometry from the optimisation process is 'pixelated' and therefore produces a rough surface finish. We imported the geometry into ZBrush[9] to smooth the surface profile and clean up the mesh. We then brought the form back into ANSYS for reanalysis and final sign-off against load capacity criteria.

3D metal printing

As proof of the process, we selected one connector to print in stainless steel. We worked closely with Addaero Inc. of New Britain, Connecticut, during the Workflow 2 stage above, who provided invaluable advice on all practical printing aspects, including overall maximum size, local detailing and geometry file accuracy requirements, among others.

The resulting printed connector is a hollow volume with a shell thickness of only 1.5mm optimised for architectural and structural performance. Printed at the same time as the connector were test coupons which are being used for tensile testing at our in-house materials lab to review for strength and ductility. This work is ongoing.

The technological development of metal 3D printing proceeds at a rapid pace. The benefits afforded by 3D printing have significant value to the AEC industry. Our vision outlined in this paper to deploy 3D metal printed components in architectural and sculptural applications takes advantage of these benefits to remove the limits of current fabrication methods and enable new structural forms. Continued development of computational tools and materials specifically designed for these applications is best served through project experience, where a goal gives focus and drives progress forward, especially with production and performance of forms previously not feasible.

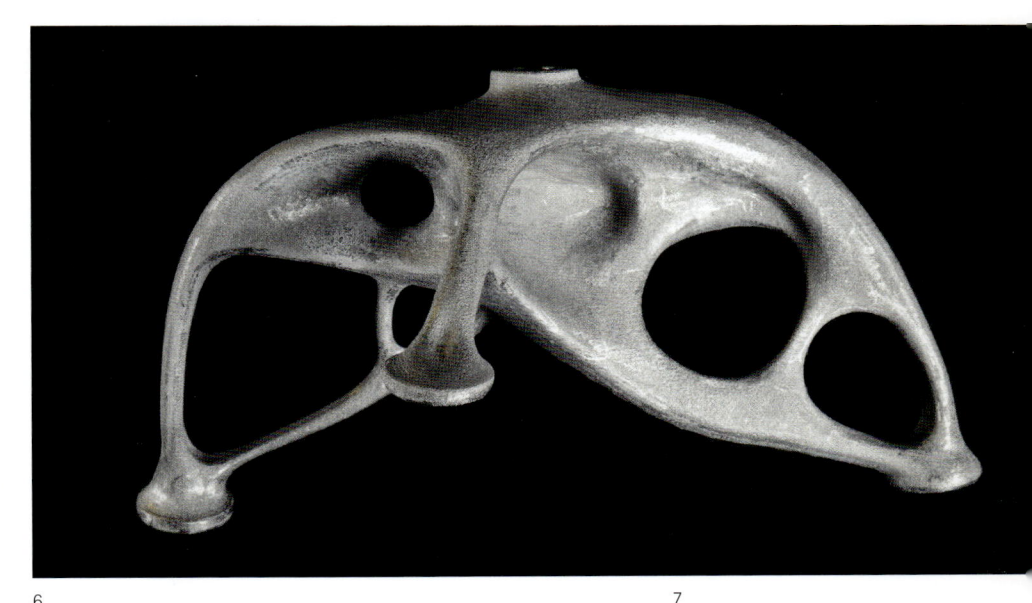

Acknowledgements

The authors wish to thank Richard Merlino of Addaero Inc. for his technical advice and production of the façade connector print.

Notes

1. See http://n-e-r-v-o-u-s.com/projects/sets/kinematics-dress/ (accessed 16 October 2016).

2. 25 April 2016: GE 3D printed metal jet engine: http://qz.com/667477/ge-fires-up-worlds-largest-commercial-jet-engine-using-3d-printed-metal-parts/.

3. Heger, F.J., Chambers, R.E. and Dietz, A.G.H., 1982, *Structural Plastic Design Manual*, Vol. 2, Federal Highway Administration.

4. Leach, N., Carlson, A., Khoshnevis, B. and Thangavelu, M., 2012, 'Robotic Construction by Contour Crafting: The Case of Lunar Construction' in *International Journal of Architectural Computing*, Vol. 10, No. 3, p.423-438.

5. Rhinoceros, Robert McNeel & Associates, version 5.

6. Grasshopper, Robert McNeel & Associates.

7. Millipede, Panagiotis Michalatos (used with permission of creator).

8. ANSYS Workbench Mechanical Enterprise, release 17.0.

9. ZBrush, Pixologic, version 4R7.

MOBILE ROBOTIC FABRICATION SYSTEM FOR FILAMENT STRUCTURES

MARIA YABLONINA / MARSHALL PRADO / EHSAN BAHARLOU / TOBIAS SCHWINN / ACHIM MENGES
Institute for Computational Design, University of Stuttgart

In the past decade, robotic fabrication in the field of architecture has developed rapidly, opening up new possibilities for architecture and design. New fabrication techniques allow the utilisation of materials like fibre composites in the field of architectural construction by employing qualities of the material that were previously not feasible. However, the equipment used for material exploration in the field is often standard industrial machines, originally designed for assembly line applications, which have scale and process limitations.

Introducing a new generation of mobile construction machines capable of operating onsite would allow expansion of the capabilities of currently developed fibre composite fabrication. This research proposes a multi-robot system of cooperative, mobile machines operating within the context of the surfaces of existing architectural environments: façades, walls, ceilings. Anchoring new tensile filament structures to these surfaces activates a new layer in the architectural environment, building upon and modifying it to current spatial requirements in real time (Fig. 1).

Using custom mobile robots

The presented project aims to expand the scope of robotic fabrication for filament and composite fibre architecture through the introduction of custom, cooperative mobile robots. Over the past decade, a significant body of work related to applying fibre composite materials to architecture and design without the need for elaborate moulds or formworks has been developed (Menges & Knippers, 2015). Simultaneously, advancement in mobile robotics and autonomous control have become more prominent in relation to design and fabrication through research projects (Jokic et al., 2014). Developments in technology and methodology in these fields allow this research to take things one step further, through the introduction of mobile collaborative robots for fibre composite fabrication. Mobile machines are directly matched to the unique affordances of fibre composites: lightweight properties of the material as well as the process of phase transition from soft filament to cured structure allow for low payload agile machines iteratively applying layers of fibre to create a structure.

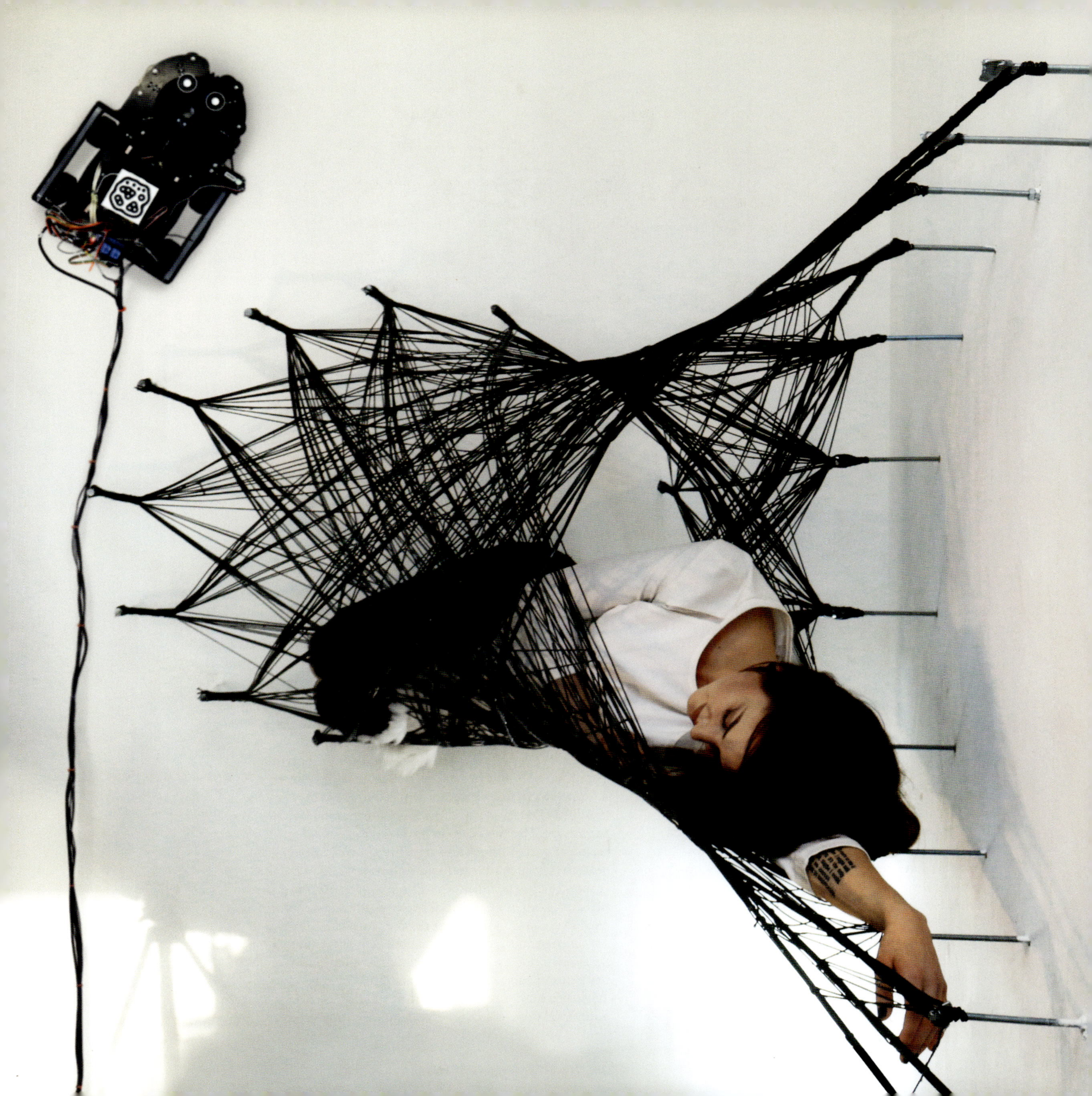

Exploring the potential for in-situ fabrication through the introduction of machines capable of operating in architectural environments would expand robotic fabrication processes beyond the constraints of the production hall. This expansion exposes the possibility of urban and interior environments as the unique framework for onsite fabrication. Multi-robot systems have the potential to provide larger solution spaces and design potentials than traditional robotic fabrication. Small machines enabled with locomotion allow fabrication in environments that are not – and could not be – equipped to house industrial-scale machines.

Significant conceptual differences between operating mobile and standing machines require a distinct change in all stages of fabrication and development, starting with design. In this work, new design (CAD) and manufacturing (CAM) processes are to be developed in order to fully take advantage of new hardware tools for construction.

A new approach to fibre composites in construction

Developments in mobile robotics and autonomous control systems allow for automation of various tasks in industrial and household applications (Novikov, 2015). Companies such as Amazon have implemented mobile robots for the automation of manual labour required at their warehouse facilities (d'Andrea, 2012). Complex locomotion systems are being developed in order to operate in dangerous and unreachable environments, such as earthquake sites (Zhang, 2007). Quadcopters have replaced complex equipment in the filming industry. Surface-climbing robots are used for the maintenance of building façades (Mahajan & Patil, 2013). In the field of digital fabrication, projects like the Aerial Construction research at the Gramazio Kohler Research group of the Eidgenössische Technische Hochschule in Zurich (Mirhan, Gramazio & Kohler, 2015) demonstrate that the application of collaborative mobile machines to construction with lightweight materials is very promising.

The integration of fibre composite materials into the architectural construction process has been a focus of exploration for designers and researchers since the late 1950s. High performance of these materials has promised a revolution in construction and design possibilities. Multiple attempts at fabrication with fibre composites at a large scale, such as the Monsanto House in California in 1957 (Phillips, 2004), influenced discourse but failed to find a foothold in the construction market. Standard fabrication techniques for fibre composites imply a serial production scale which became undesirable in a "society that increasingly valued individualism" (Knippers & Menges, 2015). The necessity of creating large complex moulds for each fabricated piece made it inefficient for the fabrication of unique elements.

1. Human-scale structure prototype fabricated using the system.

2. Series of mobile robotic prototypes for the mobile robotic fabrication system for filament structures.

3. Exploded diagram of the final robotic prototype.

Images: Maria Yablonina, Institute for Computational Design, University of Stuttgart, 2015.

Today, we see a new approach to fibre composites. A body of work developed at the Institute for Computational Design (ICD) and Institute of Building Structures and Structural Design (ITKE) at the University of Stuttgart suggests a new way of building with fibre composites using industrial robots. Through iterations of research pavilions (Menges & Knippers, 2015), coreless filament winding (Prado et al., 2014) and integrated formwork (Vasey et al., 2015) methods have been developed. These methods embrace the quality of filament materials as objects of infinite length and introduce techniques that reduce necessary moulds down to cheap and reusable formwork through continuous winding strategies. The fabrication strategies that a robot provides allow the creation of complex geometries without requiring a solid mould.

A unique property of fibre filament material is its virtually infinite length. The material can span large and small distances, which means it can work at both local and global design scales within the same system. The latter enables its use in various contexts, including furniture, interior spaces and global architectural applications.

Tensile filament systems require anchoring to solid formwork to be stable. Using the existing surfaces of an architectural environment instead of constructing new ones would create a new layer of architectural complexity in existing habitats. An industrial arm, designed to work on a production line with car-sized objects, does not provide this level of scalability and flexibility of environment interaction.

Constructing a system for complex environments

The first stage of research focuses on conceptualising and developing a locomotion system that would suit our research goals. As the aim is to develop a system for constructing complex shapes in three-dimensional space, a simple wheeled robot would not be efficient. The system needs to provide functionality for operating in complex environments, and for converting the façades, walls and ceilings of architectural surroundings into fabrication anchor surfaces. Alongside the development of mechanical locomotion solutions, software for control and real-time process analysis are required for performance in unstructured environments.

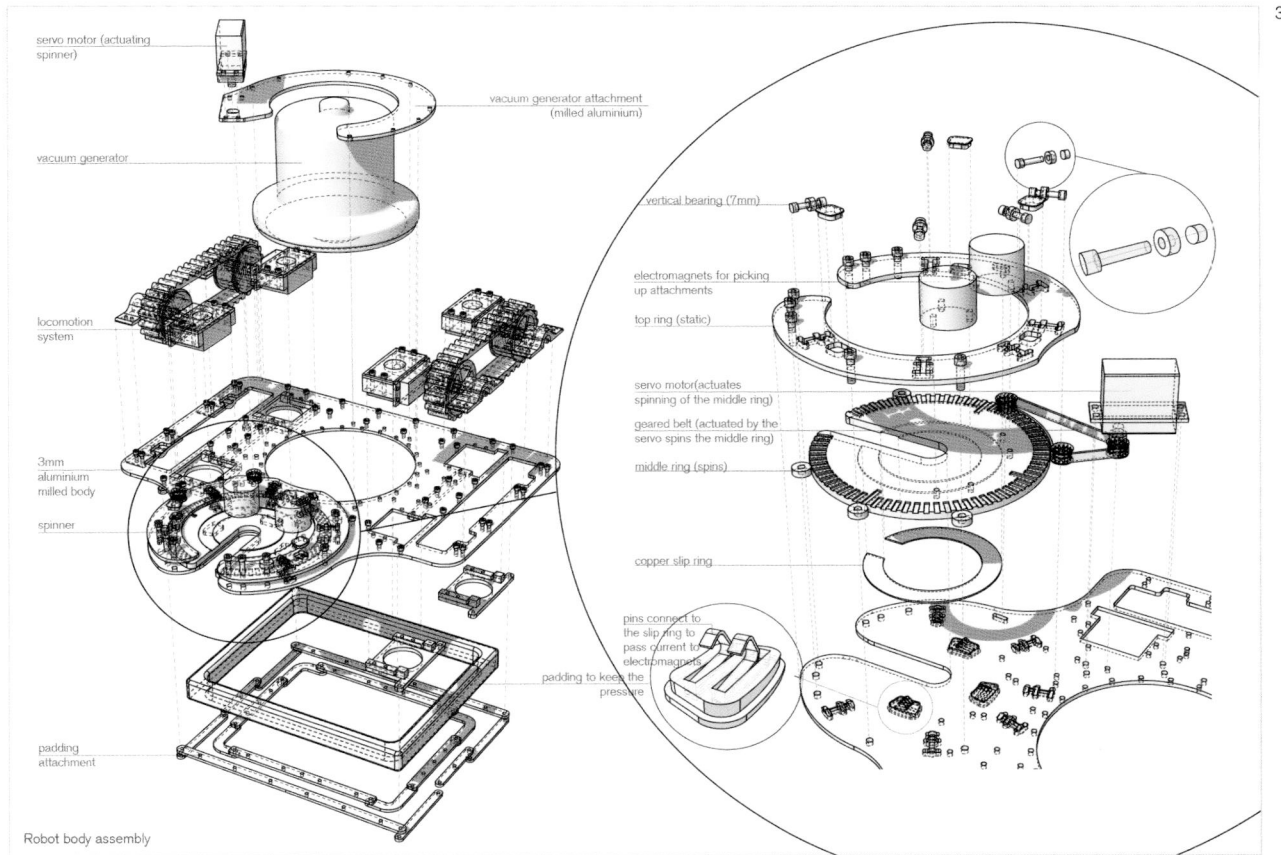

Robot body assembly

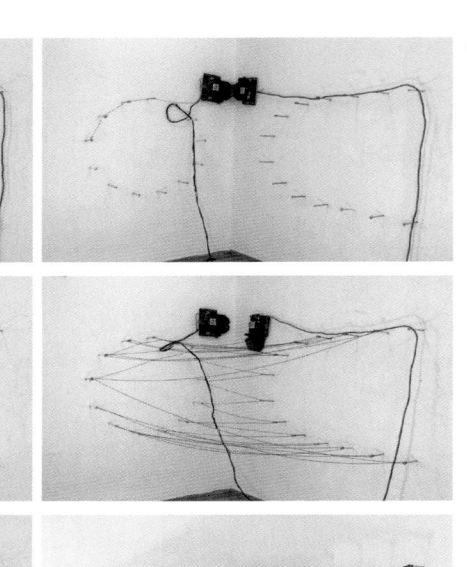

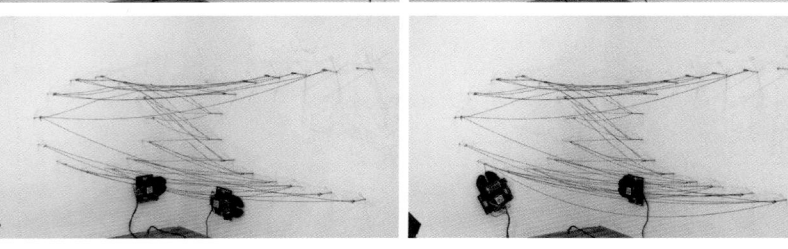

Once the locomotion system is developed, additional features of the machine need to be conceptualised. How exactly does the material interact with the environment, and what functions does the machine require in order to be able to perform the interaction? A solution for transforming an existing architectural environment into a formwork for a tensile structure needs to be developed: anchor points and an attachment mechanism.

A fabrication process involving multiple robotic units requires the development of an interaction system between the independent machines on both hardware and software levels. This ties directly into the way the machines are controlled. A strategy involving perception, actuation and localisation is required.

Developing and controlling the robots

The proposed hardware system consists of multiple robotic units of the same design enabled with various types of actuation and sensor in order to perform the fabrication process. In the concept development stage, functions and prototypes were developed iteratively (Fig. 2). The first step was to test out basic locomotion and control systems in order to explore the possibilities they offered. Initial prototypes were simple wheeled machines, with a focus on exploring methods of navigation and control. In later iterations, the need to navigate three-dimensional spaces arose. In order for a system to operate in 3D environments of human habitation, a wall-climbing prototype was developed. This prototype was based on a wheeled wall-climbing robot (Dethe & Jaju, 2014) (Fig. 3) that uses vacuum pressure to adhere itself to the surface. A centrally located vacuum motor provides enough force for the machine to carry approximately 10kg in addition to its own weight. Four independently controlled actuated wheels allow for the robot to accelerate, steer and rotate in place. Controlling each wheel with an independent motor provides more force for situations of high payload and creates a smaller turn radius for increased manoeuvrability.

In order for the machines to navigate in unpredictable environments, a control system capable of localisation and real-time path correction was also developed. Since mobile system movements are hard to predict due to an inability to directly relate motor movements to the actual distance travelled (Gil, Reinoso, Fernandez & Vicente, 2006), a feedback loop for local vector correction is required. Visual sensors (cameras) and the control unit are positioned externally to the bodies of the robots, allowing the system to capture the whole fabrication space simultaneously, process the data and send commands back to the machines. Fiducial markers (Bencina & Kaltenbrunner, 2005) placed on the robots

4. Structure building process. Image: Maria Yablonina, Institute for Computational Design, University of Stuttgart, 2015.

and perceived by the cameras provide constant feedback of each unit's position with a tolerance of 10-20mm. At every iteration, before a control signal is sent to the motors, the current acceleration vector is compared to the desired one in order to calculate a trajectory correction. For global navigation, a variation of an A* pathfinding algorithm (Hart, Nilsson & Raphael, 1968) is employed. Once obstacles and restricted areas are defined by the user on a global fabrication site map, the algorithm defines the most efficient path for the machine to move along.

A custom robotic effector was developed to efficiently attach the material to anchors and pass the thread bobbin between robots. The mechanism allows the bobbin to be wrapped around slender anchor hooks. It is actuated with a single motor through a set of gears (Fig. 3). A material bobbin is mounted onto a circular rotating plate with a slit on one side. As the robot approaches an anchoring hook, the rotational element is placed into the capturing position so that the hook slots into its centre. Actuating the motor causes the bobbin to spin around the anchor, wrapping the thread around it. Each robot is equipped with a set of electromagnets that allow each of them to pass or receive the fibre bobbin to or from other robotic units.

This application of the system is developed specifically to operate in an interior environment where anchoring surfaces are approximately at a 90° angle to each other. Each surface of the room (floor, walls, ceiling) is inhabited by one robot and has an external camera capturing it. Surfaces are manually equipped with anchors prior to the robotic fabrication process. Machines navigate the surfaces, attaching the thread to anchor points in a predetermined sequence. Each wrapping routine is followed by a passing routine where the material bobbin is passed from one machine to another, in order to span the thread in the three-dimensional space between surfaces.

Once the fabrication process begins, all of the robot movements are choreographed autonomously. However, a safety mechanism can be implemented. Whenever the operator spots a problem or a mistake, the system can be switched into troubleshooting mode and the robots can be operated manually from a pendant. This switch between autonomous and manual control can be made at any time during the process and allows smooth continuation from the previous point thereafter.

Global geometry, size and position of anchor surfaces, number of anchors and the sequence in which they are connected are defined by the user prior to fabrication.

Once the software receives the information, it computes the working space, location of the anchors and movement sequence for each robot.

Assessing the basic functions of the system

This system has been successfully tested in a scenario of interior environment fabrication process with two surface climbers spanning a simple human-scale structure made of nylon thread between two anchor surfaces (Fig. 4). Throughout the test, the machines successfully performed locomotion, interaction and anchoring. The fabricated prototype has been designed to test basic functions of the system rather than to explore design possibilities (Fig. 5). The result is a 2.5m-long and 0.5m-diameter doubly-curved hollow fibre structure capable of supporting a human. It consists of 35 layers of thread anchored to 26 anchors. The total count of passes is 455 and the total length of thread used is approximately 800m. The winding process took approximately 50 hours.

This proposed mobile robotic system is therefore successful in working with filament materials in conditions of onsite fabrication. While these machines cannot compete with industrial robots in payload and precision, they open up the possibility of building entirely new structures that would be impossible otherwise. The ability to interact with onsite environments as well as the potential for various scales of fabrication make this process extremely useful for in-situ interior and urban-scale fabrication.

Increasing the number of machines involved in the process could allow more complex multi-surface areas to be utilised, as well as increasing the speed of production. The currently existing constraint of 90° surface orientation can also be avoided through the modification of the effector hardware. Simultaneously upgrading current hardware could in turn make the system far more efficient.

The vacuum motors being utilised have a high power demand, which makes it necessary to supply power via a cable. Using more efficient vacuum motors along with powerful batteries would allow the machines to be wireless and thus to move with more freedom during fabrication. Once the robots can manoeuvre between previously laid fibres without the risk of entangling the power cable, complex fibre interactions, where subsequent fibre layers shape the previous ones into a new condition, can be achieved.

Potential for new design and construction techniques

Having proven the feasibility of the proposed system, further research is required in order to achieve a more robust fabrication strategy and to explore new design and construction potentials. Further development of current design software would allow the creation of more performative fibre patterns and structural composite spaces. Embedding tools for editing winding syntax and anchor placement would allow for planning the output to a finer degree of detail. Introducing elements such as openings, branches and space dividers would be a possible next step.

Potentially the system could occupy the external surfaces of urban environments, using building façades as formwork. The architecture that would be created is then a parasitic structure (Melis, 2004) growing on existing architectural environments, using its input as a design driver and as a formwork for the structure that is then created after. One can imagine structures being created in an urban context without human interference by autonomous machines, a space that is created where and when it is needed and disassembled once no longer relevant.

The proposed system, alongside other research into robotic applications in architecture, lays the foundation for a broader variety of task-specific machines for construction. Building a larger library of tools, including industrial arms and CNC tools as well as custom-built mobile machines and effectors, would further expand the possibilities of architecture. One could imagine a modular robotic platform consisting of various machines, where each performs a custom task, compensating for the limitations of others. This can be envisioned as a universal multi-material construction system where machines and tools can be added and removed in response to the specific requirements of a fabrication task.

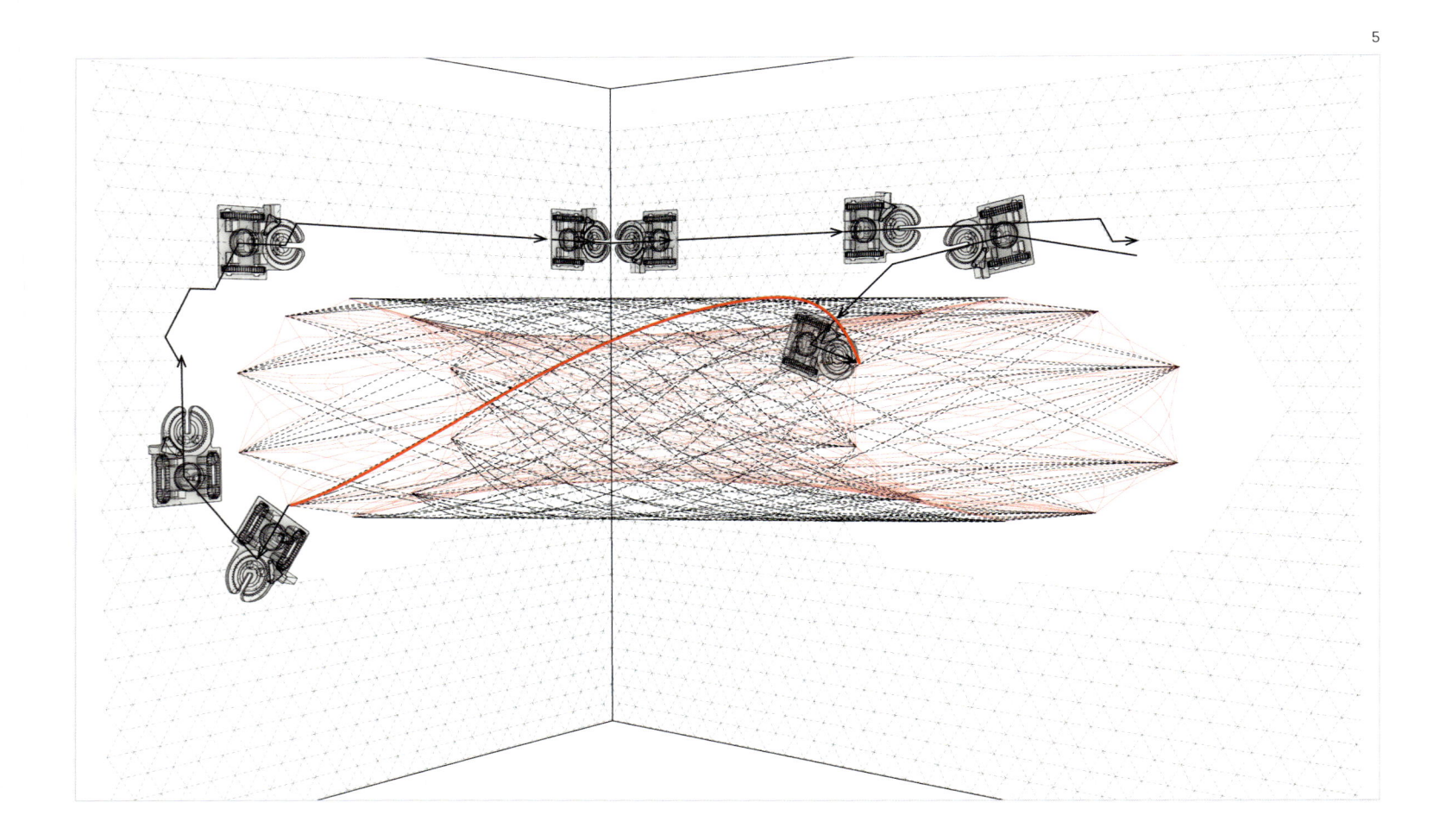

5

References

Menges, A. and Knippers, J., 2015, 'Fibrous Tectonics' in *Architectural Design*, Vol. 85, No. 5, Wiley, London, p.40-47 (ISBN 978-11118878378; doi: 10.1002/ad.1952).

Jokic, S., Novikov, P., Maggs, S., Sadaan, D., Jin, S. and Nan, C., 2014, *Robotic Positioning Device for Three-dimensional Printing*, Institute for Advanced Architecture of Catalonia, Barcelona, Spain, available at https://arxiv.org/ftp/arxiv/papers/1406/1406.3400.pdf (accessed 10 September 2016).

Novikov, P., 2015, *Architectural Robots: The Shape of the Robots that will Shape your Home*, available at http://robohub.org/architectural-robotics-shape-of-the-robots-that-will-shape-your-home-in-dom-reconfigurable-spaces/ (accessed 1 September 2016).

d'Andrea, R., 2012, guest editorial, 'A Revolution in the Warehouse: A Retrospective on Kiva Systems and the Grand Challenges Ahead' in *IEEE Transactions on Automation Science and Engineering*, Vol.9, No. 4, p.638-639.

Zhang, H., 2007, *Climbing & Walking Robots Towards New Applications*, I-Tech Education and Publishing (ISBN 978-3-902613-16-5).

Mahajan, R.G. and Patil, S.M., 2013, 'Development of Wall Climbing Robot for Cleaning Application' in *International Journal of Emerging Technology and Advanced Engineering*, Vol. 3, Issue 5, May 2013 (ISSN 2250-2459, ISO 9001:2008).

Mirjan, A., Gramazio, F. and Kohler, M., 2014, 'Building with Flying Robots' in *Fabricate 2014*, Zurich, p.267-271.

Phillips, S., 2004, 'Plastics: Monsanto Home of the Future, In Colomina, Beatriz' in *Cold War Hothouses Inventing Postwar Culture, from Cockpit to Playboy*, New York, Princeton Architectural Press, p.102.

Knippers, J. and Menges, A., 2015, 'Fibres Rethought – Towards Novel Constructional Articulation' in *Detail – Review of Architecture*, 15, p.21-23.

Prado, M., Dörstelmann, M., Schwinn, T., Menges, A. and Knippers, J., 2014, 'Coreless Filament Winding: Robotically Fabricated Fiber Composite Building Components' in McGee, W. and Ponce de Leon, M. (eds.), *Proceedings of the Robots in Architecture Conference 2014*, University of Michigan, p.275-289 (ISBN 9783319046624).

Vasey, L., Baharlou, E., Dörstelmann, M., Koslowski, V., Prado, M., Schieber, G., Menges, A. and Knippers, J., 2015, 'Behavioral Design and Adaptive Robotic Fabrication of a Fiber Composite Compression Shell with Pneumatic Formwork' in Combs, L. and Perry, C. (eds.), *Computational Ecologies: Design in the Anthropocene, Proceedings of the 35th Annual Conference of the Association for Computer-Aided Design in Architecture (ACADIA)*, University of Cincinnati, Cincinnati OH, p.297-309 (ISBN 978-0-69253-726-8).

Dethe, R.D. and Jaju, S.B., 2014, 'Developments in Wall-Climbing Robots: A Review' in *International Journal of Engineering Research and General Science*, Vol. 2, Issue 3, ISSN 2091-2730.

Gil, A., Reinoso, O., Fernandez, C. and Vicente, M.A., 2006, *Simultaneous Localization and Mapping in Unmodified Environments Using Stereo Vision*, University of Freiburg Department of Computer Science, available at http://ais.informatik.uni-freiburg.de/publications/papers/gil06icinco.pdf (accessed 3 September 2016).

Bencina, R. and Kaltenbrunner, M., 2005, *Design and Evolution of Fiducials for the ReacTIVision System*, Music Technology Group, Audiovisual Institute Universitat Pompeu Fabra, Barcelona, Spain, available at http://modin.yuri.at/publications/reactivision_3rditeration2005.pcf (accessed 5 September 2016).

Hart, P.E., Nilsson, N.J. and Raphael, B., 1968, 'A Formal Basis for the Heuristic Determination of Minimum Cost Paths' in *IEEE Transactions on Systems Science and Cybernetics SSC4*, 4(2), p.100-107, doi:10.1109/TSSC.1968.300136.

Melis, L., 2004, *Parasite Paradise: A Manifesto for Temporary Architecture and Flexible Urbanism*, NAi Publishers, Rotterdam, Netherlands (ISBN-10: 9056623303).

5. Diagram of robot choreography for a single filament layer. Image: Maria Yablonina, Institute for Computational Design, University of Stuttgart, 2015.

THE SMART TAKES FROM THE STRONG
3D PRINTING STAY-IN-PLACE FORMWORK FOR CONCRETE SLAB CONSTRUCTION

MANIA AGHAEI MEIBODI / MATHIAS BERNHARD / ANDREI JIPA / BENJAMIN DILLENBURGER
ETH Zurich

The wider aim of this research is to explore the architectural potential of additive manufacturing (AM) for prefabricating large-scale building components. It investigates the use of AM for producing building components with highly detailed and complex geometry, reducing material use and facilitating the integration of technical infrastructure.

In order to achieve this, the concept of stay-in-place 3D printed formwork is introduced. AM is employed to produce sandstone formworks for casting concrete in any shape, regardless of geometric complexity. This approach explores the synergy between the geometric flexibility of 3D printing sand formworks and the structural capacity of concrete. It allows the production of composite components with properties superior to either individual material.

This new fabrication method is demonstrated and evaluated with two large-scale 1:1 ceiling slab prototypes (Figs. 1 and 2), which are described in this paper.

Large-scale binder jetting technology in architecture

3D printing, or additive manufacturing, refers to the process of producing artefacts by successively adding material using a computer numeric control (CNC) system. A digital 3D model of an artefact is created and sliced along a vertical axis. The data about each slice is then translated and fed to a 3D printing machine, and the machine creates the artefact by building up material layer by layer.

There are a few different types of AM technological process. In the context of architecture, the interest lies in the AM processes that enable the production of large artefacts onsite and prefabricated components offsite. This research focuses on binder jetting for prefabrication (Fig. 3). Binder jetting is an AM process in which a liquid bonding agent is selectively dropped on thin layers of powder material to bind it.

Several characteristics of binder jetting make it interesting for prefabrication in architecture. Due to the nature of the process, binder jetting can theoretically

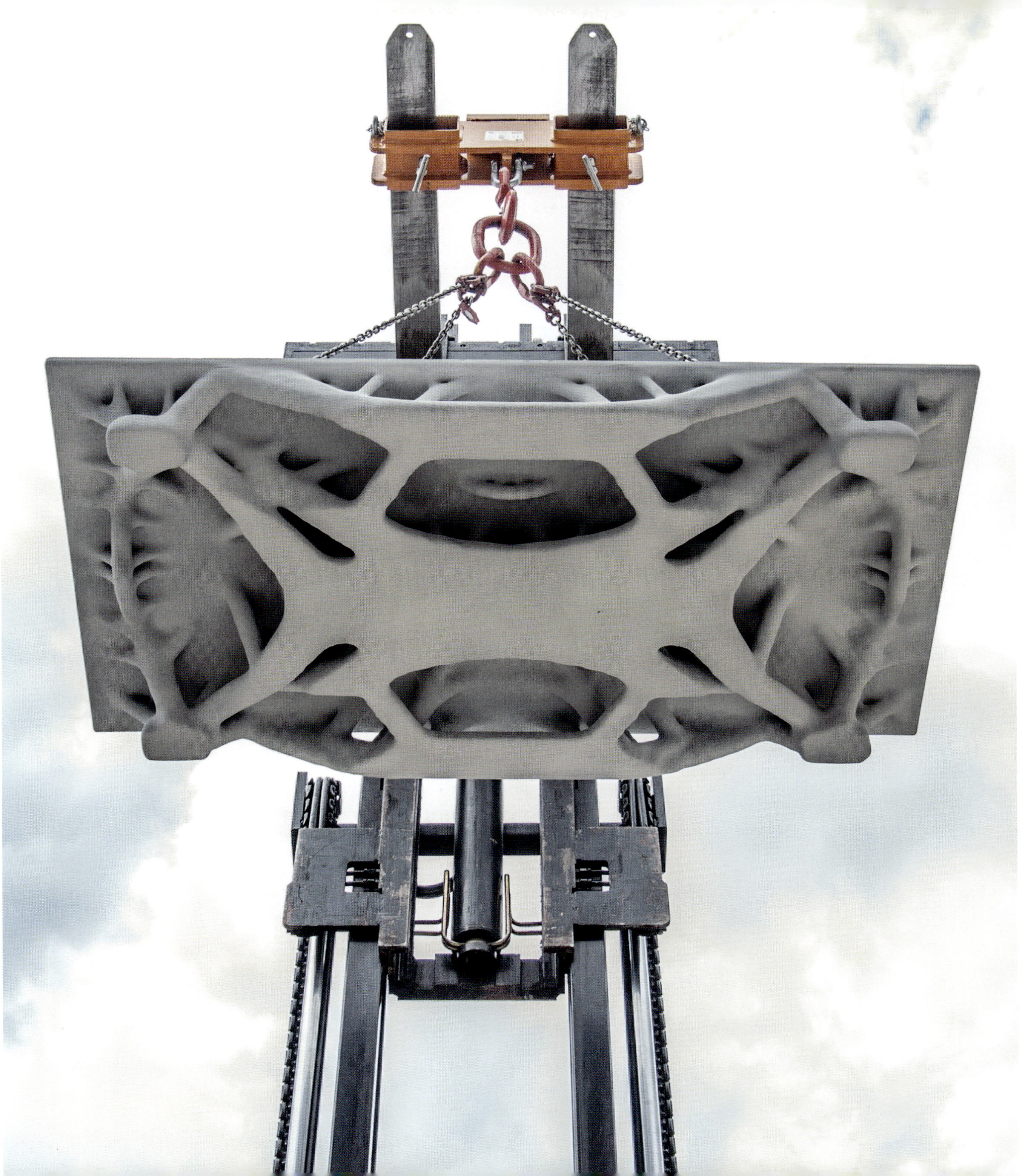

be used with any powder material that can be bonded (cement, plastics, ceramic, metals, sand, sugar, plaster, etc; Rael & San Fratello, 2011). Moreover, this process has the advantage that, within a set bounding box, increasing geometric complexity results neither in longer production time nor in higher cost. Complex cantilevering forms and even interior structures can be 3D printed without auxiliary support, because the powder-bed itself performs this function. Lastly, there are a number of larger-scale facilities that use binder jetting technology to produce large-scale artefacts. An example is the D-shape system by Enrico Dini (Dini, 2009). This is one of the largest 3D printers in the world, but unfortunately this system only reaches a limited resolution. This resolution depends on the grain size of the powder, the layer height and the resolution of the print head. In contrast, there are industrial 3D sand printers that can produce parts that are both large and highly detailed. Currently, they are used by the foundry industry to produce moulds for metal casting. These moulds can be printed at a very high resolution, in the range of a tenth of a millimetre, and at a maximum volume of 8m³.

The project *Digital Grotesque* by Dillenburger and Hansmeyer (2013) demonstrated the potential of 3D printing sand for the fabrication of highly detailed freeform components in architecture, yet the use of 3D sand printing in architecture has barely begun to reach its potential. One reason for this is that large-scale 3D printed sand parts are too weak to operate as a building material – the bending strength of 3D printed sandstone is very low. As a result, the current applications are limited to building components which are mostly under compression.

The advantages of 3D printed sandstone

The central question of this research is how to use the unique advantages of 3D printed sandstone and overcome its limitations in order to enable the fabrication of large-scale building components. The research introduces and examines the concept of stay-in-place 3D printed sandstone formworks as a solution that combines the geometric flexibility of 3D printing sandstone and the structural capacity of concrete (Fig. 4). Specifically, the following questions are investigated:

- How do concrete and 3D printed sandstone interface? To answer this question, the fabrication constraints of 3D printed formwork and the performance and efficiency (functional, structural, material) of the resulting load-bearing building components are investigated.

1. Prototype B, displaying an intricate tubular topology designed to reduce weight.

2. Prototype A, designed to reduce weight through the use of a ribbed substructure.

3. Sand binder jetting with a large industrial 3D printer.

4. Composite building element with load-bearing capacity. Physical testing of the integrity of a ceiling prototype.

- What is the impact of this new fabrication process and geometric freedom on the design of architectural components? Can this approach facilitate the fabrication of fully integrative building components with reduced material?

One reason to search for new ways to fabricate complex forms with fewer constraints is that doing so allows us to reduce material use through the optimised design of components: wall thickness can be adapted and undercuts, microstructures and complex branching topologies can be fabricated.

With its excellent geometric flexibility – recesses, undercuts, internal voids and tubular structures are possible – 3D printed sandstone formwork lends itself well to the production of such complex architectural elements. The main means of demonstrating the feasibility of this construction method in this research is the production of two large-scale 1:1 slab prototypes. The two prototypes investigated forms which were found by computational strategies (e.g. topology optimisation). The target objective of the optimisation was to reduce material use and efficiently distribute the remaining material in order to maximise the slab's strength.

Prototype A (Figs. 1 and 5) is a slab designed for a load case with three supports in the centre. This slab folds into a hierarchy of ribs that give stability to the large cantilevering areas. Prototype B addresses a load case of four perimetral support points (Fig. 2 and 4). It features a sophisticated topology of tubular elements branching in three dimensions. The amount of concrete contained within (50 litres) corresponds to a solid slab a mere 3cm thick.

To produce the large prototypes, the following steps were taken:

- Compression and bending tests of combinations of different types of powders and binders.
- Structural tests of different concrete mixtures considered for potential combination with sand-print.
- Rheology studies of casting concrete in sand-printed formworks of different geometries to derive a formal vocabulary as a design guideline (Fig. 6).
- Exploration of various computational design strategies to optimise the use of the chosen fabrication method with respect to the structural limitations of the material.

Because its main use is casting moulds for metal, relatively little was known about the structural properties of 3D printed sandstone. A series of tests was therefore initiated to measure its resistance to compression and bending forces. The tests showed that 3D printed sandstone has reasonably good resistance to compression, but is brittle when exposed to bending forces. Below is the list of parameters involved in the compression and bending tests:

Parameters of the compression tests:
- Size of the specimens: 50 x 50 x 50mm.
- Binders used: phenolic and furanic resin, with and without epoxy surface infiltration.
- Spatial orientation in the printer bed: X, Y and Z.
- Number of specimens per combination: 3.
- Total number of specimens: 36.

Parameters of the bending tests:
- Size of the specimens: 250 x 50 x 50mm.
- Three-point bending, supports at 200mm distance, central point load.
- Same binders, orientation and number of specimens as the compression tests (36 specimens in total).

The compression and bending tests were also applied for parts with different types and binders – as the table on page 215 shows, the difference between binder types is only marginal, apart from the bending strength of infiltrated parts. This is because the sand is less densified during printing, and heat curing vaporises more of the liquid. As a result, more resin infiltrates the part. As expected, additional infiltration hardens the part significantly and increases its strength.

The behaviour of 3D printed sandstone in combination with ultra-high performance fibre-reinforced concrete (UHPFRC) was investigated in partnership with the

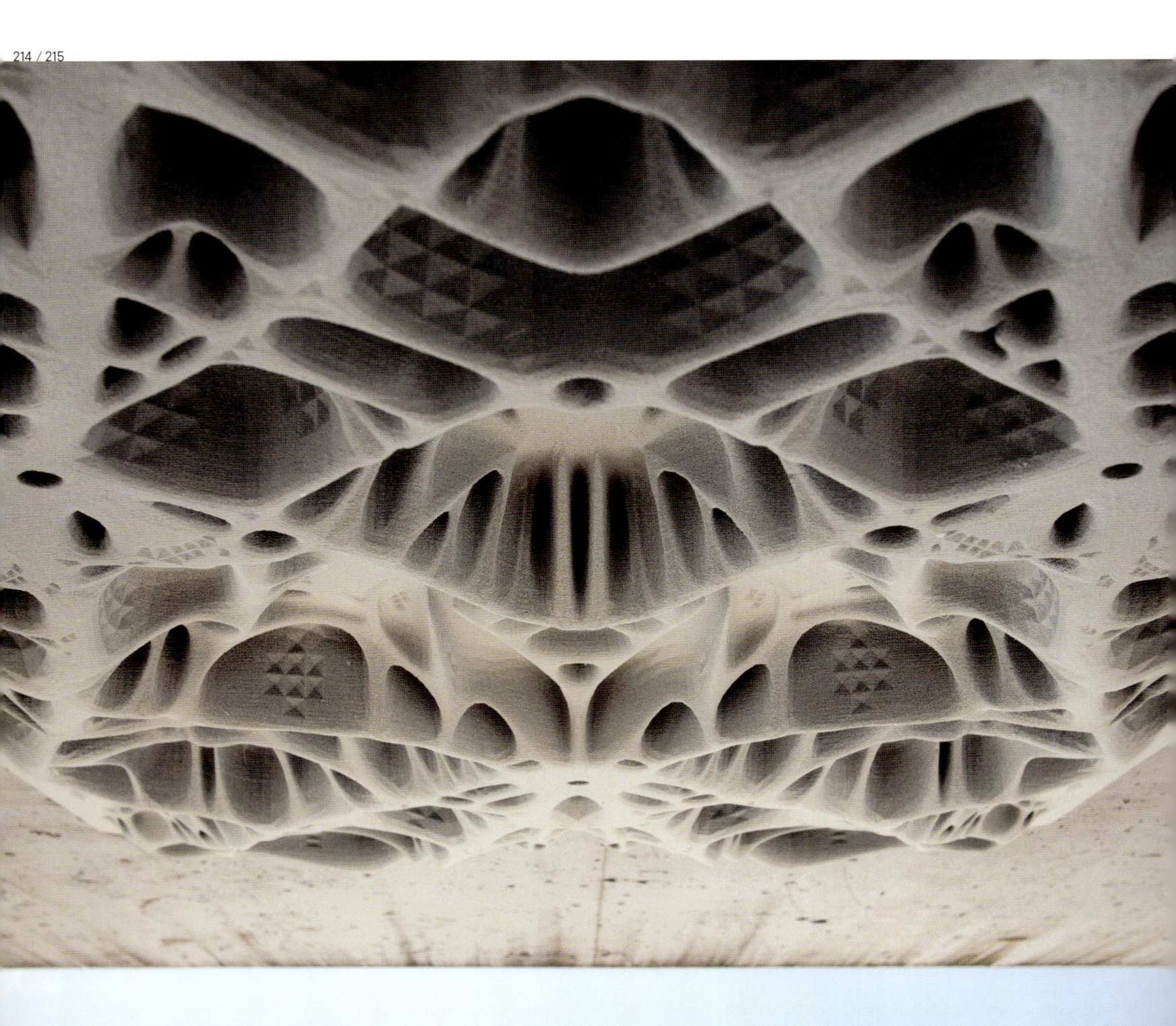

5

	Phenolic binder (PDB)		Furanic binder	
	Without infiltration	With infiltration	Without infiltration	With infiltration
Compression strength [MPa]	8.56	12.32	8.46	12.80
Bending strength [MPa]	2.95	8.85	2.96	6.49

Table 1. Load-bearing capacities of 3D printed sandstone in megapascals (MPa).

group for Physical Chemistry of Building Materials (PCBM, D-BAUG, ETH Zurich), with the following four main intentions (Fig. 7):

- Develop a concrete recipe with adequate admixtures that has the desired rheological properties.
- Adjust the length and content of the steel fibre reinforcement to achieve ductile behaviour while maintaining the ability to cast in narrow channels.
- Understand the impact of the porosity and sorptivity of the 3D printed sandstone formwork (how do the capillary absorption and transmission of water of the 3D printed sandstone influence the hardening of the concrete?).
- Mechanically test the bond between the two materials as a composite.

The details and results of the study are documented in '3D Sand-Printed High Performance Fibre-Reinforced Concrete Hybrid Structures' (Stutz, Montague de Taisne, 2016).

From a design perspective, an important finding of this thesis project is a series of formal guidelines. According to these, cavities and tubular structures in the formwork can be dimensioned in relation to both the length and the volumetric content of the fibres in the concrete mixture. These guidelines informed the design of the two prototypes in terms of dimensioning and controlling rheological aspects with regard to the concrete casting process. Moreover, both prototypes exploit the entire size (180 x 100cm) of the Ex-One S-MAX 3D printer bed.

Formwork production

Production of formworks with a high degree of detailing and precise geometric features for large concrete components is very challenging – and sometimes impossible – if using other formwork fabrication methods such as robotic wire-cutting, 3- and 5-axis CNC milling and fabric formworks. The described 1:1 slab prototypes show how 3D printing can facilitate the fabrication of such formworks.

3D printing is particularly suitable for producing stay-in-place formwork. This is because the bond between the sandstone formwork and the UHPFRC is very durable. Mechanically removing a 9mm-thick layer of 3D printed sand completely requires pressures greater than 3,000atm with a water jet. Removable temporary formwork is possible (and was successfully tested in another project) but requires a coating treatment of the formwork which closes the pores to prevent the concrete from percolating through the sandstone formwork. The geometry of the formwork and the minimum dimensions of its hollow features were dictated by the constraints of the fabrication processes, post-processing of the 3D printed formwork and the rheological properties of the concrete mix.

Parameters related to 3D printing sand

The post-processing involved removing loose sand from and infiltrating the outer surface of 3D printed formworks. Thus the geometry and diameter of the hollow features had to be designed in such a way as to facilitate removal of the loose sand (Fig. 8).

The thinness of the 3D printed formwork as it relates to the fabrication process was also studied. This dimension was tested from 6 to 10mm, and thinner walls were found to be unstable during the removal of loose sand (due to erosion from compressed air jets or vacuuming) as well as during casting (as hydrostatic pressure built up in deeper channels and penetrated the thin formwork walls).

At 1.8m^2, the overall size of the components also approached a limit in terms of both the manipulation of the formwork and the stability of the 3D printed piece. While smaller parts can increase the complexity of the assembly, they are easier to handle. Therefore the dimensioning of the parts is always a trade-off between weight, number of connections and logistical factors.

The tests revealed the fact that the friable nature of the 3D printed sandstone needs to be carefully considered, especially when scaling up the manufacturing process and fabricating components in larger volumes. A strategy

5. Detail of prototype A, showing the precise details of a finely ribbed substructure.

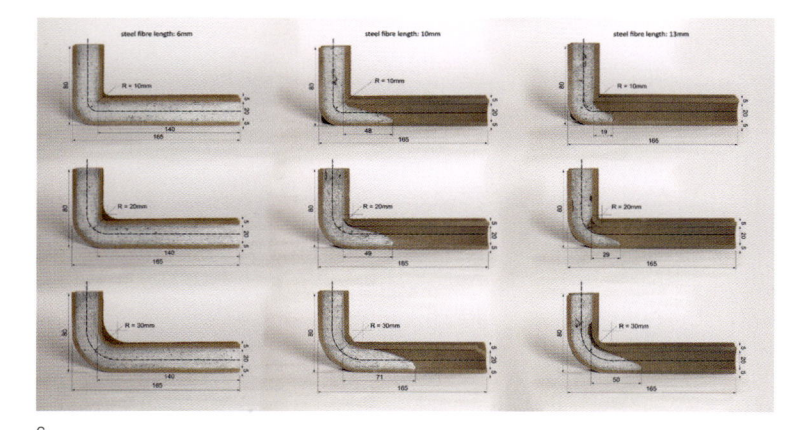

to avoid damaging the formwork before casting by integrating a protective bed of unbonded sand contained within a closed 3D printed box that also provided auxiliary support during casting was successfully tested (Fig. 9).

Parameters related to concrete

The specific post-processing operations of the 3D printed parts (i.e. vacuuming loose sand, infiltrating the outer surface of the formworks) and the rheological properties of concrete dictated minimum dimensions for the hollow features. UHPFRC mixes work well with 3D printed channels with diameters as low as 20mm and bending radii of 10mm. For features below these minimum dimensions, the stay-in-place sandstone formwork can take the role of an ornamental exposed surface that does not necessarily transfer all the details to the cast concrete inside.

A full complement of structural tests is scheduled for the next stage of the research, but the empirical tests performed so far by applying a 2,500KN/m² distributed load on a concrete component with an average concrete thickness of 30mm were encouraging. The indication is that material savings of up to 70 percent are achievable.

Successful printing of composite building components

The proposed method advances the idea of using 3D printing as an indirect fabrication method for producing composite building components with elaborate geometry. Potential applications are in the realm of one-of-a-kind, non-standard building components rather than that of mass production. While further tests are necessary to conclusively quantify the advantages of this fabrication process in comparison to others, the prototypes have shown that the method is feasible and has significant potential for application in architecture at a larger scale. For applications of this method to larger-scale building components, such as entire ceilings, structures would need to be assembled from multiple parts prefabricated in the proposed way. To prove that large spans and cantilevers are achievable, further research has to address the following challenges:

- Reinforcement considerations: steel fibre reinforcement was enough for the smaller prototypes, but in order to increase the structural spanning capabilities, traditional reinforcement bars or pre-stressing strategies must be considered. Again demonstrating its suitability, 3D printing can be used to fabricate guiding features for the precise integration of reinforcement.

- Additional functionality: as a consequence of the durability of the concrete-sandstone bond, the 3D printed formwork is ideally suited to stay in place and host additional functions. Acoustic surface treatment, heat transfer-regulating geometry and detailed ornamentation are possible, as is the integration of enclosures for mechanical and electrical services. This opens up the possibility of fabricating smart, integrative building components.
- Fabrication process development: up to this point, the research has relied on commercially available generic 3D printers. Nevertheless, this research hints at certain improvements to the technology that would benefit this specific application, such as new powder and binder combinations and the integration of post-processing.
- Digital design tool: the findings from all the experiments are to be compiled in a computational design tool specifically dedicated to the design for indirect binder jetted fabrication. This application will incorporate relevant design constraints and optimisation procedures.

The results suggest that indirect fabrication approaches can be generalised to other types of 3D printing technologies. The solution relies on a hybrid fabrication process in which a precious smart material is used minimally, only where necessary, and relies on another strong material to perform structurally. Digital fabrication is used to produce a minor proportion of the final product, but has a major impact on its performance and behaviour.

6. Development of a lexicon of formal constraints from rheology studies.

7. Cutting a sample to investigate concrete rheology.

8. Post-processing of a 3D printed sandstone formwork to prepare it for casting concrete.

9. The prototype illustrates the integration of a protective bed of unbonded sand contained within a closed 3D printed box.

Acknowledgments

We would like to thank a number of partners and collaborators whose dedication helped us to realise the projects described in this paper. They are: Prof. Dr. Robert Flatt, Nicolas Ruffray and Dr. Timothy Wangler, Physical Chemistry of Building Materials (PCBM/IFB/D-BAUG/ETH Zurich). Heinz Richner and Andi Reusser (Concrete Lab D-BAUG). Felix Stutz and Neil Montague de Taisne (Bachelor's thesis, Engineering). Hyunchul Kwon, Victoria Fard, Nicholas Hoban, Michael Thoma and Philippe Steiner (prototypes and photo-documentation). Christen Guss AG (production partner).

This research is supported by the NCCR Digital Fabrication, funded by the Swiss National Science Foundation (NCCR Digital Fabrication Agreement #51NF40-141853).

References

Dillenburger, B. and Hansmeyer, M., July 2013, 'The Resolution of Architecture in the Digital Age', in *International Conference on Computer-Aided Architectural Design Futures*, Springer Berlin Heidelberg, p.347-357.

Lim, S., Buswell, R.A., Le, T.T., Austin, S.A., Gibb, A.G. and Thorpe, T., 2012, 'Developments in Construction-scale Additive Manufacturing Processes' in *Automation in Construction*, 21, p.262-268.

Stutz, F. and Montague de Taisne, N., 2016, '3D Sand-Printed High Performance Fibre-Reinforced Concrete Hybrid Structures', supervised by Nicolas Ruffray and Mathias Bernhard, Zurich.

Rael, R. and San Fratello, V., 2011, 'Developing Concrete Polymer Building Components for 3D Printing'.

Dini, E., 2009, D-shape, Monolite UK Ltd. http://www.d-shape.com/cose.htm

PROCESS CHAIN FOR THE ROBOTIC CONTROLLED PRODUCTION OF NON-STANDARD, DOUBLE-CURVED, FIBRE-REINFORCED CONCRETE PANELS WITH AN ADAPTIVE MOULD

HENDRIK LINDEMANN / JÖRG PETRI / STEFAN NEUDECKER / HARALD KLOFT
Institut für Tragwerksentwurf, TU Graz

New developments in digital workflow

Research at ITE aims to bring computational design, digital fabrication and new materials together. The main interest in a so-called digital workflow is to develop innovative and resource-efficient building components, building systems and fabrication processes.

In past decades, we have lost structural intelligence for economic reasons; today, we mostly realise buildings of simple geometries with high-mass structural elements. By using the potential of digital planning and digital fabrication, we will in the future be able to design innovative customised structures that will be efficient economically as well as in terms of resources.

In fact, the main research focus at ITE is based on the potential of ultra-high performance concrete (UHPC) for new kinds of sustainable structures and architectures. The enormous compressive strength of this material promises a large reduction of structural material for future buildings without loss of performance, as well as the fabrication of innovative lightweight concrete structures. Conventional fabrication technologies in concrete industries do not work with this high-tech material, as geometries have to become much more complex in order to exploit this material's potential. As concrete is a mono-material, and more importantly a complex compound system with graded properties of different components, new ways of fabrication have to be taken into consideration. Conventional casting technologies cannot be used for customised concrete structures, as they are limited to planar geometries with high element thickness. New optimised materials and developments in digital fabrication are opportunities to rethink the production and design of reinforced concrete structures. They can overcome geometrical limitations and lead to a completely new design space for the use of concrete (DBZ, 2016).

So, for instance, future load-bending effects of building elements can be taken into consideration at the planning stage and can be compensated by an individual adaptation of the element curvatures using shell effects. In the fabrication of non-standard concrete structures, currently mostly customised one-way formwork is used.

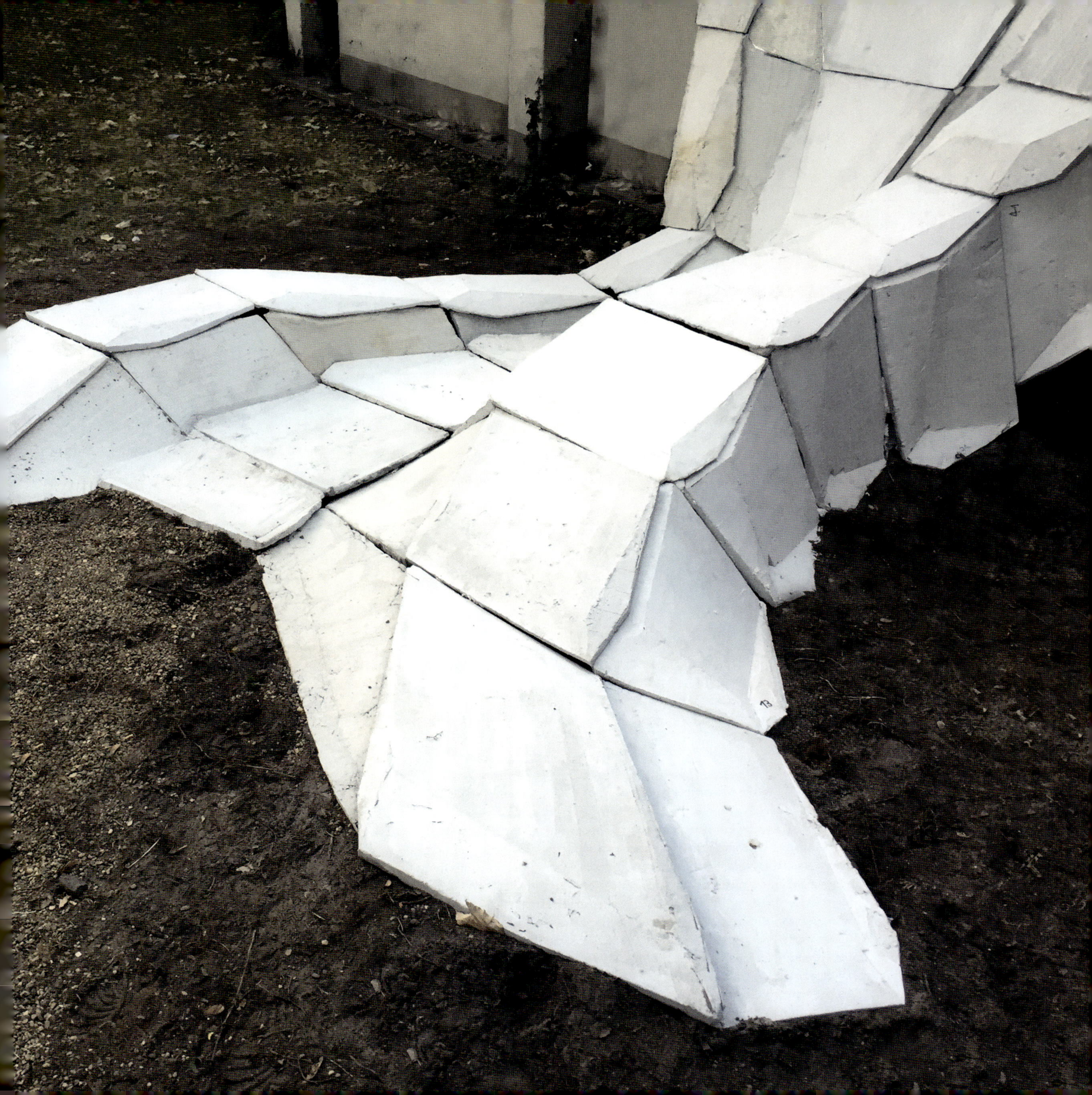

1. Finished DBF-studio mock-up (close-up).

2. Preparing the robot's movement.

3. The Digital Building Fabrication Laboratory at the ITE.

4. Overlapping shingles

5. Final fibre pattern.

6. Starting the reinforcement process.

This fabrication technology is very cost-intensive and slow and produces a lot of waste, while the reinforcement process is very complex and produced elements are limited to a large material thickness. This paper presents a method to build up customised double-curved fibre-reinforced concrete panels in a very short time without creating any waste of formwork material. The process is sequenced in several fabrication steps and involves a robot for human-machine interaction (HMI). The method is still in its testing phase, but the present results already show significant potential with regard to conventional casting techniques.

To implement the described research aims, the ITE has created a format called 'DBF-studio' where students and researchers learn how to design with materiality. The participants can directly combine data processing with fabrication techniques together with the use of specific material. Programming, design and construction become one as they learn how each element relates to the other. It is a design approach that uses bottom-up rather than top-down thinking. The use of computer programmes like Rhino and Grasshopper enables the participants to communicate easily with the advanced fabrication environments. The DBF-studio has become a key element for the Institute's workflow, a platform to push an existing idea, to verify a thesis or simply to generate a variety of possible directions.

In this paper, the robotic controlled production of non-standard double-curved fibre-reinforced concrete panels by using an adaptive mould needed to be investigated further in order to create a coherent process chain of production. The described research idea deals with a lot of parameters and constraints that have to be taken into account. To automate these complex and manifold fabrication processes, they need to be divided into several process steps and tackled independently. This strategy enables a process without complex programming, the use of sensors or complex adaptation systems. In addition, the collaboration of man and machine is a key factor insofar as it creates a very powerful combination, in which both machine and human can act to their strengths. The robot is unbeatable in its accuracy and humans are able to make flexible decisions. This creates a completely new workflow, in which physical results may be unexpected but may also lead to a new appreciation of the process and its formal traces.

The topic of advanced moulding systems for complex concrete elements has been previously addressed in different research initiatives and projects. In this respect, we would like to mention the TailorCrete project 'Industrial Technologies for Tailor-Made Concrete Structures' (ETH, 2009-13), where the research team used an adaptive metal mould that stabilised a wax cast to create precise surface geometry and form individual concrete elements. Within this context, the team from the Arnhem Station project in the Netherlands presented 'Design to Installation of a Free-form Roof Cladding with a Flexible Mold – Building the Public Transport Terminal at Arnhem' at the IASS in 2015 (UNStudio, 2015). The information derived from the digital model of each roof panel was projected on a metal mould to give the height values for the 21 adjustable pins that would individually calibrate the surface to produce the concrete roof panel. In addition, two directly related projects need to be mentioned, too – the 'Robotic Clay Molding' (ETH Zürich, 2012) and the 'Prozedurale Landschaften 2' (ETH Zurich, 2011) workshops where, together with the research team, students developed different robotically controlled methods for the surface treatment of a clay mould and sand mould for the surface treatment of concrete elements.

The next steps

How are we able to take the referenced research projects a step further? How is it possible to create a full process chain for the production of a double-curved fibre-reinforced freeform concrete panel in a reusable mould? This paper investigates the interdependent relationship between production and design, in this case a coherent unit that needs to be evaluated as one. With which tools and with which specific materials are we able to produce a double-curved fibre-reinforced freeform concrete panel in a sustainable manner with the least amount of material?

The envisioned production process results from a link between the functional and structural criteria of a freeform concrete element and its design. What does the process chain look like and what impact does it have on the architectural appearance of the final element? Is it possible to express new design ideas through technology and through the choice of specific materials? What is the impact of these factors on the final appearance? Can a production process be the driver of an overall new architectural expression?

The ICD/ITKE Research Pavilion 2012 team of Achim Menges and Jan Knippers in Stuttgart have had a great impact on the development of the research outlined in this paper, specifically within the context of robotically placed reinforcement fibre strategies.

The process of making

A preliminary method for the making of individual double-curved reinforced concrete elements was developed by the research team at ITE. The method is divided into six individual steps which all involve a 6-axis robot in combination with different end effectors combined with human interactions. The base mould material is a special mixture of sand, oil and potato starch.

At the beginning, a formwork will be created by the robot pushing the sand compound from the centre to the edges of the sand mould. This creates a rough approximation of the desired surface. Additional sand or resulting sand heaps can be added or removed by hand. In a previous step of the process chain, the deviations of the actual surface and the planed surface were tested. In the next phase, a 3D camera scans the sand surface and compares the data with the digital model of the planed surface. By evaluating the areas with too much sand and the areas with not enough sand, the geometry can be iteratively corrected and the sand compound placed in the exact shape required. By scooping from the low to high areas, a rough mould is prepared. The results were satisfying, but the digital process was too complicated and therefore too time-consuming in comparison with the manual workflow.

After this first rearranging of the sand compound is finished, a pneumatic cylinder connected to the robot is developed to stamp the surface and compress the sand mixture. Depending on the tool head and the surface curvature, the tool prints a so-called digital pattern into the mould which can be adjusted by the frequency of the up-and-down movement of the tool, the robot movement speed and the angle of the end effector according to the surface. The pneumatic cylinder minimises the robot movement and adds a significant amount of speed to the process. In preliminary tests, the cylinder acted as an independent tool and performed its up-and-down movement without being coordinated to the robot's movement. Later, the movement of the cylinder and the robot were connected so that the robot stopped for a second whenever the cylinder compressed the sand. This helped to create a more accurate surface finish by not scooping sand with the extended tool.

After finishing the sand mould, a specific amount of concrete is applied manually. Following this, the tool is changed in order to enable the robot to rearrange the fluid concrete material into the desired shape. This distribution process can be a simple offset of the mould surface, which results in a constant thickness. Additionally, the concrete can be adjusted to the structural requirements by locally varying the thickness of the concrete panel.

In a subsequent step, the end effector is changed to a specially constructed adaptive fibre placer which pulls and rolls the selected glass or carbon fibres through the fresh concrete, enabling the laying of fibre bundles in specific areas according to the structural needs of each individual element. With the help of FEA analysis tools like ANSYS, these areas can be easily located within a specific individual piece but also within the context of an overall assembled structure.

In earlier tests, patches of prefabricated standard fibre meshes were placed in the fresh concrete to reinforce specific areas. This method was only successfully applied to planar geometries because the meshes were not flexible enough to adapt to curvature or complex geometries. Whenever the curvature was getting too strong, the patches were folding back into planar mode and could not reinforce the concrete in the right location. Alternative prefabrication strategies of double-curved reinforcement cages were evaluated in an earlier DBF-studio, described in the IASS 2016 paper 'DBF-studio – Evaluation and Development of Research Topics through the Application of Advanced Fabrication Technologies' (ITE, 2016).

The newly developed method shows a lot of potential for placing the fibres, especially for double-curved surfaces which allow the fibres to stay exactly in place where they need to be (Fig. 5). As a corollary benefit, we see that this process integrates seamlessly in the production chain to create thin, structurally optimised concrete elements.

In the final step of the process, the placed fibres are covered with a second layer of concrete (Fig. 6) in the same manner. The overall set-up of the production line is a computer linked by Ethernet to a UR 5 robot running Rhino and Grasshopper. The plug-in to control the robot, developed at the ITE, enables easy access to the machinery (Figs. 2 and 3). Different multi-curved panels can be produced with this method. Patterns of twisted surfaces came out especially well, showing that this technique leaves traces of the process that can be used as a material and aesthetic expression of the fabrication method.

Moving forward

The latest DBF-studio combined the experience of the 2014 and 2015 teams working with HMI robots (IASS 2016, 'DBF-studio – Evaluation and Development of Research Topics through the Application of Advanced Fabrication Technologies'). The goal was to build a complete structure out of several unique fibre-reinforced concrete shingles (Figs. 1 and 7). The shingle design consists of four surfaces: two trapezoids, and two rhombuses on the side. The trapezoids can be twisted (double-curved) in contrast to the rhombuses, which are designed to always form a planar surface. The planar surfaces overlap with the neighbouring planar surface. Due to the convex-concave section of the design, the overlap acts as an interlocking element between the shingles.

To keep the elements positioned in a longitudinal direction, every single piece needed to have an additional nose, almost like the detail of a classic roof shingle. In order to create the shingle nose, two methods were tested. The first one was to remove sand from the mould on the desired edge and compress the gap in the mould again.

7. Finished DBF-studio mock-up.

The cavity could then be filled with concrete and worked as an additionally moulded form added to the shingle in the form of an extended nose. The second method was the result of the casting process. By scooping leftover material to the sides, a thin rim started to pile up against the mould edge. By applying extra material and programming an extra offset to the edge, the nose of the shingle was created without any extra mould. The idea of the mock-up was to evaluate the full potential of this method and to explore its limits with students. By producing the shingles either with the front side in the sand or the back side in the sand, both processes (creating the mould and filling it) can be displayed, showing their individual formal expressions and aesthetic value (Fig. 4).

With a well-connected production chain, 64 individual concrete elements were produced within 10 days. The results were quite convincing, although there is still a lot of potential to improve the method. It allowed us to produce a wide range of geometrical possibilities. Double-curved, single-curved and planar elements with sharp edges with a continuous thickness could be produced and could even vary and be fine-tuned according to the structural needs of each single piece. Reviewing the results of the mock-up shingles and study objects produced earlier shows the potential of the process. Compared to earlier investigations with more curvature, the shingle geometry does not differ enough from an actual planar surface and therefore the process traces appear like an irritation on an almost even surface rather than a tolerated process trace. Either this can be optimised in further post-production steps, or the nature of the surface has to become an integrated part of the design and the process logic to form the optimal geometry in terms of stiffness, material behaviour and production method.

New steps for the production chain

The promising results of the latest mock-up encourage the transfer of the production chain to the new Digital Building Fabrication Laboratory (DBFL), which was introduced at the beginning of 2017 by the ITE. A scaling-up has to go hand-in-hand with a full atomisation of the process. The previously tested process steps, such as the 3D scanning of the surface, could become valuable additions to the process chain. An atomised fill-up could result in an improvement of production speed and improved accuracy of the concrete elements. An important issue will be the surface finish. Integrating the design of the surface appearance, as a result of a flexible process and production technique, will lead to new perceptions of the material and its manufacturing and will give the designer a layer expression.

Traditional moulding techniques in concrete industries have led us to a limit in form and in geometric complexity. They have been optimised to perfectly replicate the design of formal architecture via top-down processes and to find a compromise between design and structural properties. By reaching the limits of material properties for creating resource-efficient and sustainable production with the new fabrication methods, as described here, the digital workflow from the early design stages to final production has to be rethought. As the optimisation processes of building elements in the fabrication stage could have a big impact on the final design, and also because architectural design ideas could not directly lead to a final building shape, recursive feedback loops have to be established in the design process. Similar to growth processes in nature, highly optimised fabrication processes in architecture as a genotype lead to a unique phenotypical shape. Following up these bottom-up principle aesthetics from these fabrication constraints becomes itself a new kind of architectural expression. One can't be done without the other. Fabrication and form have to be in balance.

References

Kloft, H., Ledderose, L., Mainka, J., Neudecker, S., Petri, J. D., 2016, DBZ: 'Hochleistungswerkstoff, Bauen mit Beton im Zeitalter digitaler Planung und Fertigung', 1 June, available at http://www.dbz.de/artikel/dbz_Hochleistungswerkstoff_Bauen_mit_Beton_im_Zeit-_alter_digitaler_Planung_und_2496142.html (accessed 28 September 2016).

TailorCrete, 2014, *TailorCrete brochure* (PDF), Denmark, available at http://www.tailorcrete.com/28625 (accessed 28 September 2016).

Schipper, R., Eigenraam, P., Grünewald, S., Soru, M., Nap, P., Van Overfeld, B., Vermeulen, J., 2008, 'Kine-Mould: Manufacturing Technology for Curved Architectural Elements in Concrete' in *Proceedings of the International Society Of Flexible Formwork (ISOFF) Symposium 2015*, available at http://repository.tudelft.nl/islandora/object/uuid:af86512f-af53-42df-ac2d-807559753621/datastream/OBJ/download (accessed 28 September 2016).

Gramazio and Kohler, 2012, *Robotic Clay Molding, Barcelona*, available at http://gramaziokohler.arch.ethz.ch/web/d/lehre/235.html (accessed 28 September 2016).

Gramazio and Kohler, 2011, *Prozedurale Landschaften 2*, available at http://gramaziokohler.arch.ethz.ch/web/d/lehre/211.html (accessed 28 September 2016).

Gramazio and Kohler, 2011, *Prozedurale Landschaften*, available at http://gramaziokohler.arch.ethz.ch/web/d/lehre/208.html (accessed 28 September 2016).

ELYTRA FILAMENT PAVILION
ROBOTIC FILAMENT WINDING FOR STRUCTURAL COMPOSITE BUILDING SYSTEMS

MARSHALL PRADO / MORITZ DÖRSTELMANN / ACHIM MENGES
Institute for Computational Design, University of Stuttgart
JAMES SOLLY / JAN KNIPPERS
Institute of Building Structures and Structural Design, University of Stuttgart

Novel design and fabrication strategies

Ongoing research conducted at the University of Stuttgart is focused on material-efficient construction through the development of novel design and fabrication strategies for fibre composite lightweight construction systems. In a long-term, bottom-up development across multiple demonstrator projects, the underlying structural principles of fibrous lightweight structures in nature have been investigated in interdisciplinary collaboration with biologists, leading to the development of building technological advancements which allow the transfer of biological lightweight construction principles into technical fibre composite structures (Menges, 2015, Dörstelmann, 2015a, Van de Kamp, 2015). A series of prototypical demonstrator projects have showcased a higher degree of material efficiency and functional integration than current building methods.

The presented project continues this line of research in a site-specific installation at the Victoria and Albert Museum in London. The project aims to further develop the previously prototypically tested processes at a larger

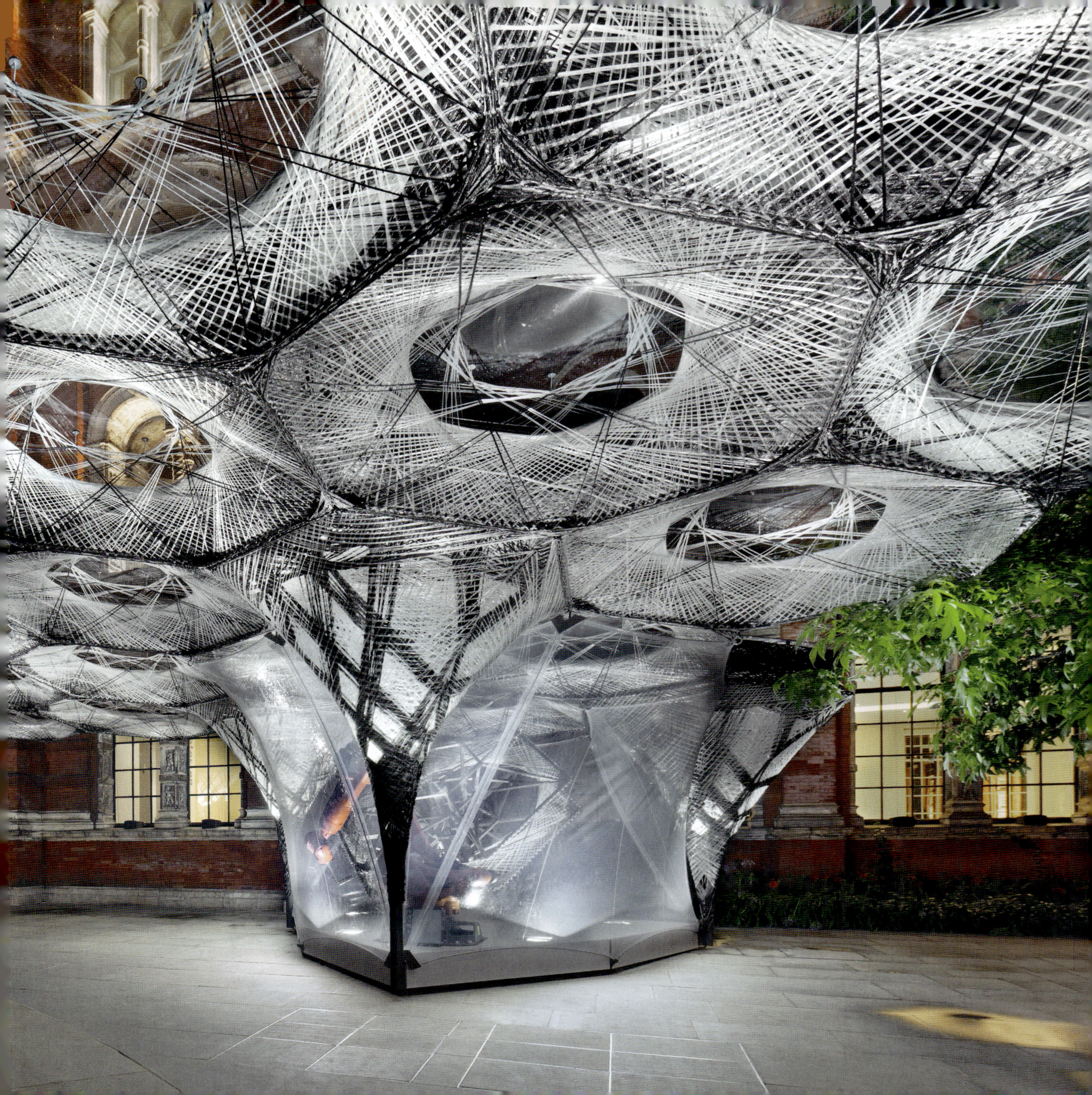

scale, with additional functional capacities being embedded and ultimately constituting a fibrous building system that is suitable for niche applications in architecture. Developments include significant advancements in coreless filament winding techniques, embedding of sensory systems into the fibre composite building parts, integration of construction detailing and interfaces to complementary building systems such as façade and ground anchoring, structural simulation methods, reconfiguration and expansion capacity based on a sensor-informed learning system in combination with a local fabrication set-up. The focus of the presented paper is the advancements in robotic fabrication methods for bespoke fibre composite parts.

Fibrous composites are versatile and structurally performative materials, useful for many architectural applications. They have been utilised in many engineering industries (e.g. aerospace and automotive) for decades, due to their high strength-to-weight ratio and high degree of formability. The use of these materials has not been fully developed with respect to architectural production, although the building sector could largely benefit from the material performance, efficiency and degree of functional integration in fibrous lightweight constructions (LeGault, 2014). Early experiments with fibre-reinforced polymer (FRP) buildings in the 1960s were unsuccessful due to the lack of appropriate design flexibility and construction methods for this group of materials. The fibrous material, which is often a hand-laid fibre-woven textile, creates a sturdy albeit homogeneous material arrangement that only takes limited advantage of the anisotropic nature of the fibres for structural efficiency. Many of the traditional composite fabrication techniques require full-scale surface moulds, which is inefficient in both material usage and cost (Weitao, 2011). This often leads to serialised production of similar parts in order to take advantage of the initial material formwork investment, which is the case in these early architectural explorations. The legacy fabrication techniques, which have not changed much in nearly nine decades, are restrictive from the perspective of both design and material performance.

Filament winding, the most efficient and cost-effective method of composite fabrication, is an alternative to hand-laid fibres that can be automated for speed and efficiency in industrial production (Peters, 2011). Fibre orientation can be controlled, which makes this process more adaptable to the structural requirements and changing boundary conditions of architectural production. Industrial processes often still require surface moulds or mandrels for geometric articulation and composite performance. Precedent work on the use of FRP in architectural construction developed at the Institute for Computational Design (ICD) and the Institute for Building Structures and Structural Design (ITKE) at the University of Stuttgart was focused on the

1. Photo of the onsite fabrication core of the Elytra Filament Pavilion. Image: © NAARO.

2. Photo of component production and connection tests. Image: © V&A Museum, London.

development of integrated design, engineering and fabrication methods that allow the harnessing of the material characteristics of fibrous materials for building construction while reducing the need for surface moulds (Dörstelmann, 2014, Waimer, 2013, Reichert, 2014). The ICD/ITKE Research Pavilion 2013-14 showcased the ability to make a dual-layered composite structural system from highly differentiated components using a 'coreless filament winding' system and reconfigurable winding frame (Prado, 2014). This project pushed the possibilities for morphological exploration and novel fabrication techniques. Two synced industrial robotic arms fitted with reconfigurable frames created a highly adaptable fabrication set-up, where geometric articulation and morphologic differentiation were key areas of investigation. This project showed high potential in both the developed dual-layered structural system and the coreless winding fabrication process.

While the previously mentioned fabrication process was capable of a high degree of morphologic freedom (useful when creating geometrically specific structures or expanding the design potential of the system), for simpler applications this process could be refined to reduce complexity and increase fabrication efficiency. The Elytra Filament Pavilion proposed an adaptive, reconfigurable construction set with structural components that could be rearranged or grow into various configurations. This resulted in a unified edge condition for each component, making a reconfigurable frame unnecessary while still allowing for unique, individualised geometries and fibre arrangements to be created. More research interest was placed on the refinement of the fabrication scenario and the performative component geometries suitable for this design implementation.

The development of a versatile fibrous building system requires further consideration in several areas beyond that of previous investigations. First, to show its applicability as an architectural system, it must be utilised at a larger architectural scale. More specifically, in the case of the Elytra Filament Pavilion, the scale required longer spans and cantilevers to test the use of the structural system in various scenarios. Furthermore, beyond purely structural considerations, a fibrous building system should be able to integrate, incorporate or interface with other building systems such as the roof enclosure, wall façade construction, floor or foundation, which are important preconditions for wider applications in the building industry.

Fibre composite building system

Small-scale pavilions have historically often served as vehicles for highlighting innovation in design, material or fabrication, while reduced programmatic, spatial and functional requirements allow a focus on specific research questions. Similar past projects, such as the ICD/ITKE Research Pavilions 2012 and 2013-14 (Knippers, 2015, Dörstelmann, 2015b), were scientific demonstrators built on the campus of the University of Stuttgart and thus did not require significant interface with other building systems. In comparison, the Elytra Filament Pavilion is formed from fibrous components interlinked with multiple other material systems. Additionally, being sited in a prominent public space, it was required to pass through the rigorous certification process required for an inhabitable architectural structure.

The key components of the pavilion building system include (from top to bottom): the makrolon cladding panels, coreless wound fibre composite cells, bolted component-to-component joints, integrated lighting and sensor systems, coreless wound fibre composite column halves, bolted component-to-frame joints, steel supporting columns, membrane-support bracketry, core-enclosing membrane, core steel frame, foundation plates and helical ground anchors. Many of these elements are common within the building industry, but as no standard interface details exist for coreless wound fibre composite components, these were developed alongside the fabrication process.

Robotic coreless filament winding allows complex spatial arrangements of filament rovings to structurally connect various points in space. The distribution and spacing of winding points not only influences the component's shape and fibre layout resolution, but is also most suitable to be used as fabrication-inherent connection detailing if equipped with aluminium or stainless steel metal sleeves. Through the nature of the robotic winding process, the fibre composite rovings are bundled around the winding points, so the metal sleeves are embedded and structurally connected to the composite material. Rather than cutting or drilling (operations often used to create mechanical connections in fibre composites), the fibre rovings remain intact and structurally uncompromised. To increase the load transfer capacity of the metal-composite interface, sleeves with structured exterior surface geometry were used. Load transfer from and to these connection points through the composite structure is enhanced as the anisotropic fibres naturally align towards the connection during the winding process.

3

The inside of the metal sleeves can be blank or threaded to enable screw and bolt connections to the various building systems mentioned above. The use and type of connector can be preprogrammed as part of the frame assembly for the specific application required.

Robotic fabrication process for coreless filament winding

The presented fabrication process utilised a refined custom production set-up as well as further developments in the coreless filament winding process. An 8-axis robotic set-up consisting of a 6-axis industrial robotic arm linked to a 2-axis turntable, which carries a multi-part fixed frame, was used for winding (Fig. 3). For offsite production, a custom resin bath and spool holder which could carry up to six carbon fibre spools simultaneously were utilised for higher production speeds, while for onsite production (Fig. 1) pre-impregnated fibre spools were used. A higher degree of integrated construction detailing was enabled by advancements in the robotic coreless filament winding techniques. A multi-stage winding process was developed, which relied on several phases of frame assembly and disassembly throughout the winding process in order to wind fibres in specific configurations (Fig. 4). With this technique, embedded connectors and structural spokes could be created to interface with a transparent shingled roof enclosure system. Through adaptations in the robotic fabrication process, the construction of a wide variety of component geometries with tailored structural performance is possible on a simplified winding frame. The differentiation in system morphology, which would not be possible with standard FRP fabrication techniques, showcases the refinement and control of the coreless winding system. The surface curvature and the size of the aperture could be controlled for each component. Expanding beyond single-surface topologies (emblematic of fabrication techniques requiring a mould) enabled a hierarchical organisation of volumetric form, including a global bilayer structure. This structural system, used in previous research, was refined to create larger diameter components with a thinner structural depth to cover more area with less volume. This made the system highly efficient, requiring less material for a larger structure.

3. Coreless winding fabrication set-up: initial multi-stage frame. Image: © V&A Museum, London, © ICD/ITKE, University of Stuttgart.

4. Diagram of multi-stage winding sequence. Image: © ICD/ITKE, University of Stuttgart.

base winding frame

winding:
lower spokes
add top frame

winding:
glass inner body
aperture reinforcement
edge reinforcement

winding:
corner reinforcement
outer body
add brackets
remove upper frame structure

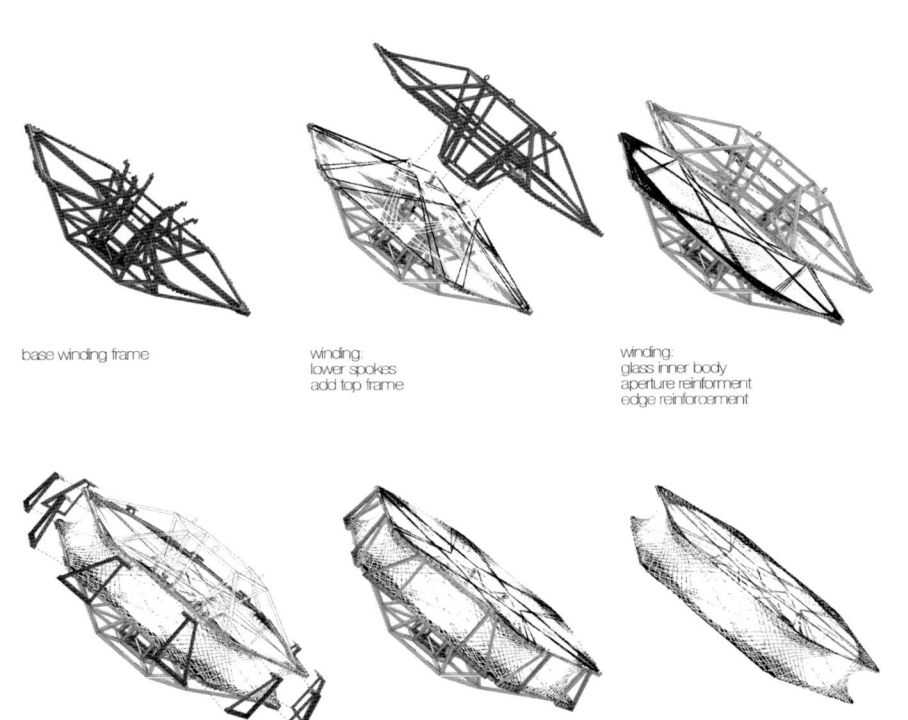

winding:
upper spokes

remove metal frame

Another evolution in the structure was the development of the closed-body reinforcement at the component scale. This created two interconnected hyperbolic surfaces which enclose a complex structural volume. With traditional mould-based fabrication techniques, these geometries would be impossible to manufacture unless the winding core remained in the finished piece or was sacrificed completely. At the material scale, coreless winding techniques were developed to control local fibre density and thus enable increased surface depth. This process, similar to three-dimensional weaving techniques, builds height from alternating dense fibre directions and a low-density, counter-directional stabilisation layer. This refinement was further demonstrated in the variation of fibre resolution, providing more control over the structural filigree.

Improved manufacturing efficiency

The refined fabrication process allowed for highly efficient production of unique roof and column components. Optimised winding times range from 4 to 9 hours per component, which were then cured and tempered overnight before being removed from the reusable winding frame. The process was highly automated and could be performed with a single robotic operator. Material handling, resin mixing and frame assembly are still manual processes in this scenario, though using industrial solutions for these could further improve manufacturing efficiency.

The Elytra Filament Pavilion cells are formed from a mix of unidirectional carbon and glass fibre roving to tailor structural efficiency. The stiffer carbon fibres provide the primary load paths within the cells, while the glass fibre creates the required geometry, distributes load and stabilises the carbon fibres. The flexibility of the fabrication set-up enables variation of the cell aperture size, changing the resulting performance of the structure. A small aperture uses more material in the top and bottom surfaces of the component, resulting in a heavier but stiffer element (Fig. 5). The cell's corners, which include connection points to its neighbours, then receive a variation in carbon material quantity based on the amount of load to be supported, with higher forces requiring greater localised stiffness and strength.

In certain critical parts of the structure, the whole cell is reinforced with a layer of carbon fibre to provide improved load transfer and strength. These strongest cells are capable of supporting a load of up to 500kg in a cantilevering condition. In earlier cellular prototypes, the free edges arising from this base geometry were susceptible to buckling issues, but this was eliminated in the final demonstrator through the closed outer body reinforcement mentioned previously (Fig. 4). The pavilion cells are therefore toroid-like beams that, when joined, create a continuous double-layer shell without free edges and with significant shear connectivity (Figs. 1, 2 and 7). Apart from geometric variation, the additive nature of coreless filament winding allows for a highly efficient use of material, placing fibres of different types only where they may be best used. With the possible variations in cell geometry and reinforcement known, a computational tool was developed to determine material placement, balancing stiffness and load distribution across the structure while achieving deflection limits (Fig. 5).

The project uses the integrative capacity of fibrous building systems to embed a sensor system that monitors visitor movements, microclimatic conditions and structural behaviour (Fig. 6). In combination with an onsite fabrication set-up, the project showcases the potential of fibrous lightweight structures to become responsive learning systems that expand and reconfigure as evolving structure and space.

Fibre optical sensors allowed for the monitoring of internal stress states of the composite structure, while thermal imaging enabled the gathering of anonymous statistics of visitor utilisation of the courtyard. Local weather data and climate simulation processes allowed predictions of local microclimatic conditions. Interpreting these data sets' interrelations allowed for reactive or proactive expansion and reconfiguration behaviours of the canopy and deriving of the respective fibre layout and fabrication data for new components. During the run of the exhibition, new components were produced at specific onsite fabrication events. The onsite fabrication set-up utilised the compactness of industrial robot arms and the fibre composite material spools. After assessing the structural capacity of the global system and local loading conditions, the new components were produced with even less material, continuing to push the boundaries of lightweight construction throughout the ongoing research process. The produced components were added during onsite reconfiguration events, resulting in two cantilevers reaching out by 5.5m and 6m from the next support, highlighting the structural performance of the implemented fibre composite building system.

The Elytra Filament Pavilion was installed in the John Madejski Garden at the Victoria and Albert Museum in London in May 2016. In its starting configuration, the fibrous canopy was constructed from 40 differentiated roof components resting on seven columns. It covered an area of 200m², which was extended to 220m² during the exhibition run. The fibre composite structure weighs only 9kg/m², while the entire canopy weighs 2.5 tonnes. The project showcases the future potential of fibrous building systems and how integrated design, engineering and fabrication strategies allow for simultaneous advancements in building technology and building culture.

Future fabrication scenarios

Future research will focus on the upscaling of building parts while maintaining the level of detail and resolution in differentiation and local adaptation. Preceding projects have shown how fabrication time can be reduced by fabricating fewer components on larger scales. This also reduces the amount of joints required and increases the fibre continuity for higher structural efficiency. Scaling up the existing fabrication scenario would not be possible due to the workspace limitations of an industrial robotic arm and transportability volume, but alternative set-ups may be used which incorporate a robotic linear axis or onsite fabrication methods using small moveable fabrication agents. Industrialisation of the winding process could require further refinement, including sensor-integrated cyber-physical winding strategies for increased automation, error correction and live adaptation of the robotic movements, or material quality control for composite durability and UV and fire resistance. Answering these questions would provide a big step towards developing a fibrous building system for architectural applications.

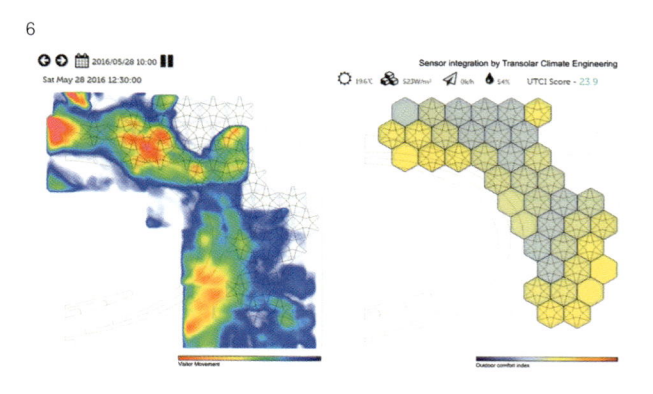

5. Diagram of structural testing of cell aperture variations.
Image: © ICD/ITKE, University of Stuttgart.

6: Diagram of sensor data visualisation: (a) visitor movements, (b) microclimatic conditions.
Image: © Transsolar KlimaEngineering.

7. Photo of canopy structure with differentiated cellular arrangement.
Image: © NAARO.

Acknowledgements

The authors would like to express their gratitude to the entire scientific development and robotic fabrication team, including students and support staff, who helped to complete this project.

Project Team

ICD – Institute for Computational Design, University of Stuttgart: Achim Menges, with Moritz Dörstelmann, Marshall Prado, Aikaterini Papadimitriou, Niccolo Dambrosio and Roberto Naboni, with support from Dylan Wood and Daniel Reist.

ITKE – Institute of Building Structures and Structural Design, University of Stuttgart: Jan Knippers, with Valentin Koslowski, James Solly and Thiemo Fildhuth.

Transsolar Climate Engineering, Stuttgart Building Technology and Climate Responsive Design, TU München: Thomas Auer, with Elmira Reisi and Boris Plotnikov.

With the support of:
Michael Preisack, Christian Arias, Pedro Giachini, Andre Kauffman, Thu Nguyen, Nikolaos Xenos, Giulio Brugnaro, Alberto Lago, Yuliya Baranovskaya, Belen Torres and IFB University of Stuttgart (Prof. P. Middendorf).

Commission:
Victoria & Albert Museum, London, 2016.

Funding:
Victoria & Albert Museum, London
University of Stuttgart
Getty Lab
Kuka Roboter GmbH + Kuka Robotics UK Ltd
SGL Carbon SE
Hexion
Covestro AG
FBGS International NV
Arnold AG
PFEIFER Seil- und Hebetechnik GmbH
Stahlbau Wendeler GmbH + Co. KG
Lange+Ritter GmbH
STILL GmbH

References

Menges, A. and Knippers, J., 2015, 'Fibrous Tectonics' in *Architectural Design*, Vol. 85, No. 5, Wiley, London, p.40-47 (ISBN 978-11118878378; DOI: 10.1002/ad.1952).

Dörstelmann, M., Knippers, J., Koslowski, V., Menges, A., Prado, M., Schieber, G. and Vasey, L., 2015a, ICD/ITKE Research Pavilion 2014–15, 'Fibre Placement on a Pneumatic Body Based on a Water Spider Web' in *Architectural Design*, 85(5), p.60-65.

Van de Kamp, T., Dörstelmann, M., Dos Santos Rolo, T., Baumbach. T., Menges, A. and Knippers, J., 2015, 'Beetle Elytra as Role Models for Lightweight Building Construction' in *Entomologie Heute*, Vol. 27, p.149-158.

LeGault, M., 2014, 'Architectural Composites: Rising to New Challenges' in *CompositesWorld*, available at http://www.compositesworld.com/articles/architectural-composites-rising-to-new-challenges.

Weitao, M., 2011, 'Cost Modelling for Manufacturing of Aerospace Composites', MSc research thesis, School of Applied Sciences, Cranfield University.

Peters, S.T. (ed), 2011, *Composite Filament Winding*, ASM International, Material Park.

Dörstelmann, M., Parascho, S., Prado, M., Menges, A. and Knippers, J., 2014, 'Integrative Computational Design Methodologies for Modular Architectural Fibre Composite Morphologies' in *Design Agency* [Proceedings of the 34th Annual Conference of the Association for Computer Aided Design in Architecture (ACADIA)], Los Angeles, p.177-188 (ISBN 9781926724478).

Waimer, F., La Magna, R., Reichert, S., Schwinn, T., Knippers, J. and Menges, A., 2013, 'Integrated Design Methods for the Simulation of Fibre-Based Structures' in Gengnagel, C., Kilian, A., Nembrini, J. and Scheurer, F. (eds.), *Rethinking Prototyping, Proceedings of the Design Modelling Symposium Berlin* 2013, Verlag der Universität der Künste Berlin, p.277-290 (ISBN 978-3-89462-243-5).

Reichert, S., Schwinn, T., La Magna, R., Waimer, F., Knippers, J. and Menges, A., 2014. 'Fibrous Structures: an Integrative Approach to Design Computation, Simulation and Fabrication for Lightweight, Glass and Carbon Fibre Composite Structures in Architecture based on Biomimetic Design Principles' in *Computer-Aided Design*, 52, p.27-39.

Prado, M., Dörstelmann, M., Schwinn, T., Menges, A. and Knippers, J., 2014, 'Coreless Filament Winding: Robotically Fabricated Fibre Composite Building Components' in McGee, W. and Ponce de Leon, M. (eds.), *Proceedings of the Robots in Architecture Conference 2014*, University of Michigan, p.275-289 (ISBN 9783319046624).

Knippers, J., La Magna, R., Menges, A., Reichert, S., Schwinn, T. and Weimar, F., 2015, ICD/ITKE Research Pavilion 2012, 'Coreless Filament Winding on the Morphological Principles of an Arthropod Exoskeleton' in *Architectural Design*, Vol. 85, No. 5, Wiley, London, p.48-53 (ISBN 978-11118878378; DOI: 10.1002/ad.1953).

Dörstelmann, M., Knippers, J., Menges, A., Parascho, S., Prado, M. and Schwinn, T., 2015b, ICD/ITKE Research Pavilion 2013-14, 'Modular Coreless Filament Winding Based on Beetle Elytra' in *Architectural Design*, 85(5), p.54-59.

4

RETHINKING CONSTRUCTIONAL LOGICS

SENSORIAL PLAYSCAPE
ADVANCED STRUCTURAL, MATERIAL AND RESPONSIVE CAPACITIES OF TEXTILE HYBRID ARCHITECTURES AS THERAPEUTIC ENVIRONMENTS FOR SOCIAL PLAY

SEAN AHLQUIST
University of Michigan, Taubman College of Architecture and Urban Planning

Computational design commonly focuses on the synchronisation of advanced manufacturing technology and material behaviour. This allows for technical specificity, or instrumentalisation, to be achieved in material, structural and architectural performance. The research discussed in this paper extends such a material-based practice by utilising aspects of sensorial experience to drive the design and engineering of material performance and architectural responsivity. This is explored as a part of the *Social Sensory Architectures* research project, through the articulation of textile hybrid structures and their application to the development of skills in motor control and social interaction for children with autism spectrum disorder (ASD). The research is developed at the University of Michigan, through a collaboration between the Taubman College of Architecture and Urban Planning, the Department of Psychiatry and the School of Kinesiology. This alignment of disciplines integrates material research with methodologies for assessment of social function and kinesthetic activity.

This interdisciplinary research is described, in this paper, through the development of the *sensoryPLAYSCAPE* prototype. The prototype, as a malleable multi-sensory architecture, seeks to unravel associations between deficiencies in motor planning and processing of sensory stimuli with limitations in social function for children with autism. Defined as a spectrum disorder, a hallmark of autism is the highly unique and specific sensory and behavioural issues related to each individual. This is captured in the commonly used phrase: "when you've met one person with autism, you've met one person with autism". Accordingly, a significant criterion for the prototype is to enable the child to instrumentalise the sensorial experience of the architecture to suit their particular preferences. The intentions are to develop skills in motor planning that will assist social functioning through collaborative play. Navigating through the tactile architecture simultaneously reinforces such physiological and social activities through the sensorial adaptability of the architecture.

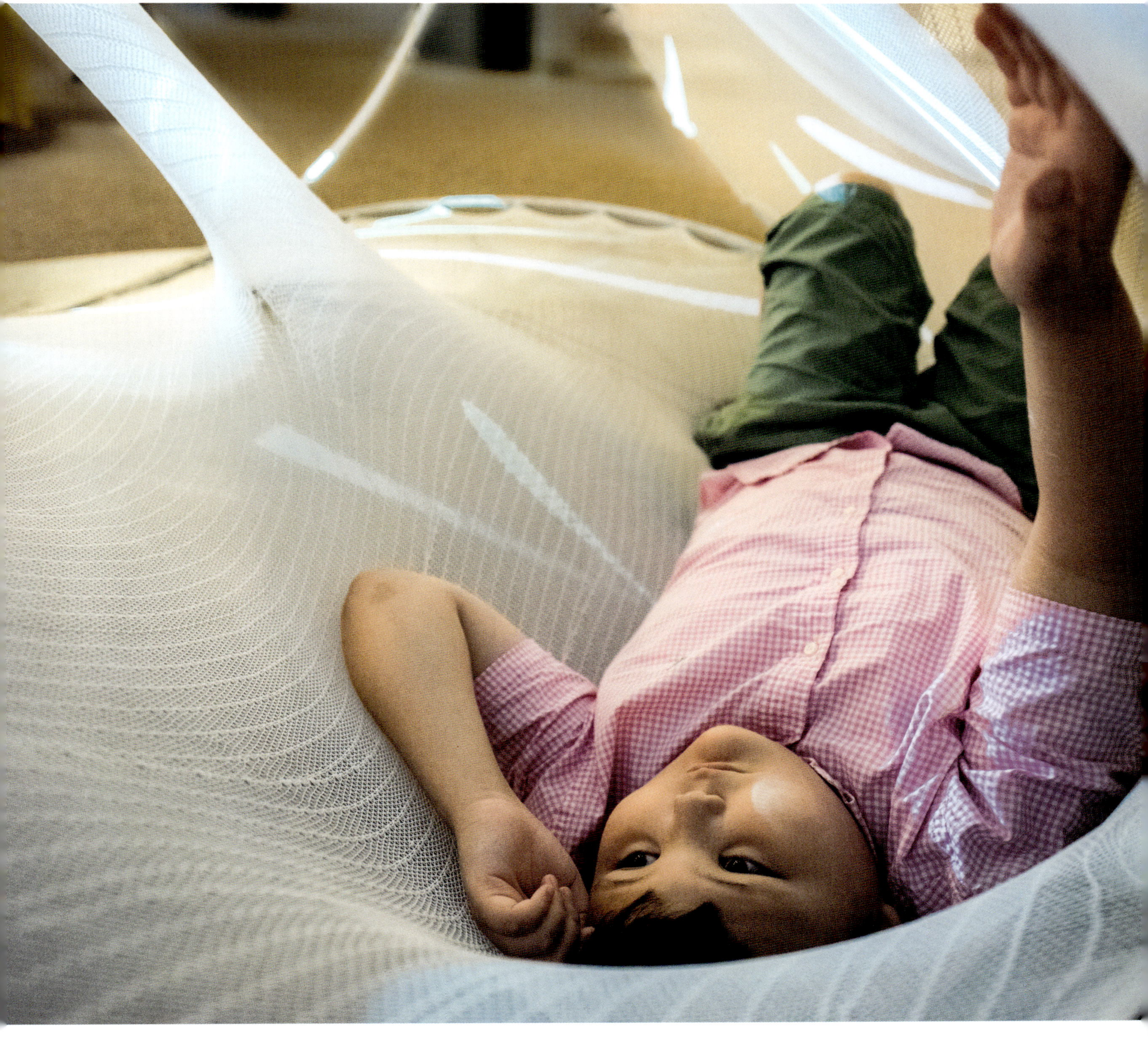

This research concentrates on tailoring the hierarchical relationships between the various multi-modal sensations triggered via the interactive textile hybrid environment. Tactile, visual and auditory stimuli are activated through physical deformation of textile surfaces and moving through the intricate spatial arrangements of the playscape prototype (Fig. 2). This requires the priority of the textile hybrid system to move beyond primarily resolving the stresses of internal material behaviour and structural forces, where minimal load-bearing reserves remain. Through advances in the use of CNC-knitting of the tactile, structural interface and lamination of bending-active glass fibre-reinforced (GFRP) beams, the highly variable and dynamic loads incurred as a structural landscape for play are enabled.

Nexus of movement and social function

Autism is a neurological disorder affecting 1 in 68 children and involving global impairments in communication, social interaction and the regulation of physical and emotional behaviour (Baio, 2014). An underlying yet prevalent factor is the inability to receive, sort and integrate sensory information related to social and environmental stimuli (Spradlin & Brady, 1999). Such ineffective means for integrating sensory information prevents the learning of adaptive, generalised behaviours and coordinated movement (Bundy et al., 2002).

Specifically, atypical tactile sensory processing is often a characteristic in children with autism (Rogers et al., 2003). The development of fine and gross motor control relies heavily on the somatosensory system, where accurate tactile and proprioceptive sensation are most critical (Cauller, 1995). Thus impairments in fine and gross motor skills are also commonplace, hindering precision for task-oriented movement, hand-eye coordination, social imitation, gait, posture and balance (Dawson & Watling, 2000). Overall, the quality of motor performance is influenced when guided feedback from the sensory system is diminished (Baranek, 2002).

As a part of the research into sensorial architectures, the primary exploration is the interconnection between the domains of movement, as driven by sensory processing, social function and communication. Touch is a primary method for rudimentary non-verbal communication, while the whole of the somatosensory system is pivotal in more nuanced interaction. Gestures and facial expressions function via feedback from stretch receptors of the skin and muscles in the hands and arms (Cascio, 2010). Abnormalities in the somatosensory system, such as for children with autism, are thus seen to correlate with reduced social attention and impairment in non-verbal communication (Foss-Feig et al., 2012). Children who experience limitations in motor skills are shown to have fewer opportunities for social interaction with peers, correlating with lower levels of physical activity (MacDonald et al., 2014). In comparison with children having speech-language impairments or learning disabilities, those with autism are approximately 50 percent less likely to be invited to social activities and 450 percent more likely never to see friends (Shattuck et al., 2011).

Environment also plays an influential and often magnified role in the socio-sensory experience. Stress for a child may emanate from a mismatch between the environment and the child's aberrant processing of its multi-modal stimuli. Research has shown that successful intervention can occur through a focus on altering environment as opposed to eliminating the atypical behaviour which results from dysfunctional sensory processing (Lovass & Smith, 2003). The performance of motor tasks has been shown to be better for children with autism when developed in a meaningful and related context (Baranek, 2002). This is a core principle of dynamic systems theory, where one domain – environment – affects another domain – movement (Ketcheson et al., 2016). Therefore

1. Pilot study of the *sensoryPLAYSCAPE* prototype at the HandsOn Museum in Ann Arbor, Michigan, for the MyTurn Event for children with autism. Image: Sean Ahlquist, University of Michigan, Peter Matthews, Michigan Photography, 2016.

2. Projected graphics and highly differentiated spatial configurations of the textile hybrid system form the multi-scalar, multi-sensory nature of the prototype. Image: Sean Ahlquist, University of Michigan, 2016.

the forming of new behaviours has to account for both domains, to ideally trigger the cascading effect of producing new opportunities for social interaction.

Sensory responsiveness in textile hybrids

The ability to formulate and execute patterns of movement, through feedback between motor commands and sensory data, is pivotal to the development of social behaviour. The relationship between movement and its sensory consequence forms an understanding of the intentions of movement and, ultimately, provides the knowledge that allows the interpretation of other people's gestures (Izawa et al., 2012). For children with autism, learning new patterns of movement is most reliant on proprioceptive feedback – sensation from muscle and joint articulation to determine position and orientation of the limbs and body. Visual stimulation has a secondary impact, meaning the non-physical stimuli can often play a less influential role (Haswell et al., 2009).

To synthesise movement and social behaviour, the multi-sensory nature of the playscape prototype is focused most heavily on its tactile qualities. This operates through multiple scales and in the instrumentalisation of elasticity at each scale. One level attends to forming skills for grading of movement, the ability to assess and execute the appropriate amount of pressure needed to complete a task. Yarn, variegated stitch structure and the calibration of tensile forces generate an increasingly magnified tactile feedback as one pushes on the surfaces to greater depths (Fig. 3). Another level of engagement corresponds to movement of the body through space and time, the proprioceptive and vestibular senses that guide orientation and pace. The calibration of the pre-stressed textiles, laminated GFRP beams and spatial arrangement generates the combined experience of localised pressure at the interaction of the body with the textile and minimised (though recognisable) deflection at the scale of the entire material system (Fig. 1). Elasticity is tailored to satisfy deeper sensations of touch and register fine and gross movements. Correlation with the visual and auditory landscape fosters continual variability and saliency.

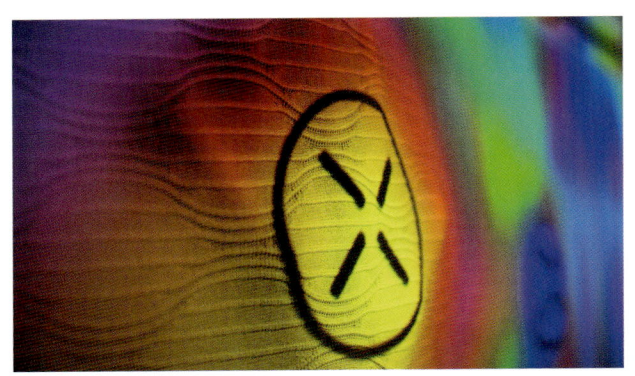

Textile hybrid sensoryPLAYSCAPE prototype

A hybrid structure denotes a system which integrates more than one fundamental structural strategy (Engel, 2007). The textile hybrid incorporates tensile form-active surfaces and boundary elements stiffened through their configuration into curved bending-active geometries to generate a structural form (Lienhard et al., 2012, Ahlquist et al., 2013). More specifically, through this research at the University of Michigan, the hybrid system is uniquely comprised of seamless CNC-knitted textiles and bending-active GFRP rods laminated into curved structural beams (Ahlquist, 2015) (Fig. 4). In the design, engineering and manufacturing of the playscape prototype, the topologies of the textile architecture and rod configurations are articulated through simulation in the Java-based springFORM software (Ahlquist et al., 2014).

Bending-active laminated GFRP beams

The active bending of the GFRP rods in a textile hybrid serves to maximise stiffness and simultaneously activate tension in the integrated textile surfaces. Traditionally, the relationship between the GFRP rod cross-section and desired stiffness is designed solely to satisfy a target geometry. Unfortunately, this leaves little in structural reserve for additional load-bearing purposes. Typically, the bending-active GFRP boundary is comprised of a single rod cross-section, meaning rods of the same cross-section are utilised throughout the entire system. This is problematic, as it clamps the scale of the entire structure (or the GFRP component within it) to its

3. *StretchCOLOR* interface, developed in Unity, generates colour based on the location and amount of pressure applied to the tactile, elastic knitted textile.

4. Textile hybrid structure, installed at Southern Illinois University, Carbondale, formed of bending-active laminated GFRP beams interconnected with form-active CNC-knitted textiles.

Images: Sean Ahlquist, University of Michigan, 2016.

5

5. Mesh topology for form-finding of the tensile surfaces in springFORM, focused on the sharp transitions between surficial and cylindrical forms.

6. Fundamental logic for CNC knitting based upon mesh topologies and forms generated in springFORM, indicating the transition from planar to tubular knitting and the shifting (transferring) while knitting of the tubular sections.

7. Detail of CNC-knitted textile, indicating tuck-stitch one-by-one knit structure, with transition between planar and tubular knitting.

Images: Sean Ahlquist, University of Michigan, 2016.

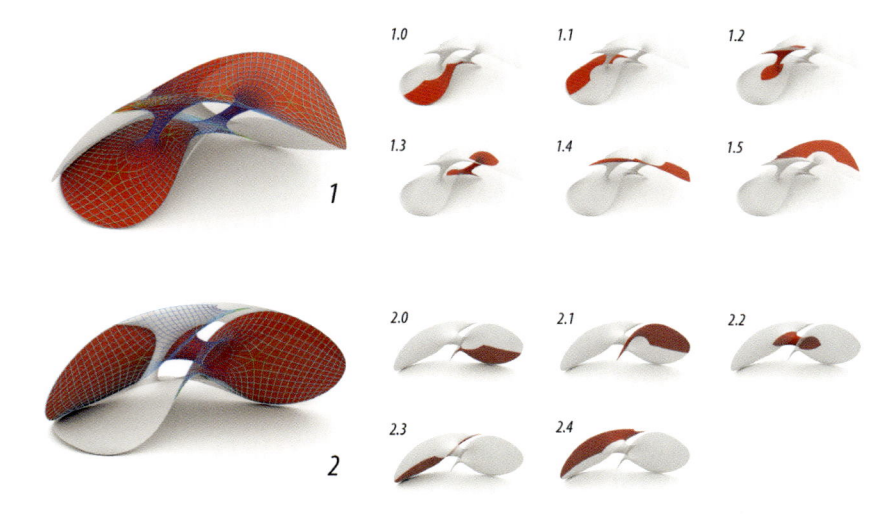

manufactured length. If serialised in a linear assembly, this produces a significant structural discontinuity from the end-to-end condition of the rods. Both scenarios are especially problematic given the erratic and unpredictable loading to be incurred as the system is deployed as an architecture for play.

In response, GFRP rods are strategically bundled and laminated in their bending-active state to form curved beams with shear-stiff connections. A critical advantage is gained in geometric freedom, where individual rods of minimal cross-section can be used to accomplish a wide range of radii, and in structural stiffness, where, once laminated, strength is increased by a factor of 10 (Ahlquist & Lienhard, 2016). The basic construction of the laminated beams for the prototype consists of GFRP bundles of two to four rods, CNC-knitted sleeving and vacuum-formed impregnation with epoxy resin. Experiments testing the composition of the sleeving indicate that a high performance polyester yarn provides enough consolidation and flow of epoxy to sufficiently bond the rods, without the introduction of gained stiffness through the sleeve itself (Ahlquist & Lienhard, 2016).

Seamless CNC-knitted textiles

Where the bending-active laminated boundary takes on more structural capacity, the integrated textiles are given a certain freedom for spatial articulation, while still contributing to overall structural stability. In this instance, the topology of the knitted structure is designed to acutely control transition and non-orthogonal orientations between surface and cylindrical geometries. Rather than a more traditional method of shaping tubular geometries, panels are merged to and from tubular forms as a part of generating singular seamless textiles. Such logic is initiated in the springFORM simulation in order to tailor the tensile forms and also follow the logic for programming the CNC-knitted textiles (Fig. 5).

CNC knitting machines are equipped with two independent but adjacent needle beds, easily allowing for tubular textiles to be manufactured by knitting across the front needle bed continuously to the back bed, returning again to the front and repeating the pattern. Shaping, or altering the number of stitches from one pass along the needle bed to the next, provides the capacity to contour the tubular form. To accomplish the dramatic transition from a surficial to tubular condition in the playscape prototype, two independent panels are knitted, each on a separate needle bed, and merged at the ends of the tubular structure (Fig. 6). To accomplish the offset between the top and bottom surfaces of the two interlocked textiles, the tubular portion is both iteratively knitted and shifted, or transferred, across the needle bed. Where it is branched from the bottom surface at one edge, the tubular structure is linked to the top surface at the opposite edge, producing seamless textiles which span across the length of the GFRP boundary.

The overall textile architecture is dictated through extrapolating geometry and relative force calculations from the springFORM model in comparison to knitted 1:1 textiles swatches. The knitted swatches utilise a nylastic (co-mingled elastic nylon and spandex core) yarn with an alternating tuck-tuck-stitch structure knitted on every other needle (referred to as one-by-one) of a 14-gauge CNC knitting machine (Fig. 7). The method of extrapolation is approximate, as the stitch structure is altered in the final textile via shaping of the overall form and manipulating the stitch length, in order to accomplish certain conditions such as achieving maximal stretch to fit across the widest dimensions of the structure. The performance of the tensioned textile surface is defined primarily by providing a high degree of elasticity, which serves as the tactile interface. Yet this is still in balance with its service to the textile hybrid system, where the CNC-knitted textiles improve the overall structural stiffness by approximately 15 percent.

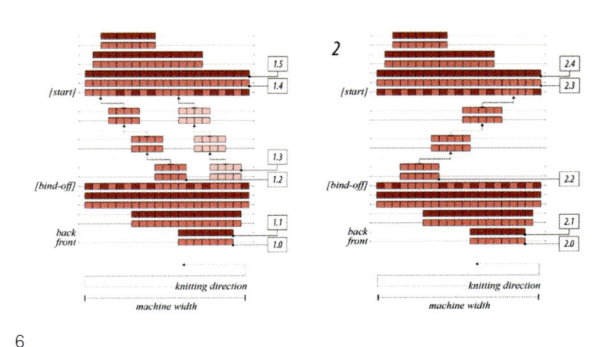

Interface architecture

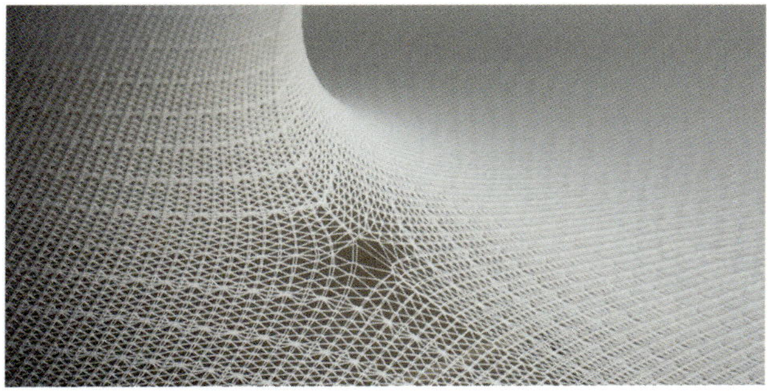

To embed visual and auditory interactivity, the prototype integrates projection, sensing via the Microsoft Kinect and interface design developed in the programming environment Unity (Fig. 8). The depth map data are extracted from the Microsoft Kinect for use in capturing the base geometry of the textile surfaces and also in identifying, through difference mapping, any alterations to the geometry based on physical interactions. To locate the point and exact depth of touch, the difference map is posterised to produce clear contours and to search out local maxima, allowing for the identification of multiple touches at any given moment.

In order to align physical space with the projected imagery, a chessboard mapping is utilised with a homographic translation. Each region of the chessboard is translated in isolation, minimising the effects of distortions from one region to the next. This facet of the algorithm is critical, as it allows for contoured surfaces to be analysed with higher accuracy through an increased resolution of the chessboard. The same method can be utilised with a lower resolution chessboard to track interactions on a two-dimensional surface.

The method of sensing functions as a standalone algorithm outputting data for location and depth of interaction with the textile surfaces. This allows for complete interchangeability between the form of the structure and the modes of visual and auditory feedback. It defines a designation between *sensoryARCHITECTURE* and *stretchINTERFACE*. For the prototype in this research, the architecture is defined as the *sensoryPLAYSCAPE*, while a series of interfaces, developed in Unity, have been employed. *StretchCOLOR* is developed as a painting tool where colour is determined by the amount of pressure being applied to the surface. *StretchANIMATE* projects pre-rendered animations across the surface of the structure based on touching the textiles at key locations. *StretchSWARM* provides more intimate interactions, where a quick touch disturbs a free-flowing school of fish, while a long touch generates an attractor for the fish to circle around, also triggering a randomised soundscape of wind chimes.

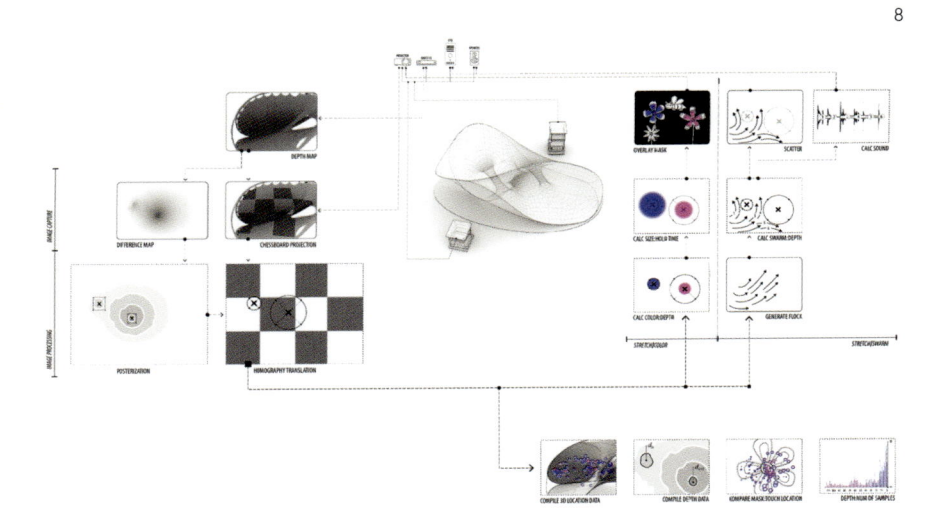

Therapeutic capacity of sensory architectures

Two primary skills are being addressed – motor planning and social function – through the *sensoryPLAYSCAPE* prototype in combination with the various software modes for multi-sensory feedback. Through an ongoing pilot study with the Spectrum Therapy Center in Ann Arbor, the *stretchCOLOR* interface is utilised to attend to the development of skills for grading of movement. Where poor signalling from the somatosensory system and lack of muscle tone may contribute to diminished nuance in motor function, the visual and physical feedback of the prototype provides compensatory data.

8. Arcnitecture of the *sensoryPLAYSCAPE* prototype, showing the interconnection between the textile hybrid structure, sensing hardware and software, interfaces programmed in Unity and output and analysis of diagnostics.
Image: Sean Ahlquist / Oliver Popadich, University of Michigan, 2016.

Projected colours, based on depth of touch, and resistance in the elastic textiles, increasing with the amount of pressure applied, provide magnified feedback for varying degrees of movement. Through iterative experience, an understanding of motor planning emerges through the child's own unique physiological and sensory processing capabilities. Data are collected through the software, capturing location, depth and frequency of touch. Motor skills are measured through a pre- and post-kinesthetic assessment using the Peabody Developmental Motor Scales and Bayley Scales of Infant and Toddler Development for measuring motor skills and identifying developmental delays.

The social component of this research is assessed through the concepts formed by the PLAY project, an early intervention program developed in Ann Arbor by Dr. Richard Solomon and focused on modes of play to encourage communication and social interaction (Solomon et al., 2007). Assessment tracks three characteristics: (i) fundamental developmental level (FDL) – milestones for emotional, social and cognitive development, (ii) sensory motor profile – the dynamics between environment, social interaction and self-regulation, and (iii) comfort zone – preferred, often non-social, activities. The intent of the PLAY project approach is to generate reciprocal interaction, or circles of communication. This is generated through following the child's lead, yet tempering activities to refrain from the isolating comfort zone while staying within their FDL. The sensory component serves dually to make the activity attractive while also providing a positive reinforcer to the back-and-forth social interactions. The *sensoryPLAYSCAPE* prototype embeds these concepts through the synthesis of variable and multi-scalar tactile qualities with modes of interaction that encourage combined play (Fig. 9).

Working with the sensorial experience to create new architectures

The research described in this paper provides the foundation for an architecture which sets the sensorial experience as the primary performative constraint by which material, spatial, visual and sonic landscapes are instrumentalised. Yet perception of space and time, in its social and environmental constituents, is largely atypical for children with autism. In response, those who engage with the architecture are given considerable agency to actively and dynamically articulate its material and immaterial natures. Performance of the prototype is defined by the measured understanding of the physiological and sociological human behaviours that occur within it. The manner in which the architecture is transformed communicates the individualised nature of the socio-sensorial experience.

9. Encouragement of social interaction through the synthesis of spatial, tactile and visual stimuli.
Image: Sean Ahlquist, University of Michigan. Photo by Gregory Wendt, Southern Illinois University, Carbondale, 2016.

Acknowledgements

This research is being developed at the University of Michigan through funding from the MCubed programme for the project 'Tactile interfaces and environments for developing motor skills and social interaction in children with autism'. The principal investigators are Prof. Sean Ahlquist, Prof. Costanza Colombi from the Department of Psychiatry and Prof. Dale Ulrich and Leah Ketcheson from the School of Kinesiology. The team consists of researchers from architecture: Oliver Popadich, Shahida Sharmin and Adam Wang; and Erin Almony and Erika Goodman from Kinesiology. Research for the engineering of the textile hybrid structures is developed in collaboration with Julian Lienhard of str.ucture. Concepts for therapies in social play and kinesthetic activity are developed in collaboration with Mary Burke of the Spectrum Therapy Center and Onna Solomon of the PLAY project.

References

Ahlquist, S., Lienhard, J., Knippers, J. and Menges, A., 2013, 'Physical and Numerical Prototyping for Integrated Bending and Form-Active Textile Hybrid Structures' in Gengnagel, C., Kilian, A., Nembrini, J. and Scheurer, F. (eds.), *Rethinking Prototyping: Proceeding of the Design Modelling Symposium*, Springer, Berlin, p.1-14.

Ahlquist, S., Kampowski, T., Oliyan, O., Menges, A. and Speck, T., 2014, 'Development of a Digital Framework for the Computation of Complex Material and Morphological Behaviour of Biological and Technological Systems' in *CAD: Special Issue on Material Ecology*, 60, p84-104.

Ahlquist, S., 2015, 'Social Sensory Architectures: Articulating Textile Hybrid Structures for Multi-sensory Responsiveness and Collaborative Play' in Perry, C. and Combs, L. (eds.), *ACADIA 2015. Computational Ecologies*, p.262-273.

Ahlquist, S. and Lienhard, J., 2016, 'Extending Geometric and Structural Capacities for Textile Hybrid Structures with Laminated GFRP Beams and CNC Knitting' in Kawaguchi, K. et al. (eds.), *Proceedings of the IASS Symposium, Spatial Structures in the 21st Century*, Tokyo, Japan.

Ashburner, J., Ziviani, J. and Rodger, S., 2008, 'Sensory Processing and Classroom Emotional, Behavioural, and Educational Outcomes in Children With Autism Spectrum Disorder' in *The American Journal of Occupational Therapy*, 62(5), p.564-573.

Baio, J., 2014, 'Prevalence of Autism Spectrum Disorder Among Children Aged 8 Years: Autism and Developmental Disabilities Monitoring Network, 11 Sites, United States, 2010' in *Morbidity and Mortality Weekly Report*, 63 (2), p.1-21.

Baranek, G., 2002, 'Efficacy of Sensory and Motor Intervention for Children with Autism' in *Journal of Autism and Developmental Disorders*, 32 (5), p.397-422.

Bundy, A.C., Lane, S.J., Murray, E.A. and Fisher, A.G., 2002, *Sensory Integration: Theory and Practice*, Philadelphia, F.A. Davis Company.

Cascio, C., 2010, 'Somatosensory Processing in Neurodevelopmental Disorders' in *Journal of Neurodevelopmental Disorders*, 2(2), p.

Cauller, L., 1995, 'Layer I of Primary Sensory Neocortex: where Top-down Converges upon Bottom-up' in *Behavioural Brain Research*, 71(1). p.63-70.

Dawson, G. and Watling, R., 2000, 'Interventions to Facilitate Auditory, Visual, and Motor Integration in Autism: A Review of the Evidence' in *Journal of Autism and Developmental Disorders*, 30(5), p.415-421.

Engel H., 2007, *Tragsysteme – Structure Systems*, fourth edition, Hatje Cantz.

Foss-Feig, J., Heacock, J. and Cascio, C., 2012, 'Tactile Responsiveness Patterns and their Association with Core Features in Autism Spectrum Disorders' in *Research in Autism Spectrum Disorders*, 6, p.337.

Gomot, M., Bernard, F.A., Davis, M.H., Belmonte, M.K., Ashwin, C., Bullmore, E.T. et al., 2006, 'Change Detection in Autism: an Auditory Event-related fMRI Study' in *NeuroImage*, 29, p.475-484.

Haswell, C.C., Izawa, J., Dowell, L.R., Mostofsky, S.H. and Shadmehr, R., 2009, 'Representation of Internal Models of Action in the Autistic Brain' in *Nature Neuroscience*, 12, p.970-972.

Izawa, J., Pekny, S., Marko, M., Haswell, C., Shadmehr, R. and Mostofsky, S., 2012, 'Motor Learning Relies on Integrated Sensory Inputs in ADHD, but Over-Selectively on Proprioception in Autism Spectrum Conditions' in *Autism Research*, 5, p.124-136.

Ketcheson, L., Hauck, J. and Ulrich, D., 2016, 'The Effect of Early Motor Skill Intervention on Motor Skills, Level of Physical Activity, and Socialization in Young Children with Autism Spectrum Disorder: A Pilot Study' in *Autism*, Epub.

Lienhard, J., Alpermann, H., Gengnagel, C. and Knippers, J., 2012, 'Active Bending; a Review on Structures where Bending is Used as a Self-formation Process' in *Proceedings of the International IASS Symposium*, p.650-657.

Lovaas, O.I. and Smith, T., 2003, 'Early and Intensive Behavioural Intervention in Autism' in Kazdin, A.E. and Weisz, J.R. (eds.), *Evidence-based Psychotherapies for Children and Adolescents*, Guilford Press (New York), p.325-340.

Macdonald, M., Lord, C. and Ulrich, D., 2014, 'Motor Skills and Calibrated Autism Severity in Young Children with Autism Spectrum Disorder' in *Adapted Physical Activity Quarterly*, 31(2), p.95-105.

Rogers, S.J., Hepburn, S. and Wehner, E., 2003, 'Parent Reports of Sensory Symptoms in Toddlers with Autism and those with Other Developmental Disorders' in *Journal of Autism and Developmental Disorders*, 33, p.631-642.

Shattuck, P.T., Orsmond, G.I., Wagner, M. and Cooper, B.P., 2011, 'Participation in Social Activities among Adolescents with an Autism Spectrum Disorder' in *PLoS One*, 6(11), p.1-9.

Spradlin, J.E. and Brady, N.C., 1999, 'Early Childhood Autism and Stimulus Control' in Ghezzi, P.M., Williams, W.L. and Carr, J.E. (eds.), *Autism: Behaviour Analytic Perspectives*, Context Press (Reno), p.49-65.

BENDING-ACTIVE PLATES PLANNING AND CONSTRUCTION

SIMON SCHLEICHER
University of California, Berkeley, College of Environmental Design (CED), Berkeley, United States of America
RICCARDO LA MAGNA / JAN KNIPPERS
University of Stuttgart, Institute of Building Structures and Structural Design (ITKE), Stuttgart, Germany

Bending-active plate structures

In 2015, researchers at the University of California, Berkeley, and the Institute of Building Structures and Structural Design (ITKE) at the University of Stuttgart collaborated with the aim of contributing to the current research on bending-active plate structures. They placed particular emphasis on the further development of the formal and structural potential of this relatively new structural system and construction principle. In general, bending-active structures are fascinating because they take advantage of large elastic deformations as a form-giving and self-stabilising strategy. Previous research has mainly focused on a bottom-up form-finding approach, in which typical characteristics of plates or strips were predefined first and the global shape of the structure resulted from the interaction of assembled parts. In contrast, the main emphasis of this work will be on demonstrating a possible top-down approach that is based on form-conversion.

For bending-active plate structures that implement form-conversion, the process starts with the design

1

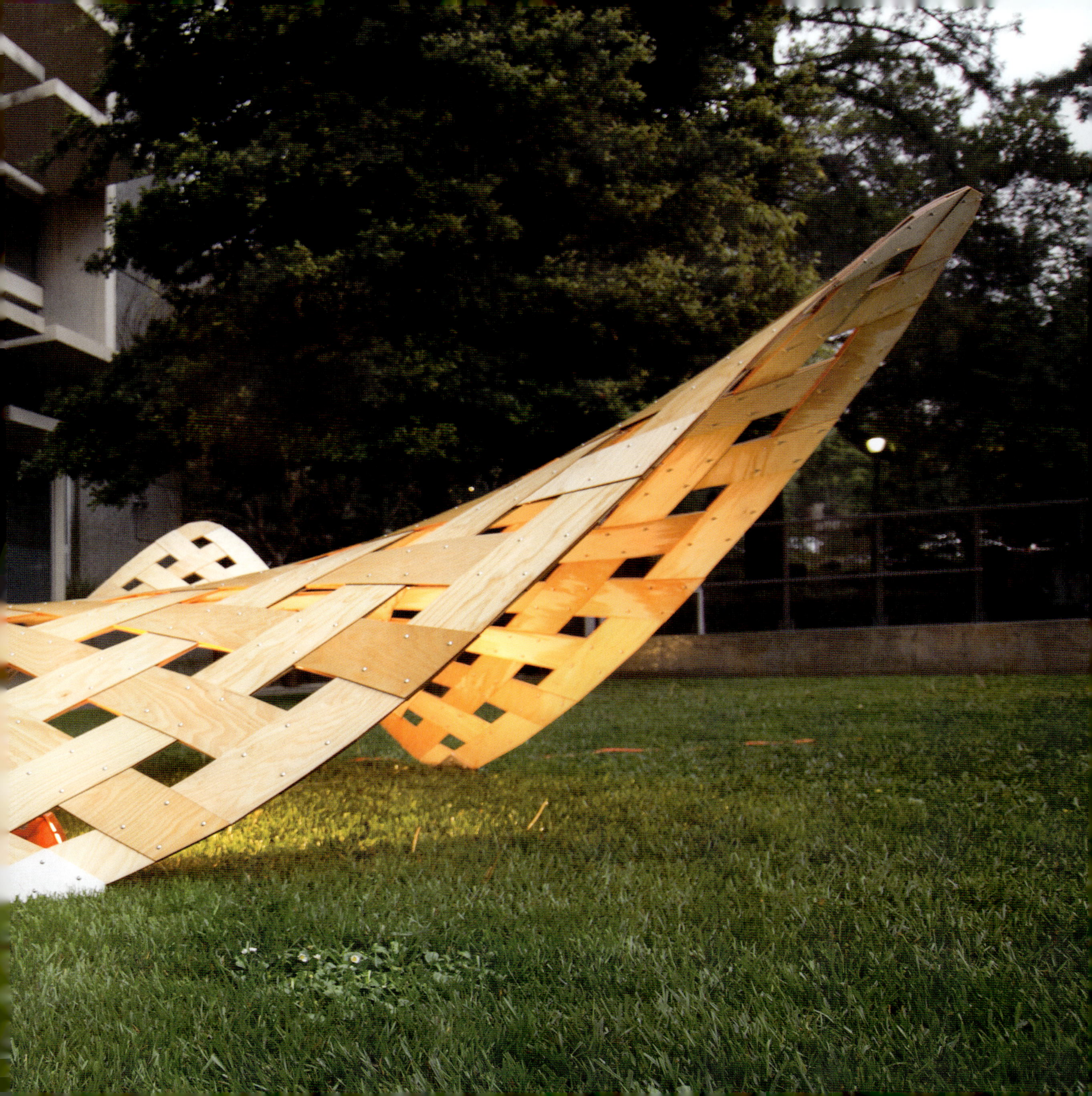

of a target shape, which is then subdivided into bespoke panels, with due consideration given to their specific geometry and structural characteristics. By attaching greater importance to the target shape, form-conversion offers several benefits and opens up a larger design space than a bottom-up form-finding. However, the key challenge remains – and essentially boils down to – the question of how to assess both the global shape and the local features of the constituent parts for structures in which geometric characteristics and material properties are inevitably linked together and similarly affect the result.

To demonstrate the potentials and challenges of the form-conversion approach, the authors will discuss this research method in general and show its feasibility in the planning of two built case studies in particular. Each structure emphasises a different aspect of this design approach. While the first case study takes advantage of translating a predefined shape into a self-supporting woven pattern, the second case study gains significant stability by translating a given form into a multi-layered shell. Finally, the means of architectural prototyping will provide proof of concept. By reflecting on how these case studies were actually constructed, the authors will give valuable insights into the opportunities and limitations of designing bending-active plate structures by form-conversion, and will hopefully spark further research in this direction.

Breakthroughs in modelling tools

In recent years, the architecture community has witnessed astonishing changes in digital design and modelling tools. With programs such as Kangaroo Physics, Karamba and SOFiSTiK, new types of versatile tool have become widely accessible, enabling the creative design of highly complex geometrical models and also allowing for the integration of real-time, physics-based simulations in common CAD environments. Thus equipped, it is nowadays possible to rapidly form-find and freely interact with particle systems, or accurately analyse and optimise structures by means of the finite element method. Due to these changes, one can now describe and evaluate the mechanical behaviour and structural

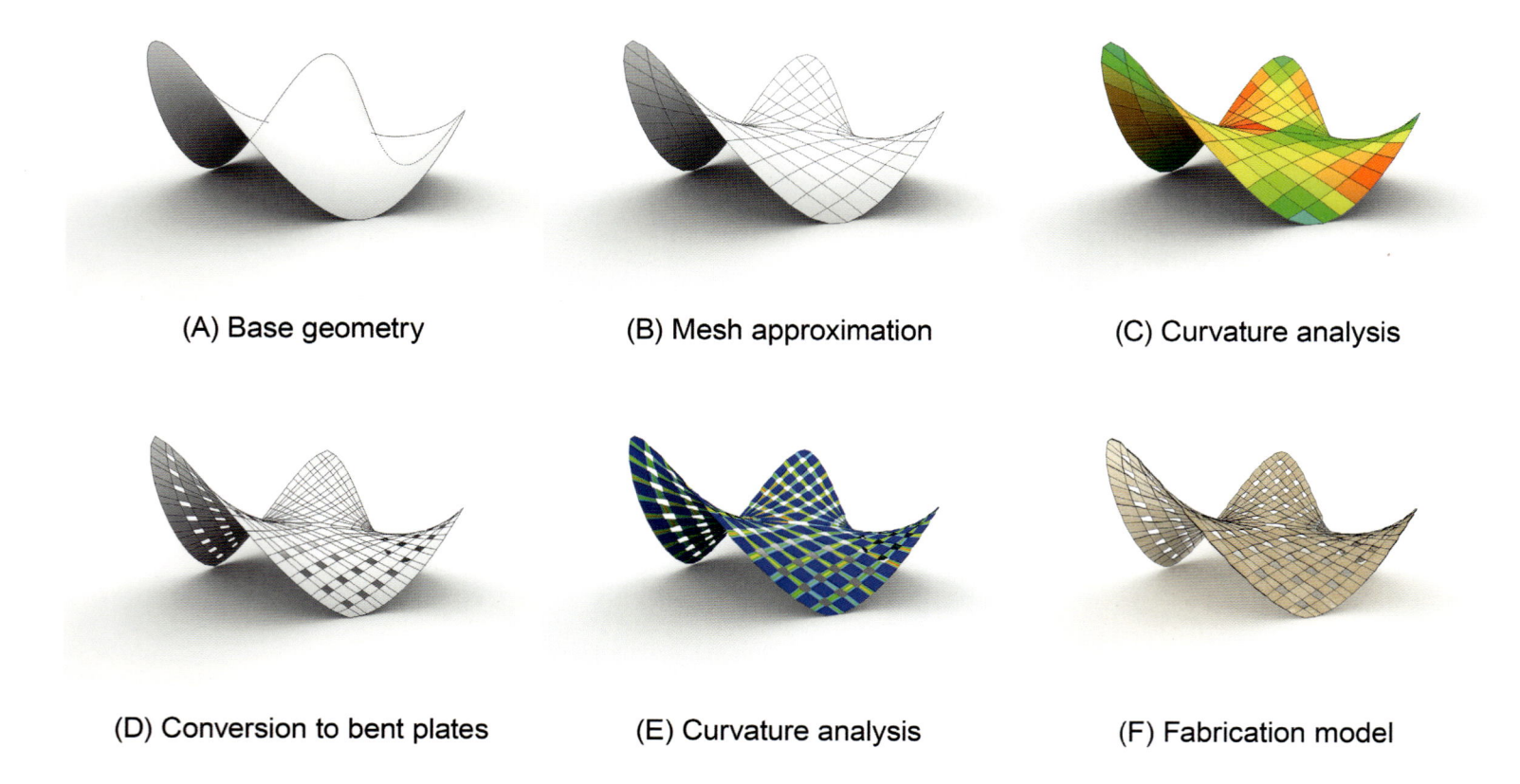

(A) Base geometry

(B) Mesh approximation

(C) Curvature analysis

(D) Conversion to bent plates

(E) Curvature analysis

(F) Fabrication model

capacity of a model under simultaneous consideration of external forces and internal material stresses. In response to these developments, architects and engineers are rediscovering a widespread interest in structural systems in which form and stress state cannot easily be predicted, but result from a delicate balance between geometry, interacting forces and material properties. In this context, bending-active structures are perfectly suited to illustrate the innovative potential that physics-based simulations can have on the design process. As a relatively new typology, bending-active structures are characterised by the clever integration of large elastic deformations of initially planar building materials in order to generate geometrically complex constructions (Knippers et al., 2011). While the conventional maxim in engineering is to limit the amount of bending, this structural system promotes the opposite approach and instead harnesses material flexibility for lightweight designs. This idea is as simple as it is versatile. It can be used, for example, as a form-giving and self-stabilising strategy in static structures, as suggested by Lienhard (2014), or be considered for the design of compliant mechanisms and kinetic structures, as shown by Schleicher (2015).

Bending-active structures can generally be divided into two main categories, which relate to the geometrical dimensions of their basic building blocks. For instance, one-dimensional systems can be built from slender rods, while two-dimensional systems employ thin plates. While extensive knowledge exists for 1D systems, with elastic gridshells as their most prominent application, plate-dominant structures have not yet received much attention and are considered more difficult to design. However, what makes this subset of bending-active structures particularly interesting is the fact that plates have a clear scale separation. They are typically very large in one dimension and progressively smaller in the other two. Their length is specified in metres, their width in centimetres and their height only in millimetres. This hierarchy makes it easier to assess the structural behaviour and accurately anticipate the plates' deformed geometry with digital simulations. Among the most prominent examples for bending-active plate structures are Buckminster Fuller's plydomes or the ICD/ITKE Research Pavilion 2010. While the first example follows a rational geometry-based approach in which the shape of a sphere is approximated with a regular tiling of identical plates (Fuller, 1959, Marks, 1973), the design of the second example integrates intensive structural simulations and takes advantage of computational mass customisation (Lienhard et al., 2012, Fleischmann et al., 2012).

3

Design space

The design space of bending-active plate structures is limited by material formability. The only shapes that can be achieved within stress limits are those that minimise the stretching of the material. For plate-like elements, these are reduced to developable surfaces: cylinders and cones. Attempting to bend a sheet of material in two directions will result in either irreversible plastic deformations or ultimately failure. Due to these constraints, designers mostly follow a bottom-up form-finding approach, which usually starts with planar sheets and recreates the bending process digitally (Lienhard et al., 2011). By using the method of ultra-elastic cables, as described by Lienhard et al. (2014), one can deform multiple plates and couple them to form complex structures in equilibrium. Depending on the simulation software used, this method can be very quick and interactive or particularly accurate and reliable regarding its results. The drawback of form-finding, however, is that the final shape and the caused stresses are often not known from the start. A designer with a certain shape in mind would therefore have to conduct multiple simulations with gradually changing parameters to approximate a target design (Schleicher et al., 2015).

1. *Berkeley Weave* installation at UC Berkeley's College of Environmental Design (CED). The ultra-thin bending-active shell is assembled out of 3mm birch plywood panels.

2. Form-conversion process and analysis of *Berkeley Weave*.

3. Architecture students at the College of Environmental Design (CED) assembling *Berkeley Weave*.

Images: Simon Schleicher.

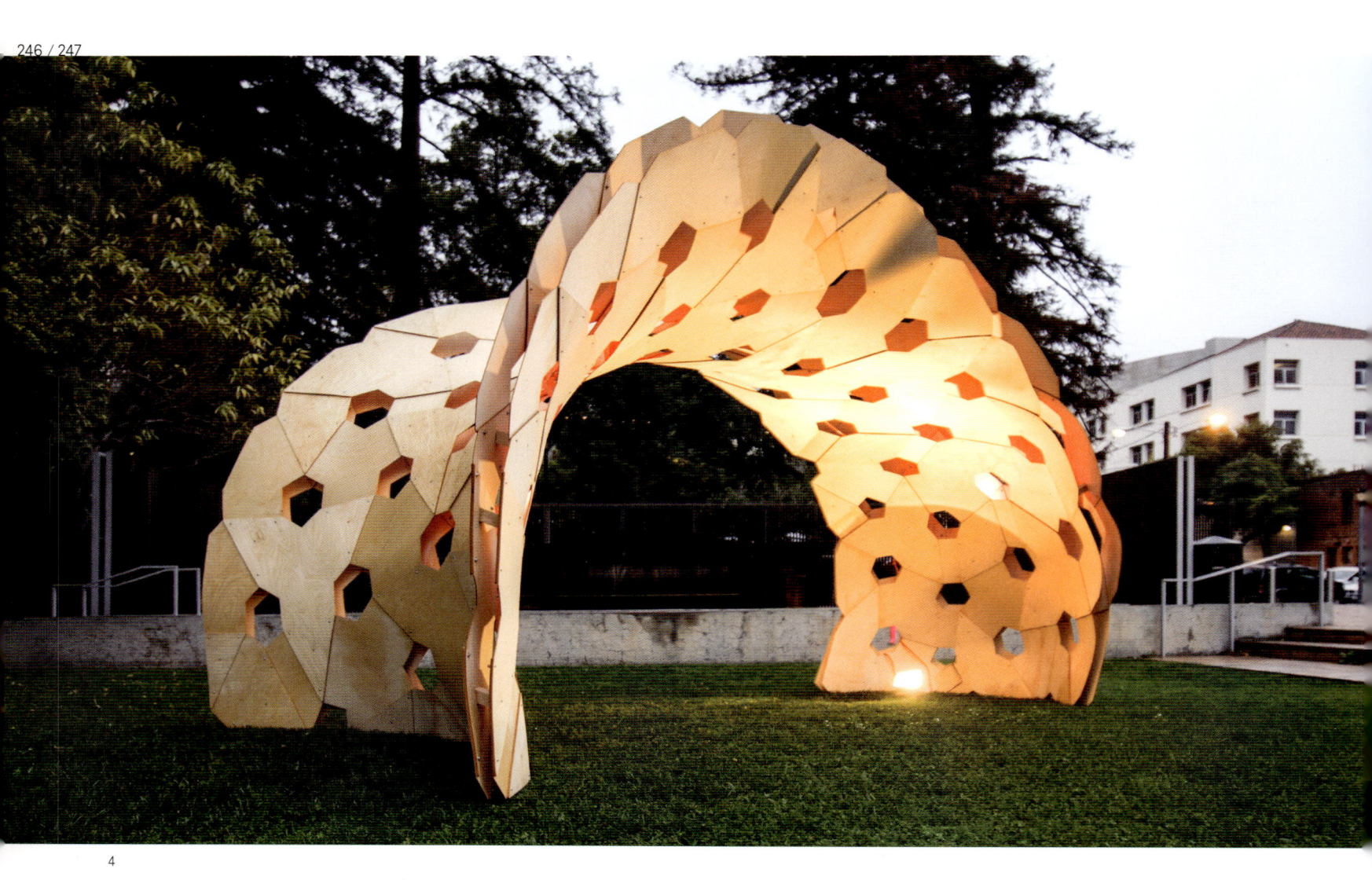

4

4. View of the *Bend9* structure assembled out of 3mm-thin birch plywood in the courtyard at UC Berkeley's College of Environmental Design (CED).

5. Assembly process of the *Bend9* pavilion at the courtyard of UC Berkeley's College of Environmental Design (CED).

Images: Simon Schleicher.

5

The constraints related to form-finding raise the burning question of whether a radically different approach could give the designer greater control over the final shape while at the same time guaranteeing that components are only bent within permissible limits. In an earlier publication, the authors introduced a different approach and coined for it the term 'form-conversion' (La Magna et al., 2016). Here, the design process is top-down and begins with a predefined target surface or mesh, which is then discretised further into smaller bent tiles based on the flexibility of the plate material used. Investigating this strategy further allows for the possibility of significantly expanding the feasible design space of bending-active structures.

Multi-directional bending

The key conceptual idea behind form-conversion is to overcome the obstacle that a plate can only be bent in one direction and will not easily be forced into double curvature without stretching or plastically deforming the material. To achieve multi-directional bending, one needs to remove material strategically and thereby free the plates from the stiffening constraints of their surroundings. As a result, one would get single-curved developable surfaces with no or very little Gaussian curvature. A similar approach was presented by Xing et al. (2011) for more complicated geometries. This principle can be integrated into the subdivision of virtually any freeform surface. In order to prove this point, the authors applied this method to the design of two exemplary case studies.

The first case study that follows a form-conversion approach is called *Berkeley Weave*. This project considers not only the effect of bending of slender strips but also their torsion. A saddle-shaped design, based on a modified Enneper surface (Fig. 2A), was chosen because of its challenging anticlastic geometry, with locally high Gaussian curvature. The subsequent conversion into a bending-active plate structure followed several steps. The first one was to approximate the surface with a quad mesh (Fig. 2B). A curvature analysis of the resulting mesh reveals that its individual faces are not planar but double-curved (Fig. 2C). The planarity of the quads, however, is an important precondition for the later assembly process. In a second step, the mesh was transformed into a four-layered weave pattern with composed strips that feature pre-drilled holes. Here, each quad was turned into a crossing of two strips in one direction, with two other strips at a 90° angle. The resulting interwoven mesh was then optimised for planarity. However, only the regions where strips overlapped were made planar, while the mesh faces between the intersections remained curved (Fig. 2D). A second curvature analysis illustrates the procedure and shows zero Gaussian curvature at the intersections of the strips, while the connecting faces are both bent and twisted (Fig. 2E).

Specific routines in the form-conversion process guaranteed that the bent zones stayed within the permissible bending radii. In the last step, this converted shape was used to generate a fabrication model that featured all the connection details and strip subdivisions (Fig. 2F). To allow for a proper connection, bolts were only placed in the planar regions between intersecting strips. Since the strips were composed of smaller segments, it was also important to control their position in the four-layered weave and the sequence of layers. A pattern was created which guaranteed that strip segments only ended in layers 2 and 3 and were clamped in between continuous strips in layers 1 and 4. A positive side effect of this weaving strategy was that the gaps between segments were never visible and the strips appeared to be made out of one piece. The resulting challenge, however, was that each segment needed a unique length and required individual positioning of the screw holes.

The second case study, called *Bend9*, showcases another take on form-conversion. This project is a multi-layered arch that spans over 5.2m and has a height of 3.5m. It was designed to prove the technical feasibility of using bending-active plates for larger load-bearing structures. In comparison to the previous case study, this project implements a different tiling pattern and explores the possibility of significantly increasing a shape's rigidity by cross-connecting distant layers with each other. To fully exploit the large deformations that plywood allows for, the thickness of the sheets had to be reduced to the minimum, leading to the radical choice of employing 3mm birch plywood. Since the resulting sheets were very flexible, additional stiffness needed to be gained by giving the global shell a peculiar geometry, which transitioned from an area of positive curvature to one of negative curvature. This pronounced double curvature provides additional stiffness and helps avoid undesirable deformation of the structure. Despite the considerable strength achieved by the shape alone, the choice of using extremely thin sheets of plywood at that scale necessitated additional reinforcement to provide further load resistance. These needs were met by a double-layered structure with two cross-connected shells. As in the previous example, the first step of the process was to convert the base geometry into a mesh pattern. In the next step, a preliminary analysis of the structure was conducted and informed the offsetting of the mesh to create a second layer. As the distance between the two layers varies to reflect the bending moment calculated from the preliminary analysis, the offset of the surfaces changes along the span of the arch. The offset reflects the stress state in the individual layers, and the distance between them increases in the critical areas to improve the global resistance of the system. The subsequent form-conversion process was once again driven by material constraints and by the permissible stress limits with respect to bending and torsion. The resulting tiling logic that was used for both layers affected the size of the members and guaranteed that each component could be bent into the specific shape required to construct the whole surface. More precisely, this was achieved by strategically placing voids into target positions of the master geometry, ensuring that the bending process could take place without prejudice for the individual components. Although initially flat, each element underwent multi-directional bending and was locked into position once it was fastened to its neighbours. The flexible 3mm plywood elements achieved consistent stiffness when jointed together, as the pavilion, although a discrete version of the initial shape, still retained substantial shell stiffness. This was validated in another finite element analysis that considered both self-weight and undesirable loading scenarios.

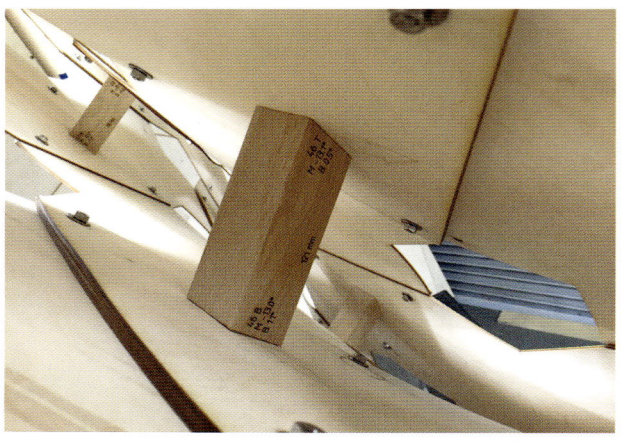

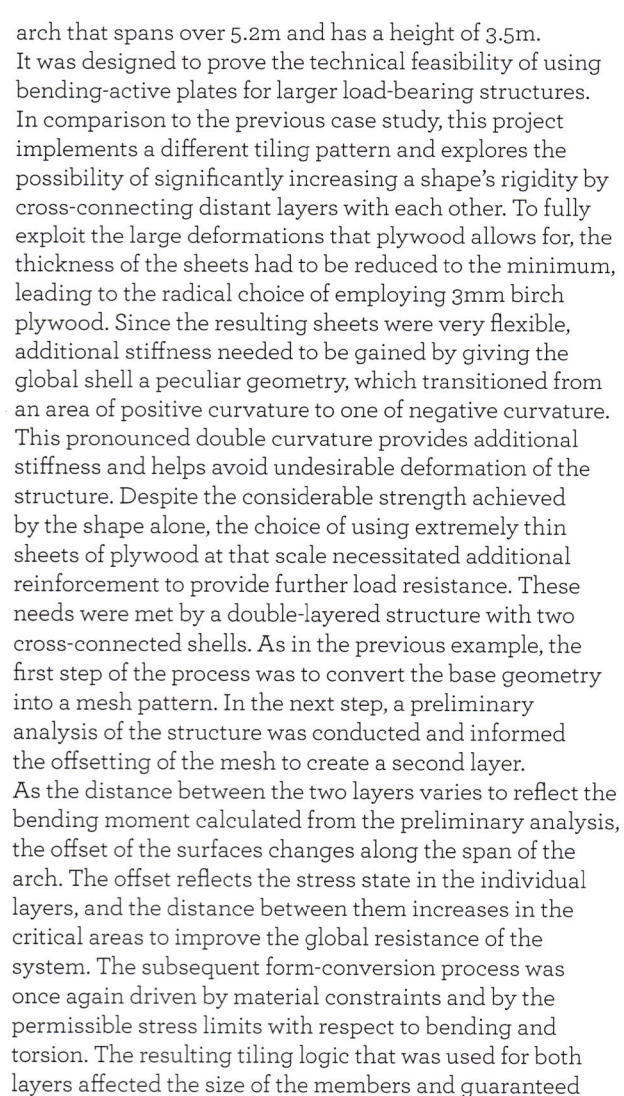

Prototyping

To evaluate these case studies and to demonstrate proof of concept, the authors referred to architectural prototyping. Constructing with the actual material is still one of the best ways to quickly validate assumptions, gain intuition about practical design issues and lay the foundations for future research.

The first case study was constructed in the dimensions of 4m x 3.5m x 1.8m and was exhibited at various occasions at UC Berkeley. The structure was assembled from 480 geometrically different plywood strips that were fastened together with 400 bolts. The material used was 3mm-thick birch plywood with a Young's modulus of $E_{mII} = 16,471 N/mm^2$ and $E_{m=\perp} 1,029 N/mm^2$. Dimensions and material specifications were employed for a finite element analysis using the software SOFiSTiK. In consideration of self-weight and stored elastic energy, the minimal bending radii in both the digital simulation and the built structure were no smaller than 0.25m and the resulting stress peaks were below 60 percent of the permissible

6. Left: detail of the elements. Right: detail of the connecting elements. Image: Riccardo La Magna.

material utilisation. The structure was assembled from the centre outwards, and during the construction process it was interesting to experience how the global stiffness increased the more elements were added and the more the structure was forced into its double-curved configuration (Fig. 3).

Similarly, the second case study was also constructed in the original scale and was shown at Autodesk's Pier 9 and at UC Berkeley. The built structure employed 196 elements unique in shape and geometry (Fig. 5). 76 square wood profiles of 4cm x 4cm were used to connect the two plywood skins (Fig. 6). Due to the varying distance between the layers, the connectors had a total of 156 exclusive compound mitres. The whole structure weighs only 160kg, a characteristic which also highlights the efficiency of the system and its potential for lightweight construction. The smooth curvature transition and the overall complexity of the shape clearly emphasise the potential of the construction logic. Furthermore, both implemented form-conversion processes can be applied to any kind of double-curved freeform surface, not only the ones presented here.

Feasibility for the future

The two case studies clearly illustrate the feasibility of form-conversion for the planning and construction of bending-active plate structures. Both structures are directly informed by the mechanical properties of the thin plywood sheets employed for the project. Their overall geometry is therefore the result of an accurate negotiation between the mechanical limits of the material and its deformation capabilities.

The assembly strategy devised for both prototypes drastically reduces fabrication complexity by resorting to exclusively planar components which make up the entire double-curved surfaces. Despite the large amount of individual geometries, the whole fabrication process was optimised by tightly nesting all the components to minimise material waste, flat-cut the elements and finally assemble the piece onsite. The very nature of the projects required a tight integration of design, simulation and assessment of the fabrication and assembly constraints. Overall, the *Bend9* pavilion and the *Berkeley Weave* installation exemplify the technical feasibility of a form-conversion process and showcase the capacity for bending-active surface structures to be employed as lightweight constructions. For ongoing research, the buildings serve as first prototypes for the further exploration of surface-like shell structures that derive their shape through elastic bending.

Acknowledgements

For the *Berkeley Weave* installation, the authors would like to thank Sean Ostro, Andrei Nejur and Rex Crabb for their support. The *Bend9* pavilion would not have been possible without the kind support of Autodesk's Pier 9 and its entire staff.

References

Richard, B.F., 1959, *Self-strutted geodesic plydome*, U.S. Patent 2,905,113.

Knippers, J., Cremers, J., Gabler, M. and Lienhard, J., 2011, *Construction Manual for Polymers + Membranes: Materials, Semi-finished Products, Form Finding, Design*, Walter de Gruyter.

La Magna, R., Schleicher, S. and Knippers, J., 2016, *Bending-Active Plates*.

Lienhard, J., Schleicher, S. and Knippers, J., 2011, 'Bending-active Structures – Research Pavilion ICD/ITKE' in *Proceedings of the International Symposium of the IABSE-IASS Symposium*.

Lienhard, J., La Magna, R. and Knippers, J., 2014, 'Form-finding Bending-active Structures with Temporary Ultra-elastic Contraction Elements' in *Mob Rapidly Assem Struct IV, 136*, p.107.

Lienhard, J., 2014, *Bending-active Structures: Form-finding Strategies using Elastic Deformation in Static and Kinetic Systems and the Structural Potentials therein*.

Marks, R.W. and Fuller, R.B., 1973, *Dymaxion World of Buckminster Fuller*, Anchor Books.

Schleicher, S., 2015, *Bio-inspired Compliant Mechanisms for Architectural Design: Transferring Bending and Folding Principles of Plant Leaves to Flexible Kinetic Structures*.

Schleicher, S., Rastetter, A., La Magna, R., Schönbrunner, A., Haberbosch, N. and Knippers, J., 2015, 'Form-Finding and Design Potentials of Bending-Active Plate Structures' in *Modelling Behaviour*, Springer International Publishing, p.53-63.

Xing, Q., Esquivel, G., Akleman, E., Chen, J. and Gross, J., August 2011, 'Band Decomposition of 2-manifold Meshes for Physical Construction of Large Structures' in *ACM SIGGRAPH 2011 Posters*, ACM, p.58.

PRECAST CONCRETE SHELLS
A STRUCTURAL CHALLENGE

STEFAN PETERS / ANDREAS TRUMMER / FELIX AMTSBERG / GERNOT PARMANN
Graz University of Technology

The primary focus of this research project is the fabrication and joining of thin-walled, double-curved prefabricated concrete elements. By using a process-based approach, many different research questions were combined into one interdisciplinary project. The material technological aspects led to the search for interesting architectural uses for ultra-high performance concrete (UHPC). It is extremely well-suited for thin-walled, double-curved prefabricated elements; however, use in highly efficient structures is only imaginable once appropriate joining methods have been developed.

Modern possibilities in digital design and manufacturing in combination with industrial robots raise questions about alternative shaping methods that could achieve a higher quality and efficiency in the production process of structural elements. These fundamental ideas were investigated over a period of three years by a team of architects, civil engineers, material engineers and mechatronic engineers. The result was a production process that covered every step from the first design ideas all the way to the final product.

In the last hundred years, concrete has greatly influenced building culture worldwide. Today it is one of the most used consumer goods. UHPC is unique due to its quasi-non-porous structure and its high compressive strength, which ranges up to $200N/mm^2$. Due to its material properties, it is ideal for use in light structures and structures which span large distances. The sophisticated processing of the raw materials into UHPC is very similar to that of standard precast concrete production. Joining precast elements using mechanical screwing systems and press-fitting the contact surfaces is extremely effective when compared to conventional methods of filling the joints with in-situ concrete. It also creates a new and different feeling for concrete structures.

The heyday of the concrete shell structure is long gone. However, it is just as relevant today that a structure which is engineered efficiently can transfer loads mainly as membrane forces. This, in turn, means that slender elements can be produced and material utilisation optimised. This is not the case with standard flexural concrete elements, and most concrete elements that are designed are flexural elements. The historical decline in

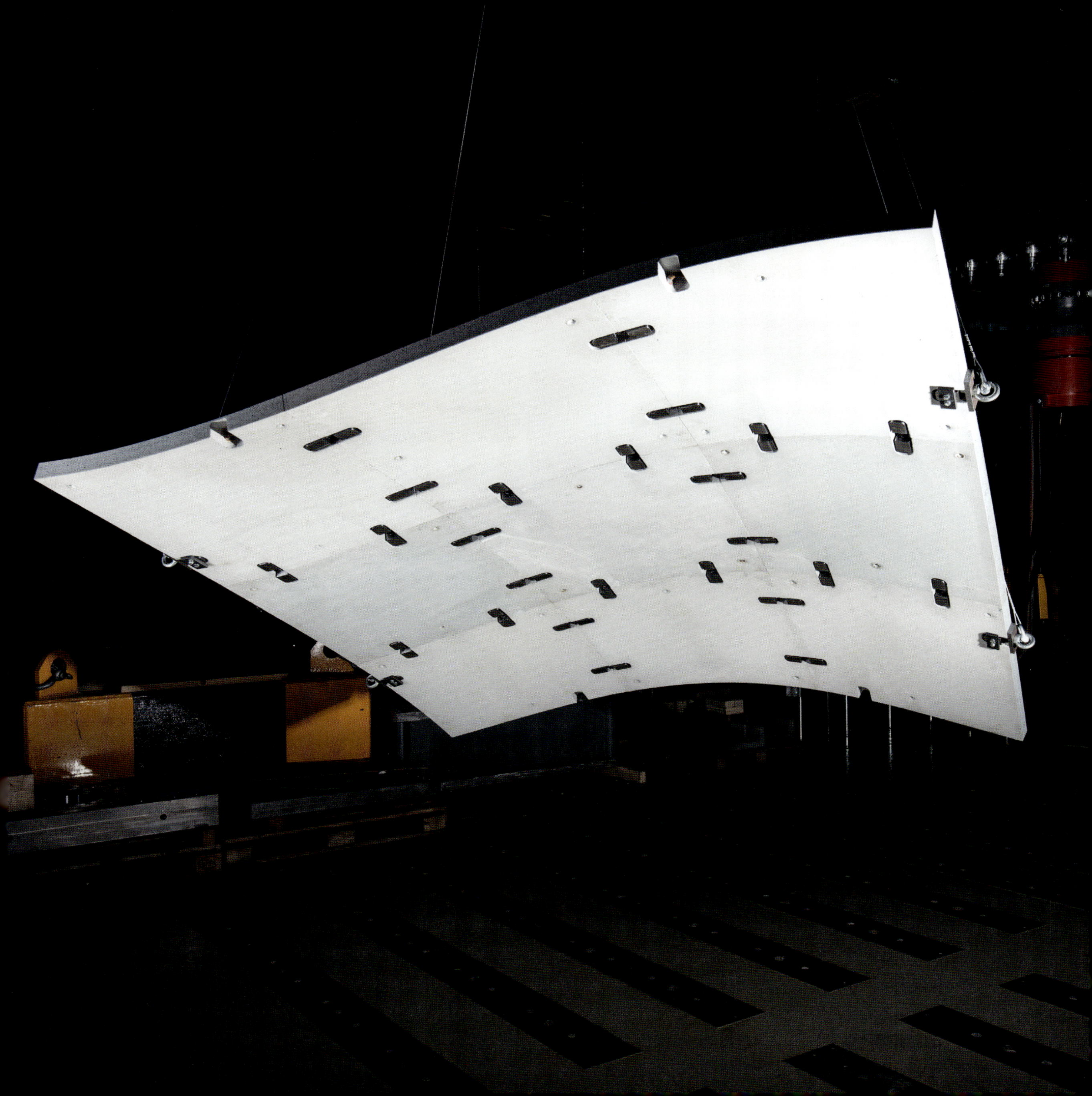

concrete shells is usually blamed on the large costs associated with their production. This is illustrated by the fact that the largest portion of Felix Candela's structures was built in the 1950s-60s, and the increase in the Mexican minimum wage was responsible for the end of this boom. The high costs involved in producing complex, time-consuming formwork compared to the costs of the cheap materials used to produce concrete are obviously unfavourable[1]. After the first patent application from Wallace Neff, a new branch of research was born. This concentrated on principles of pneumatic formwork – the most well-known of these being the BiniShells[2]. Pier Luigi Nervi (1891–1979) suggested an alternative. After he founded his building company in 1920, he developed a building system based on semi-precast panels which were supported by a falsework and finally finished with in-situ concrete[3]. These three examples show that examining the building method is the key to a re-evaluation of concrete shell production. These three different approaches show that there is a connection between the history of shell constructions and the search for an efficient production method.

Research context

Because of the rapid developments within architecture in digital fabrication and robotic production, questions regarding the efficient production of double-curved elements are increasingly in the spotlight. Numerous questions have been posed, with solutions and strategies varying considerably. One such project was TailorCrete. This involved pouring the concrete onsite into a milled foam formwork. A part of the project introduced a variable moulding table based on adjustable pixels and an elastomer mat. The formwork was then created using wax and the parts were cast conventionally[4]. The steel rebars were then bent and welded automatically using robots[5]. A similar method was also developed by the ADAPA, which made it possible to create double-curved shell elements. This method was also based on a flexible membrane, which was shaped using adjustable pins[6]. The PhD thesis 'Double-Curved Precast Concrete Elements' presented a variable moulding table and a complementary concrete mix whereby the shape was adjusted after the initial setting time. This meant that no countering formwork was necessary. This project concentrated on the properties of the concrete[7]. The problem of joining precast concrete parts is usually solved, in the same manner as in Nervi's structures, by pouring concrete into the joints onsite. The project 'Lokale Lasteinleitung... mit Implantaten in Bauteilen aus ultra-hochfestem Beton' proposed a steel connector for thin concrete elements. These are suitable for tension, compression and shear forces[8].

Digital prefabrication for concrete shells

Taking current research aspects into account, the following goals were defined. Concrete shell structures should not be cast individually using large, complex, onsite form- or falsework, but constructed by joining elements that have been accurately prefabricated. This requires the double-curved surfaces to be divided into a number of individual elements. The dimensions of these elements are based on the boundary conditions of the laboratory where they are produced, as well as the possibilities for transportation.

If it can be assumed that the structure is a freeform one without any type of symmetry, a large number of irregular elements will be produced and few, if any, of them will be identical. As soon as the formwork cannot be produced using flat panels, the question of alternative production methods is even more relevant. The structuring of this question was based on the production chain, from the first concepts through to the final joining of the elements. The main aim was to design a flexible formwork which could be controlled by a robot and would be robust enough to survive in a prefab concrete factory. The requirements of the concrete element, including the carbon fibre reinforcement grids and steel fibres, called for the expertise of concrete technologists.

It was also necessary to consider alternative joining techniques for these slender prefabricated ultra-high performance concrete elements. The conventional joining method, such as that used by Nervi, involving filling the

1. Final mock-up and steel connectors.

2. Pixel field prior to moulding.

3. Completed half-moulds before closing.

3

4

4. Finished joint surfaces.

5. Evaluation of the deviation from digital model in one of the fabricated elements.

joints with in-situ concrete, would not do the aesthetics or the material properties of the UHPC justice. Much can be taken from the methods of historical stonecutters, who built vaults where the forces were transferred through the contact surfaces. Compared to these historical vaults, however, stability was not provided by the element being thick or extremely heavy (which is advantageous for a press-fit). Instead, this was replaced by a mechanical press-fit on the contact surfaces, which is common in concrete construction. This method requires the contact surfaces to be extremely precise and to have a high-grade finish. Due to the requirements of the contact joint and the precision involved, it was necessary to document the deviation from the planned geometry constantly. This method resulted in continual measurements, as well as suggestions for sensor-controlled iterative processing cycles for both the settings of the variable moulding table and the grinding of the joint surfaces.

Process-based design

A fictive hall construction, a sort of case study, was therefore devised on which the research approaches for design and implementation could be tested. The questions of how the surface should be divided and the size of the elements were investigated by using the freeformed, wave-like roof structure of the fictitious hall. All the building elements that were analysed and all the information derived were related to parts of this construction. This information was also linked to the digital model of the structure. The aim of this project was not to optimise the model for specific external forces. The goal was rather to calibrate and define the boundary conditions of maximum curvature for the moulding table. This meant that both the design process and the production process could be developed for one specific exemplary design. Setting limits for the manufacturing processes for larger and smaller objects followed in other projects. By using parametric construction tools, it was possible to keep the information consistent for all members of the team, in every phase of the project. The extremely clear separation and focus of the development of the joining system and the moulding table made it possible to work independently with clearly defined interfaces.

Conventionally, the formwork for casting double-curved concrete elements is milled from extruded polystyrene, painted and then sanded. The disadvantage of this method is that every element needs two forms which are then no longer required and have to be discarded after a single use.

By evaluating the results obtained from experiences in other projects, not only does this method consume large amounts of resources but it is also not very economical. For example, to achieve a fair-faced concrete, long milling times are necessary and therefore the cost of manufacture increases dramatically. This is why a variable moulding table was favoured in this project, making it possible to produce different double-curved surfaces simply. From the very beginning, one of the primary goals was to create a simple, robust tool which had a long life expectancy and was appropriate for use in a precast concrete factory environment without breaking.

Two different moulding tables were investigated: a so-called pin field and a so-called pixel field. Both of these can be controlled or adjusted by an external industrial robot. The robot can be used for other parts of the production process, as it is separate and not fixed to the moulding tables. The pin field has a formable surface connected to joint-mounted heads. These heads are connected to the pins, which are evenly distributed across an orthogonal field. The double curvature is then produced by moving the pins along their longitudinal axis and deforming the surface.

On the other hand, the pixel field is made up of a number of plastic rods, each with a square cross-section, which can be slid along their longitudinal axis (Fig. 2). In this case, the industrial robot pushes the plastic rods into the correct position for the final precast concrete element shape. The pixels are then fixed and used as the basis for the elastic mat. When considering the fastest reuse of the pixel field, as well as the separation of the concrete casting process from the moulding table, it became clear that an additional step was necessary: taking a negative form made of quartz sand. Here, a layer of bonded sand was put on the elastomer mat and compacted, as is usual in casting techniques. This has many advantages: the sand adopts its shape quickly and therefore only needs to be on the pixel field briefly. The quality of the surface is also very high. According to what is known today, there is hope that with this bonded sand a formwork material has been found which expands the possibilities for fair-faced concrete formwork (Fig. 3). A UHPC concrete with steel fibres from Dykerhoff was used, with Nanodur Compound 5941 binding material. This was combined with two layers of carbon fibre grid mats. In this project, spacers were developed which could be clamped between the two sand forms. They held the carbon fibre reinforcement mats (CFRP) 5mm away from the surface as precisely as possible. After the concrete was poured and set, the edges were ground in a wet state. The connecting edges are extremely complex. They are spatially curved, stripe-like surfaces. An essential requirement for this step is the system of three points which are always in the same position relative to each other, which are integrated into the panels. These points are the interface between the reference points in the CAD/CAM files and the real plate. This makes it possible for the plate and edges to be spatially positioned correctly over and over again. At the momentary stage of development, it is necessary to remove 5-10mm from each joint surface. Approximately 1mm can be removed in each processing stage when using a water-cooled, diamond-tipped grinding bit (Fig. 4).

The joints of the nine plates were press-fitted using the specially developed screw connection. The bent rebars, which were anchored into the cross-section of the concrete plate, transferred the tension forces from pre-tensioned screw connections into the concrete and the contact surfaces were then pressed together[9] (Fig. 1). The calculation of the reinforcement and the design of the screw connectors were carried out using finite element software.

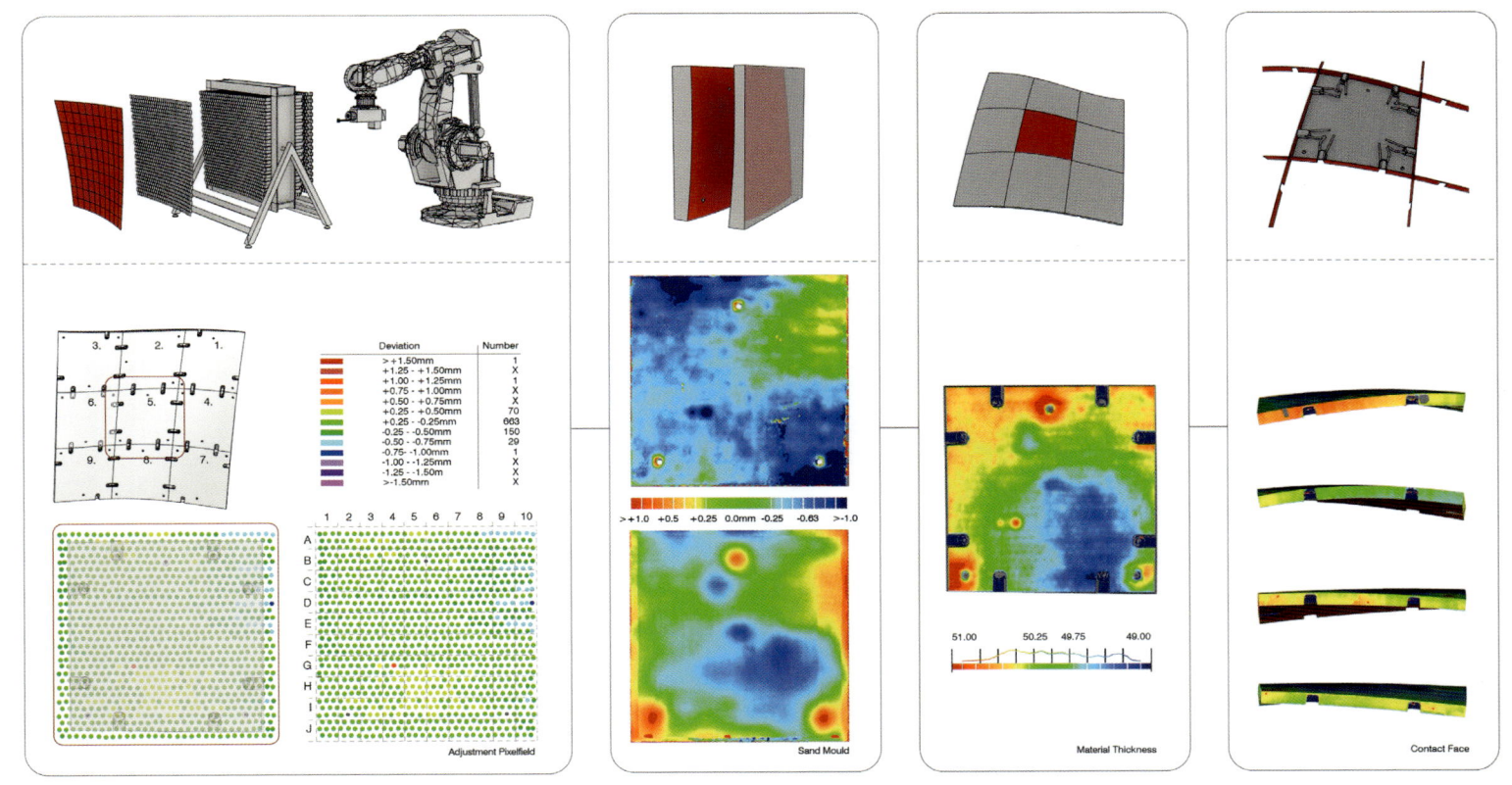

6

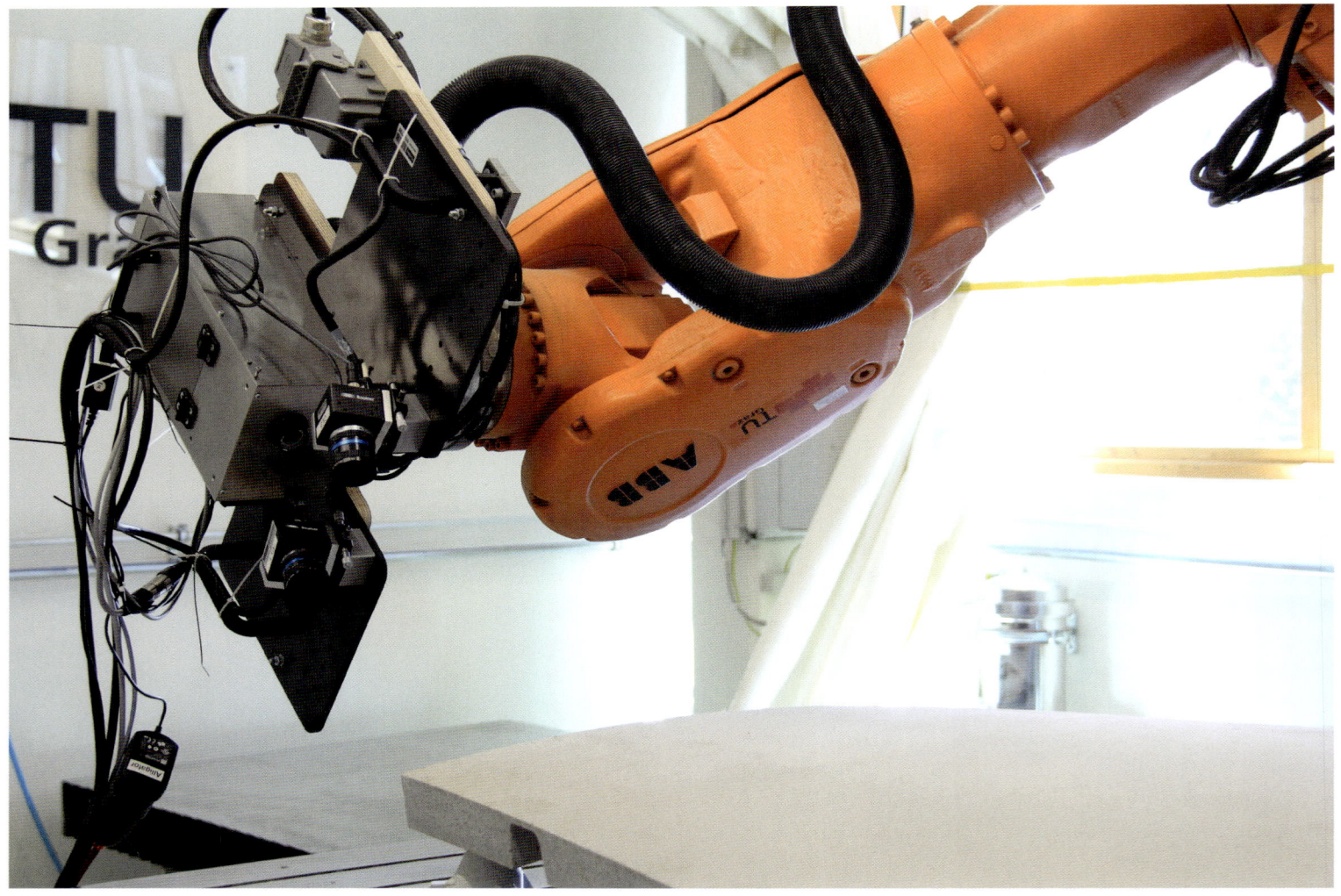

Sensor-based evaluation

An important step in the development of the manufacturing process was the observation of the different manufacturing steps. The digital workflow process enables safe and accurate production. It is, however, interrupted by several intermediary steps. Firstly, this means that the two steps which are carried out by the industrial robot are at the beginning and the end of the production process. Secondly, the digital processes themselves can also deviate from the desired output. This deviation can also go beyond the defined accuracy of the industrial robot.

The industrial robot that is used for both the adjustment of the pixels in the pixel field and the grinding of the contact surfaces can carry out production steps with an accuracy of ±0.25mm. To determine the cause of the size and shape deviations of the surfaces, every single step was recorded using measurement technology and checked: from the production and the robotic adjustment of the moulding table to creating the sand mould, all the way up to the final grinding of the joining surface (Fig. 5). The evaluation of the information showed that there were two possible reasons for the deviations, both of which can be controlled and automated using sensory technology.

1. **Adjusting the pixels**
   Setting up the moulding table by adjusting the pixels worked well. After the moulding table had been correctly adjusted, 80% of the pixels were within ±0.25mm of the planned position. The tolerance of ±0.50mm was only exceeded by pixels around the edge. The source of these larger displacements is that

6. Highly precise robot-driven optical sensor system.

some pixels move their neighbouring pixel with them even though they have already been adjusted. To be able to detect these discrepancies automatically and put the pixels back in their correct position, the tip of the robot tool was coupled with a prototype sensor. An extra routine was also added to the adjustment script, which checked the position of the neighbouring pixel after adjustment to make sure that it had not been unintentionally moved. If so, it was also readjusted.

2. **Tool/component interaction**
   Formatting the concrete panels using wet-state grinding is very dependent on the tool/component interaction. The combination of UHPC and steel fibres leads to wear on the tool. Within just one processing stage where the plate is reduced to an acceptable size within the tolerance range, the tool experiences significant wear. The wear on the tool is also dependent on the amount and direction of the steel fibres and therefore it cannot be estimated beforehand. A high-precision measuring device was installed on the robot, which checked the results after the processing step and decided if further processing steps should be carried out to correct discrepancies (Fig. 6).

The tool mentioned above for pixel adjustment makes it possible to control large numbers of pixels easily. During the pixel adjustment, the decision as to whether to proceed or go back and readjust – and the iterative process of grinding, measuring and regrinding – are not particularly typical manufacturing cycles, but they could help to develop new production concepts in the fields of civil engineering and architecture.

Collaboration between experts

Because the project introduced here was extremely broad, it was necessary for a number of different experts to work together on it. A process-oriented approach and the exemplary processing of the linked case study showed the method to be successful. By including digital manufacturing methods and robotic technology, it led to a usable, variable moulding table for flexible shapes. New standards were set for high surface quality and formability by using sand with a binding agent as a formwork. The quality of the grinding using an industry robot makes it possible for small factories to produce precise, prefabricated concrete elements. The newly developed joining system makes installing prefabricated concrete comparable to glass construction. This method shows great economic potential that validates it for future use. A practical case study, which the company Max Bögl is presently carrying out, should show that this method can be used for large format, ultra-thin prefabricated concrete elements. The case study is a slender roof construction made from four 10m-long, 2m-wide and 6cm-thick double-curved prefabricated concrete elements. This should also show that the technical innovations described will also find their way into the construction industry. Being able to build light constructions out of concrete and reduce the amount of formwork will be the key to success.

Notes

1. Michel, M. and Knaack, U., 2014, 'Grundlagen zur Entwicklung adaptiver Schalungssysteme für frei geformte Betonschalen und Wände' in *Bautechnik*, 91(12), p.845-853.

2. Sobek, W., 1987, 'Auf pneumatisch gestützten Schalungen hergestellte Betonschalen', Stuttgart, p.9.

3. Pier Luigi Nervi, cited in Herrmann Ruhle, 'Wie wurden Schalen gebaut? Ein erlebter Ruckblick', in *Arcus* 18 (1992), p.32-49, esp. p.42.

4. Willmann, J., Kohler, M. and Gramazio, F., 2013, 'TailorCrete', in *The Robotic Touch: How Robots Change Architecture*, Park Books, p.216-223.

5. Cortsen, J., Oesterle, S., Sølvason, D. and Stehling, H., 2013, 'From Digital Design to Automated Production' in *Rob | Arch 2012*, Springer Vienna, p.149-154.

6. Adapa, 'Adaptive Moulds', 2016, available at http://adapa.dk/products/adaptive-moulds (accessed 2 October 2016).

7. Schipper, H.R., 2015, 'Double-curved Precast Concrete Elements: Research into Technical Viability of the Flexible Mould Method', diss. TU Delft, Delft University of Technology.

8. Sobek, W. and Mittelstädt, J., 2014, 'Introduction of Compressive, Tensile and Shear Forces into Elements made of Ultra-High Performance Concrete by the Use of Implant', in Schmidt, S. (ed.), *Nachhaltiges Bauen mit ultra-hochfestem Beton: Ergebnisse des Schwerpunktprogrammes 1182: Sustainable Building with Ultra-high Performance Concrete*, Kassel University Press, p.643-659.

9. Santner, G., 2016, 'Fügetechnik im UHPC', Schalenbau, Graz University of Technology (90000).

FROM LAMINATION TO ASSEMBLY
MODELLING THE SEINE MUSICALE

HANNO STEHLING / FABIAN SCHEURER
Design-to-Production
JEAN ROULIER
Lignocam SA
HÉLORI GEGLO / MATHIAS HOFMANN
Hess Timber

The Seine Musicale by Shigeru Ban (formerly known as Cité Musicale) is envisioned as the flagship project for the urban renewal attempt of the Île Seguin in the west of Paris. Built in place of a former Renault manufacturing plant, the complex will host various concert and rehearsal spaces. The egg-shaped auditorium features a doubly-curved timber structure consisting of 1,300 individual glue-laminated and CNC-machined beam segments, as well as a secondary structure formed by 3,300 individual timber pieces supporting the hexagonal and triangular façade elements.

For fabrication and assembly of both timber structures, a fully parametric 3D CAD model was implemented, detailed down to the last screw and containing both the raw and final geometries of all timber elements. This model was the central node in the digital planning process. It was the origin of fabrication data for lamination and CNC milling of all timber pieces, acted as the basis for structural calculations and was used to simulate assembly situations throughout the whole structure.

This paper gives an overview of the digital planning and fabrication process of the primary timber structure of the Seine Musicale. The second part describes how Woodpecker, the timber fabrication plug-in for the parametric modelling environment Grasshopper, was further developed in this context.

Topology and detailing

The primary timber structure is a hexagonal grid consisting of 15 horizontal rings and 86 diagonals running around the egg-shaped building. Structurally, the rings are formed by up to 24m-long segments (Fig. 2), acting as tension or compression rings in the lower or upper building parts respectively. The diagonals are segmented into shorter pieces of 4-5m in length (Fig. 4), always spanning from one ring to the next. The whole structure rests on supports at the lowermost and uppermost rings with no additional support points in between.

In terms of detailing, there was a requirement by the architects to use as little steel as possible within the

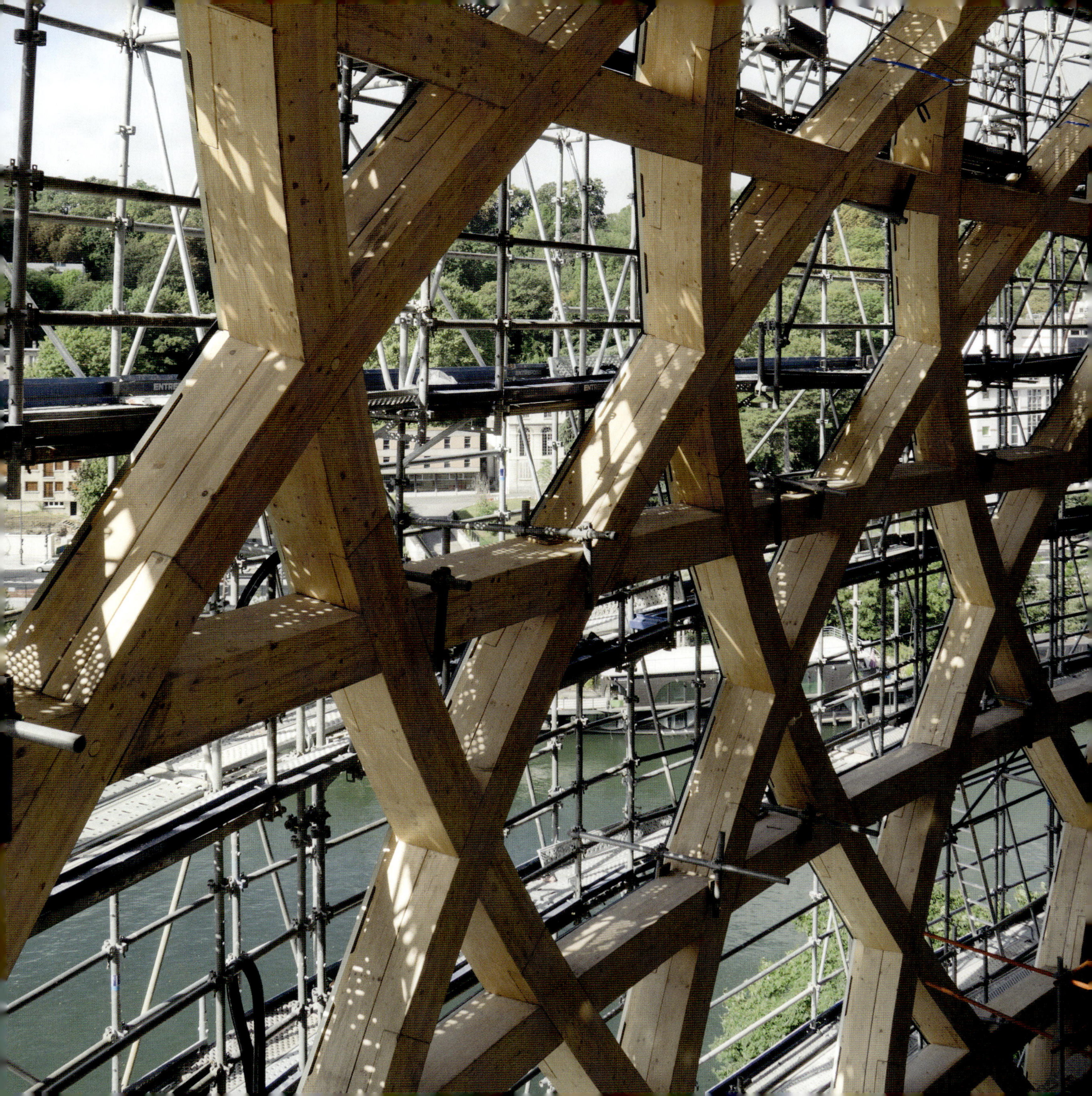

timber structure. All the cross joints, as well as the longitudinal joints of the compression rings, were designed as lap joints, which is a traditional timber detail. Screws are taking lateral forces and beech dowels assure precise positioning. The ring/diagonal crossings also act as longitudinal joints for the diagonals. For the longitudinal joints of the tension rings, a splice joint was developed, featuring toothed inlays CNC-cut from beech plywood (Fig. 3).

Describing the structural properties of these details in depth would exceed the scope of this paper. However, for freeform projects, the purely geometric properties are equally important, namely to ensure the assemblability of all pieces (see F. Scheurer, H. Stehling, F. Tschümperlin, 2013, 'Design for Assembly – Digital Prefabrication of Complex Timber Structures', *Beyond the Limits of Man, Proceedings of the IASS 2013 Symposium*).

Assembly

Traditional lap joints have only one degree of freedom, meaning that there is exactly one possible assembly direction ('from above' in respect to the joint plane). With curved beam segments spanning over multiple crossings, many lap joints with different directions have to be engaged at the same time, blocking assembly altogether. This problem has to be solved in every freeform project, with solutions highly dependent on the respective geometric properties.

In case of the Seine Musicale, assembly was solved by slightly skewing the lap joint side faces depending on individual assembly directions for every beam segment.

The diagonal segments were pre-assembled into X-shaped elements. Onsite, these elements had to be mounted by engaging two lap joints at the same time, leading to a pairwise assembly direction for these joints.

For the rings, the assembly was defined as a circular movement rather than a linear translation. With this concept, the four to eleven lap joints of each segment could be engaged one after the other, rather than all at the same time.

Notably, the 'toothed splice joint' helped a lot in easing assembly, as it features a wide range of possible assembly directions. This is in contrast to a more conventional connection with slots, steel plates and steel dowels, which would have limited assembly direction to the plane of the slots/plates.

Assembly of every single segment was simulated in the 3D CAD model in order to detect and solve collisions and other issues blocking assembly.

Lamination

Beam segments for structures like the one discussed are usually CNC-milled from a mixture of straight, single-curved and double-curved glue-laminated timber blanks. The decision of which type of blank to use is a trade-off between structural strength, material cut-off and lamination costs.

For the Seine Musicale, a special constraint for the primary structure was that all timber beams be fabricated with the timber fibres exactly following the final geometry, in order

1. The timber structure viewed from the inside.

2. Ring beam segments in different stages of pre-assembly.

3. The longitudinal tension joint features toothed beech plywood inlays instead of steel plates. Next to the structural properties, the main advantage over more conventional details is its great freedom in terms of assembly direction: the joint can be engaged within the opening angle of the teeth along the beam (alpha), from vertical to horizontal crosswise (beta).

4. Diagonal segments during finishing after milling.

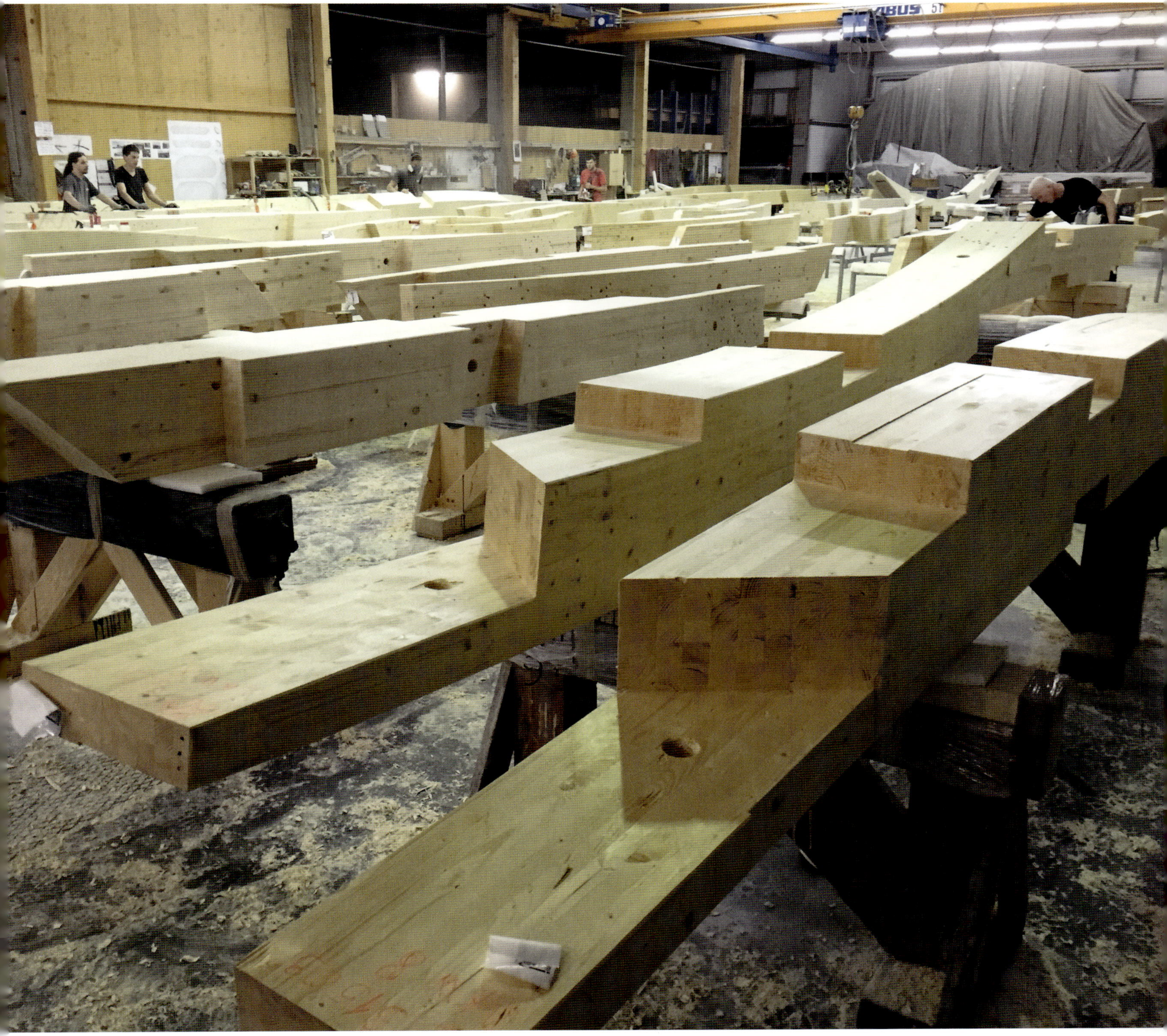

to reach a flawless appearance without any visibly cut glue seams. As the final geometry of all pieces was double-curved, this meant that all glue-laminated blanks had to be double-curved, too.

More than 1,200 pieces were laminated from stick lamellas with a cross-section of only 32 x 40mm. Thus the typical piece consisted of about 110 lamellas which had to be precisely placed in the press bed, which itself had to be adjusted to the desired shape of every single piece. To streamline this process, a simulation of the press bed was implemented in the parametric 3D CAD model, permitting export of data sheets and drawings for press settings and quality control. Due to the number of pieces, two different kinds of press beds were used, requiring different variants of setting data.

For some of the longest ring segments, lamination from stick lamellas was not feasible. Instead, these were produced with a more conventional two-step approach: straight planks are laminated into a single-curved beam on a conventional large-scale press bed for single curvature. The beam is then cut into strips crosswise to the lamination direction, resulting in single-curved plank lamellas. A second single-curved lamination process then yields a double-curved result.

To ensure precise placement of the up to 24m-long beam segments in the CNC milling machine, despite lacking any planar face as reference, positioning points were defined in the 3D CAD model and exported along with the lamination data. These points were defined based on press bed positions and thus could be marked on the pieces during lamination. As data for CNC milling were later generated from the same model, the positioning points could be referenced again and related to physical support points in the CNC milling machine.

This process allowed for the minimum blank oversize to be no more than 10mm per side, which was necessary to meet the criterion of not cutting through the first lamella during CNC milling. In addition to the aesthetic quality, the small oversize helped to save material, which in turn sped up both the lamination and milling processes.

CNC milling

The interface from the CAD model to the CNC machine is the critical point in any digital fabrication process (see H. Stehling, F. Scheurer, J. Roulier, 2014. 'Bridging the Gap from CAD to CAM', *FABRICATE - Proceedings of the International Conference.* Zurich: gta Verlag). While parametric modelling enables the definition of thousands of individual components through the same set of rules, in fabrication every piece becomes a physical instance which has to be laminated, machined, post-processed, transported and finally assembled. In conventional processes, this is mirrored on the software side, where every piece is individually prepared for CNC milling based on a CAD model showing the desired result in full detail. To streamline this process, a set of BTL (Building Transfer Language, see www.design2machine.com/btl) files was exported for every piece. Described in more detail in the aforementioned *FABRICATE 2014* paper, BTL allows the definition of fabrication operations based on geometry, not machine features. So BTL does not remove the individual machining preparation of every single piece, but brings it to a level where already defined operations can be batch processed instead of trying to define operations based on a piece of volumetric geometry.

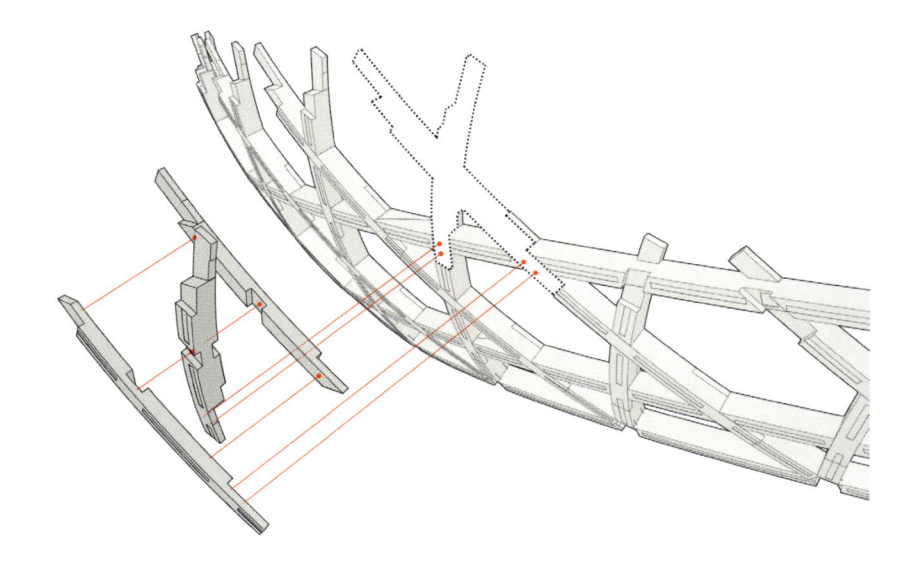

This process can be described as optimal in terms of quality control, as the systematic layout and parametric origin of the BTL data prevent individual mistakes during machining preparation, while every piece is still looked at by an experienced operator, who spots possible problems in exceptional geometric situations that might otherwise have been overlooked.

Especially for complex details like the beech-toothed splice joint, close collaboration between the parametric

5

5. Assembly concept for the diagonals. To facilitate the central crossing, one of the legs forming the X has to be subdivided into two layers.

6. The erected timber structure viewed from the outside. The secondary structure forming the transition to the (not yet mounted) façade elements can be seen on top of the main beam segments.

modeller (knowing all the details and their geometric range) and the fabrication operator (knowing the machine and its capabilities) is necessary in order to ensure an efficient process in fabrication preparation. In the case of the Seine Musicale, several test iterations were run until a satisfying data set was achieved, and improvements to the BTL layout were made even after production had already started.

Conclusion

For the timber structure of the Seine Musicale, a highly integrated digital fabrication process incorporating lamination and CNC milling has been set up. High demands in terms of aesthetics and detailing lead to innovation in the fields of lamination and connection details. Assemblability is one of the key aspects (if not *the* key aspect) in freeform projects and has to be taken into account as early as possible. The interface into fabrication can rely on established exchange formats and processes, but has to be further developed to meet the specific needs of each project. Optimally, a balance between automatisation and manual control is found.

Erection of the timber structure of the Seine Musicale finished in summer 2016. At the time of writing, the façade is being installed. The scheduled opening date of the building is April 2017.

Addendum

Keeping it state-of-the-art – update on Woodpecker, the timber CAD/CAM interface for Grasshopper

The BTL has proven itself as a very suitable CAD/CAM interface format in many freeform timber projects, such as the D1 Tower Canopies (Innovarchi, Dubai 2015), the French Pavilion at the Expo in Milan (X-Tu, Milan 2015) or the 'Haus des Brotes' ('House of Bread') (Coop Himmelb(l)au, Asten 2016).

Originally a side-product of project-specific implementations, a BTL export plug-in for Grasshopper was released by the authors in 2014 (see www.food4rhino.com/project/woodpecker). This plug-in features the most generic operations and allows the generation of BTL files including 5-axis contours directly from Grasshopper. Since then, development has focused on projects such as the ones mentioned above, which were notably not done in Grasshopper but used the same BTL export code. In spring 2017, the first major update is being released as Woodpecker Version 2. As well as supporting a wider variety of BTL operations and a series of bug fixes and other improvements, the plug-in will allow the export of BTLX. BTLX is XML-based and is meant to be a successor to the aged ASCII-based BTL format. Version 1.0 was released in 2015 and is gradually being adopted.

Woodpecker remains free for educational purposes.

SCALING ARCHITECTURAL ROBOTICS CONSTRUCTION OF THE KIRK KAPITAL HEADQUARTERS

ASBJØRN SØNDERGAARD / JELLE FERINGA
Odico Formwork Robotics

At FABRICATE 2011, the authors of this article encountered two new research trajectories (Dombernowsky, 2011, Verde, 2011), on, respectively, the design of topologically optimised concrete structures and hot-wire-cutting of expanded polystyrene (EPS) construction elements. Over lunch, the potential for a synthesis was gauged. In the years that followed, the intense collaboration that ensued resulted in a number of projects and articles (McGee, 2013, Feringa, 2014, Søndergaard, 2016). The industrial merit of the approaches explored paved the way to further develop these at an industrial scale, leading to the founding of Odico Formwork Robotics in the spring of 2012 (Søndergaard, 2014). At Odico, the challenges faced when deploying and building with robotics at scale are addressed. Over the years, a range of novel fabrication processes have been developed in an industrial context.

Are quantity and quality mutually inclusive?

Automation is often discussed in the framework of efficiency – of increasing productivity at lower labour costs. This is to say that robotics is discussed in a quantitative framework, rather than a qualitative one. The potential quality that robotics has to offer the building industry is central to its further development. Architectural robotics has been enthusiastically embraced by the design-led research community, exploring specific traits of machining processes for their intrinsic or tectonic potential. The cultivation of new manufacturing aesthetics, precipitated by the new degrees of freedom and material control offered by digital machining, has been a central motif over the past decade. Performance is rarely addressed, especially in direct quantitative terms.

So far, the literature lacks an accepted methodology and criteria to assess and contrast the relative merits of various existing technologies. Within internal technology research and development at Odico, quantity and quality represent the axes on which the merits of methods are plotted. The following criteria serve as guidelines to gauge the pertinence of technology:

- Transferability – does the approach translate across multiple applications, disciplines or material systems?

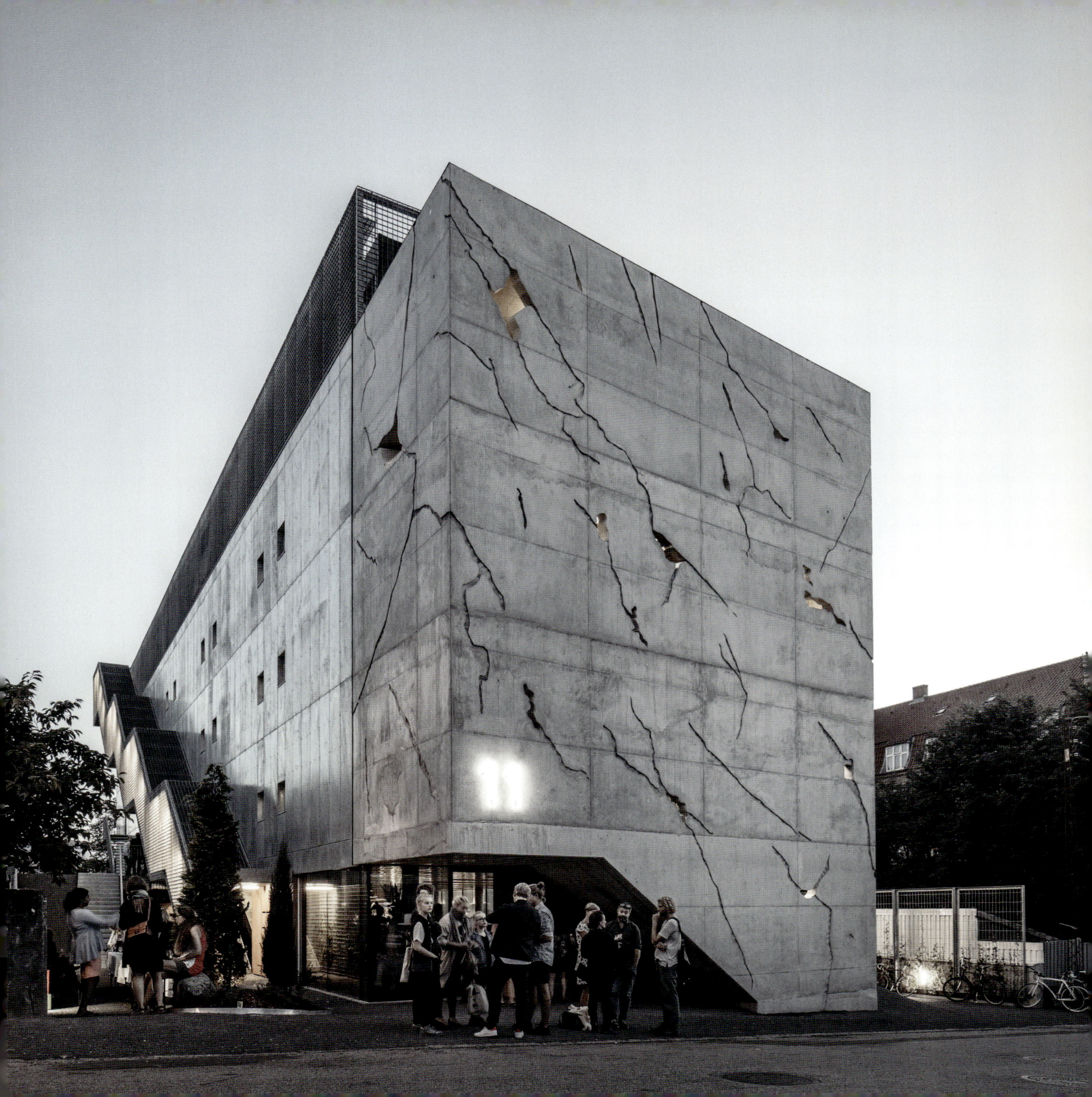

2

3

- Performance – does the approach offer a faster or more effective manner of producing results, compared to existing methods?
- Degrees of freedom – does the approach under consideration enable new opportunities in design, either by relaxing existing production constraints or by offering a way to explore previously uncharted design space?

Within these frameworks, quantitative and qualitative propositions are complementary – not conflicting – attributes of underpinning principles, with potential for large-scale impact in construction. Considerable attention has been directed within Odico to exploring the implications of one such technical approach and its derivatives – robotic hot-wire-cutting (RHWC) of expanded polystyrene (EPS) formwork for concrete casting. The following paper outlines the central developments within this effort.

Scaling production – Kirk Kapital Headquarters

The insight that underpinned the founding of Odico was that RHWC of expanded polystyrene formwork for advanced concrete casting could offer transformative advantages when deployed at industrial scale. When founding the company, the respective research projects by the authors of this paper allowed for the comparison of efficiencies between robotic CNC milling versus hot-wire-cutting of EPS moulds, which found that RHWC reduced machining times at a factor of between 10- and 100-fold (McGee, 2012). This finding is particularly relevant in achieving feasible scalability within construction manufacturing, where the throughput of large material volumes is a central concern.

For robotic fabrication of such volumes, machining time replaces labour as the key cost factor and hence is a primary focus. While robotic CNC milling has long proven its versatility, its mechanical principle of incremental material subtraction is inherently slow and thus not suited to scale economically beyond the exclusivity of high-profile construction projects. As such, the capacity of RHWC to cut through large volumes of expanded foams at significantly lower processing times, while resulting in high-smoothness casting surfaces, can yield considerable cost reductions in formwork manufacturing.

Odico set out to engage the construction market for early-stage adoption and to mature the technology through input from the commercial pilot production. Production began at small-scale installations, with the objective of working towards construction-sized productions. These early efforts initiated a continuous cycle between the ongoing technology R&D and its commercial implementation. The experience gained in production informed the development of the technology required to meet industrial ambitions. Conversely, the production pipeline provides a continuous testbed for further advances in new technology that might be considered tangential to the objective of reaching an industrial scale in production.

This milestone was reached in 2013 when Odico Formwork Robotics received the commission to produce over 4,500m² of bespoke formwork for the Kirk Kapital Headquarters (KKHQ) in Vejle, Denmark. KKHQ is a six-storey office complex designed by the Berlin-based Studio Olafur Eliasson and is architecturally Scandinavia's most ambitious office building. The project represents an international first in that it applies architectural RHWC for the production of critical load-bearing concrete structures (Fig. 2).

The design comprises four intersecting cylindrical perimeter walls, which rise out of the harbour basin. With a height of 32.3m, the cylindrical walls are interspersed with 19 intersecting hyperbolic paraboloid void walls, spanning vertically across all storeys. With dimensions varying from 7.4 x 2.8 x 5.2m to 4.2 x 3.2 x 5.2m, the volume of formwork to be produced would surmount 70-110m³ per storey section.

While building a test mock-up, traditional wooden moulds were contrasted with the EPS moulds supplied by Odico. The EPS mould stayed more true to form under casting pressures, while a relatively low-density EPS material was selected for the test where the traditional formwork dealt with deformation. Through this critical finding, Odico obtained a vote of confidence from the building contractor Jorton to go ahead and produce the formwork for the project.

The formwork system developed for the project entailed three primary variants. First, an in-situ prefabrication workflow, where polystyrene mould parts were inserted into a rectangular timber scaffolding box. This procedure was applied for onsite prefabrication of curved wall segments, which were subsequently hoisted into position (Fig. 3).

A second workflow was established for in-situ casting of lower-level hyperbolic walls. Here, the formwork was designed as a 110m³ solid foam plug, orthogonally segmented relative to the size of the standard foam stock dimensions of 1,200 x 1,550 x 2,400mm. To minimise the volume of the plug, a single timber insert structure was produced and repeatedly used in all topologically similar cases, achieving minimisation of material as well as providing auxiliary support against the casting pressure, while imposing few geometric constraints on the formwork design itself. The final formwork system application was designed for parabolic endwalls to intersect the cylindrical perimeter walls. In this case, rolled steel repetitive-use formwork was used to create the main wall geometry, while foam inserts were used as vertical plugs to achieve the parabolic opening. This ability to utilise RHWC seamlessly within the existing casting workflows was decisive in adopting the process for the project.

The above workflow required the organisation, design and manufacturing of around 3,800 unique RHWC formwork units. With the design not developed with the RHWC approach in mind, aspects of fabrication had not been a concern in the design and engineering development. As such, a considerable post-rationalisation effort was required. In order to segment the building to patterns that fitted stock material, a semi-automated CAD workflow was developed in McNeel Grasshopper and GH Python. While the project in principle would have sustained a shared, central BIM model, at the time the IFC 4[1] specification, which allows for NURBS[2] surfaces, was not available throughout the involved digital chain. The ability to exchange geometry in NURBS was a hard requirement, given the sophisticated ruled geometry of the project.

As a result, the model was sourced from a number of CAD platforms; and with the lack of software supporting IFC 4 at the time, this effectively disrupted a fluent interchange of modelling data, which compromised the geometric integrity of the model. As a result, a substantial effort in geometry pre-processing and optimisation of formwork design was required. Since then, Odico has provided support for IFC 4 for its offline robotics platform, PyRAPID (Feringa, 2015).

Taking on a recently founded start-up to deliver a central feature of the project – the production of over 130 truckloads of unique formwork for realising the most prestigious office building in Scandinavian construction history – was a risk offset by the disruptive properties of the robotic process. This enabled Odico Formwork Robotics to deliver at a considerably lower price point while handling all aspects involved with a small team.

1. Robotic hot-wire-cut façade patterns and apertures for the in-situ cast concrete façade of the Sonnesgade 11 mixed use complex, Aarhus, by Sleth Architects.
Image: © Rasmus Hjortshøj, COAST.

2. Construction site overview: the robotically hot-wire-cut formwork is used in combination with standardised timber and steel modules to resist casting pressure, applied for creating the hyperbolic void openings of the cylindrical main walls.
Image: Courtesy Kirk Property A/S.

3. Onsite prefabrication using a standard, rectangular scaffolding for formwork support against bespoke EPS infills.

Industry engages

Following the KKHQ project, Odico Formwork Robotics has seen a rapid expansion, completing over 200 projects in the four years since its formation, including several high profile commissions in the United Arabic Emirates, the United Kingdom, Norway and Denmark.

One recent example was the design and production of EPS foam guides for the manufacturing of 2,000 uniquely bent aluminium profiles, targeted at the doubly-curved glass façades of Opus Dubai, an iconic premium hotel resort designed by Zaha Hadid Architects in Dubai, UAE. In this case, enabled by the geometric coherence of the design scheme, a complete automation of the workflow was established. This enabled the entirety of profile geometries to generate mould design and resulting robot code in a single batch operation. This optimisation allowed for an increase of output from 80 unique units per 24 hours to 200-300 units, helping to accelerate the production schedule.

High volume applications, such as stairs, panels and structural components – as well as advanced infrastructural developments, where formwork expenditure represents a significant cost factor – form a testbed for the demonstration of the combined effects of the hot-wire machining speeds, the degrees of freedom offered in robotic control and the cost-effective EPS material. Indeed, this represents a viable pathway for a dramatic offsetting of costs in industrial concrete production.

The production of hot-wire-cut moulds for the KKHQ main structure corroborates that RHWC can act as a cost-effective method for the production of complex concrete moulds, applicable to a wide range of construction uses.

In such cases, as is typical for the majority of Odico's production, the primary threshold is the successful demonstration that the technology can be effectively applied to designs that did not anticipate the use of advanced robotic fabrication – or RHWC specifically – in the conceptual design phase.

Conversely, a growing interest from design partners in exploring the inherent vocabulary of RHWC concrete production is starting to complement these initial efforts. The architectural capacity of robotically controlled wire-cutting is being investigated as a constitutive

4. BladeRunner concrete panel demonstrator designed by 3XN Architects / GXN Innovation. The demonstrator is part of a series of explorations of design assemblies, seeking to capitalise on the aesthetic opportunities within the constraints of the hot-wire-cutting process.

5. The benches as installed in the Winton Gallery, December 2016. Image: © Luke Hayes.

6. Production UPHC prototype of bench B5 for the Science Museum, Winton Gallery of Mathematics, design Zaha Hadid Computation and Design Group.

premise for design, with the outline of capitalising on the specific degrees of freedom offered by the process, while maintaining the cost advantage demonstrated in practical applications (Fig. 4).

One of the first to address this potential in a commercial context, Zaha Hadid Computation and Design Group engaged with Odico to develop process-specific designs within various applications. An initial outcome of this effort was the design for 14 unique UHPC benches for the Winton Gallery of Mathematics at the Science Museum, London. For this project, a design scheme was developed within the constraints of wire-cutting moulds, enabling Odico to offer a production scheme favorable over existing fabrication approaches. The resulting benches were produced as 35mm high performance concrete shells surrounding a lightweight foam core (Figs. 5 and 6).

While robotically controlled hot-wire-cutting of concrete formwork offers a distinct solution space in which novel design vocabularies can be explored, the mechanical concept per se can be extended across several domains of material processing and motion types. This line of thinking constitutes an important exploration within Odico's internal development efforts. Over the course of four tooling prototypes, robotic abrasive wire-cutting (RAWC) has been developed and implemented within Odico's production.

While subjected to the same geometric and motion constraints as RHWC, abrasive wire-sawing enables the processing of hard materials such as marble (Feringa 2014), timber, non-flammable foams and ice (Fig. 9). This in turn facilitates a conceptual shift from producing the intermediate product of formwork designs to the architectural component itself.

Adjacent to this strand of development, Odico recently began to explore the domain of ceramic brick fabrication. In collaboration with Strøjer Tegl, a leading Danish producer of ceramic bricks, a robotic system was devised for production of bespoke tile designs. Early work on the topic (e.g. Adreano, 2012) indicated the architectural potential for bespoke ceramic tiles. Odico explored a different mechanical approach for processing the clay material due to the density of the clay utilised. By the development of an oscillating end effector, in which forward and quick lateral movement of a wire is combined, a rapid manufacturing process was devised, paving the way for rapid production while directly integrating with Strøjer's manufacturing process. As such, the installation enables the production of uniquely designed tiles. This quality was explored shortly after the initiation of the facility for an interior wall cladding of Odense Theater by Creo Arkitekter A/S, emulating the undulating motion of the theatre curtain.

Double-curved formwork – blade cutting

Odico tendered in a consortium for the production of the formwork of the Waalbrug bridge extension project by Zwarts and Jansma Architecten. The design required many thousand square metres of double-curved formwork. The constraint of double curvature could not be met in a satisfactory way using ruled surface rationalisation and hot-wire-cutting, so that approach was dismissed in favour of timber formwork, which meant that Odico did not participate in the realisation of the project. However, the tender did inspire an idea: by bending a blade, double curvature could be closely approximated. This method allows the production of moulds at a new level of scale and efficiency, enabling the realisation of large-scale double-curved concrete structures. The cross-disciplinary research project *BladeRunner* was formulated, and two more years of development culminated in a patented technology where unique double-curved formwork no longer incurs an unreasonable cost penalty (Søndergaard, 2016, Brander, 2016). This method is now under preparation for pilot production (Figs. 7 and 8), with expected construction-scale roll-out over the course of 2017.

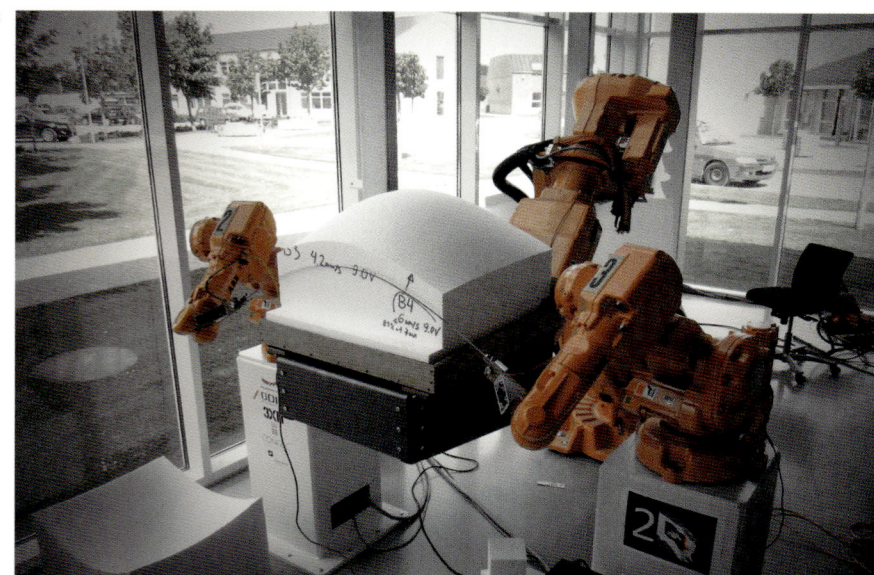

What technology wants

The past decade has seen the genesis of a range of specific robotic construction technologies and process concepts – some of which hold promise for adoption in construction. Thanks to this accumulation of academic efforts, momentum is building. The critical test is whether architectural robotics can scale beyond the lab to the construction site and become a commercially sustainable industry, possibly breaking the current technological stasis.

In *What Technology Wants*, Kevin Kelly offers a compelling perspective on the forces that drive technology: "The second great force pushing evolution on its immense journey is positive constraints that channel evolutionary innovation in certain directions. In tandem with the constraints of physical laws outlined above, the extropy of self-organisation steers evolution along a trajectory. While these internal inertias are immensely important in biological evolution, they are even more consequential in technological evolution. In fact, in the technium, self-generated positive constraints are more than half the story; they are the main event" (Kelly, 2011).

In the context of advanced architectural fabrication, we may characterise these positive constraints as methods and techniques that are tangential to the demands of a progressive architecture, having the capacity to scale architectural artefacts of a novel character, while coincidentally challenging the price point at which these can be delivered.

Due to the inertia of the building industry, there is still ample time to learn from other industries, especially when the former concepts of work and industry are changing. Considering the efficiencies of the vast, highly automated production lines in the automotive industry, automotive entrepreneur Elon Musk said: "The biggest epiphany I've had this year is that what really matters is the machine that builds the machine – the factory."[3] He noted that an increase in output orders of magnitude

9

7. Early doubly-curved hot-blade demonstrator produced at the Robarch 2016 Workshop, 'SuperForm', by the BladeRunner Research consortium and workshop participants.

8. Experimental multi-robot cell deploying 3 ABB multi-move manipulators for for production of doubly-curved geometries via sweeping of a flexible, heated blade along a surface.

9. Robotic abrasive wire-sawing of a 2-tonne ice block, excavated from the Torne River, 200km north of the Arctic Circle near Kiruna, Sweden.

greater than today's levels was to be expected through the redirection of creative engineering efforts towards this target, rather than through the product itself.

Does the same hold true for construction? Could a shift in design orientation from the object to the 'machine that builds the house' trigger unprecedented architectural innovation? Ironically, in the construction industry, the field of architectural conservation may offer us an insight. With the processing of natural stone becoming highly automated, has its manual handling evolved to become a punitive task, reminiscent of the image of Howard Roark working in the granite quarry in *The Fountainhead*?

Architectural conservation is an area where novel fabrication methods involving robotics have been adopted early, resulting in industry-wide acceptance. Companies that have not invested in the past two decades in CNC or robotic fabrication will today or in the near future no longer be able to compete, given the cost of labour and the efficiencies gained by automation. The Sagrada Familia has been architectural conservation's most enterprising project, and its expected completion date has been brought nearer by embracing robotic fabrication (Burry, 2008). Today, the architectural merits of the past are being (re)built with state-of-the-art technology. Scanning sculptures and reproducing stone elements has become a default approach, as the recent recreation of Palmyra's Arch underscores[4].

The challenges faced by large-scale automation in construction could be the call to disrupt the present order:

"Not alone have the older forms of technics served to constrain the development of the neotechnic economy, but the new inventions and devices have been frequently used to maintain, renew and stabilise the structure of the old order… Paleotechnic purposes with neotechnic means: that is the obvious characteristic of the present order" (Mumford, 2010).

References

Andreani, S., Bechthold, M., Castillo, J., Jyoti, A. and King, N., 2012, 'Flowing Matter: Robotic Fabrication of Complex Ceramic Systems' in *Proceedings of ISARC2012 – International Symposium on Automation and Robotics in Construction*, Eindhoven.

Brander, D. et al., 2016, 'Designing for Hot-Blade Cutting: Geometric Approaches for High-Speed Manufacturing of Doubly-Curved Architectural Surfaces' in Adriansseen, S., Kohler M., Gramazio, F., Menges, A. and Pauly, P. (eds.), *Advances In Architectural Geometry 2016*, Zurich, VDF Hochschulverlag AG, p.306-327.

Burry, M., Armengol, J. and Tomlow, T., 2008, *Gaudi Unseen: Completing the Sagrada Familia*, Berlin, Jovis VERLAG GmbH.

Dombernowsky, P. and Søndergaard, A., 2011, 'Unikabeton Prototype' in Glynn, R. and Sheil, B. (eds.), *Fabricate: Making Digital Architecture*, Waterloo, Riverside Architectural Press, p.56-61.

Feringa, J. and Søndergaard, A., 2014, 'Fabricating Architectural Volume – Stereotomic Investigations in Robotic Craft' in Kohler, M. and Gramazio, F. (ed.), *Fabricate: Negotiating Design and Making*, Zürich, GTA Verlag, p.44-51.

Feringa, J. and Krijnen, T., 2015, 'BIM and Robotic Manufacturing – Towards a Seamless Integration of Modeling and Manufacturing' in *Proceedings of the International Association for Shell and Spatial Structures (IASS) Symposium*, 17-20 August 2015, Amsterdam, The Netherlands, Technische Universiteit Eindhoven, p.1-11.

Kelly, K., 2011, *What Technology Wants*, London, Penguin Books Ltd, p.119.

McGee, W., Feringa, J. and Søndergaard, A., 2013, 'Processes For An Architecture Of Volume: Robotic Hot-Wire Cutting' in Brell-Cockan, S. and Braumann, J. (eds.), *Rob I Arch 2012: Robotic Fabrication in Art, Architecture & Design*, Vienna, Springer Verlag AG, p.62-72.

Mumford, L., 2010, *Technics & Civilization*, Chicago, University of Chicago Press, p.266.

Verde, M., Hosale, M. and Feringa, J., 2011, 'Investigations in Design & Fabrication at Hyperbody' in Glynn, R. and Sheil, B. (eds.), *Fabricate: Making Digital Architecture*, Waterloo, Riverside Architectural Press, p.106-109.

Søndergaard, A., 2014, 'Odico Formwork Robotics' in Gramazio, F. and Kohler, M. (eds.), *Made by Robots: Challenging Architecture at the Large Scale*, *Architectural Design*, May-June 2014, Profile no. 229, London, Wiley Academy, p.66-67.

Søndergaard, A. et al., 2016, 'Robotic Hot-Blade Cutting' in Reinhardt, D., Saunders, R. and Burry, J., *Robotic Fabrication in Architecture, Art and Design 2016*, Vienna, Springer International Publishing, p.150-164.

Notes

1. "To support the best way to exchange rich geometry-preserving parameters, the resulting schema includes several additional geometry types, such as advanced B-rep (NURBS), faceted B-rep and surface models, constructed solid geometry (CSG) and advanced sweeps, including tapering and presentation styles, such as colours and textures, which can be added to these geometries." http://www.buildingsmart-tech.org/specifications/ifc-view-definition/ifc4-design-transfer-view/ifc4-dtv-objectives.

2. Non-uniform rational B-splines.

3. https://www.ycombinator.com/future/elon/.

4. https://theconversation.com/should-we-3d-print-a-new-palmyra-57014.

MPAVILION 2015

AL_A

Responding to climate and landscape

MPavilion is a unique architecture commission and design event for Melbourne, Australia.

A new temporary pavilion is commissioned each year from a leading international architect by the Naomi Milgrom Foundation.

Each structure takes shape in the downtown oasis of Queen Victoria Gardens to accommodate a free programme of talks, workshops, performances and installations from October to February. Building on unexpected collaborations, MPavilion is a catalyst and a meeting place – an intriguing form, a temporary landmark, a spontaneous detour, a starting point and a base to explore design's role in the creative city.

At the conclusion of its lifespan in Queen Victoria Gardens, the pavilion is demounted and gifted to the City of Melbourne for reassembly in a permanent location to create an enduring legacy.

The brief was an opportunity for a structure that responds to its climate and landscape, exploiting the temporary nature of the pavilion form and producing a design that speaks in response to the weather.

Rooting the pavilion in its parkland setting, the vision for MPavilion was to create the sensation of a forest canopy, with beautiful dappled light where visitors could see the sun and the sky – a dreamy atmosphere that could inspire a diverse programme of events for four months.

The design was driven by an ambition not only to integrate the pavilion with its parkscape environment but also to involve the wind, and sometimes the rain, as part of the experience. And so the structure needed to balance a degree of flexibility in its response to the atmosphere with subtle movements, with sufficient stability to safely host thousands of visitors over the summer. The pavilion would be a celebration of those natural shelters where people come together: an exceptionally light, open structure that sits gently on the land while affording protection from the unpredictable weather of Melbourne.

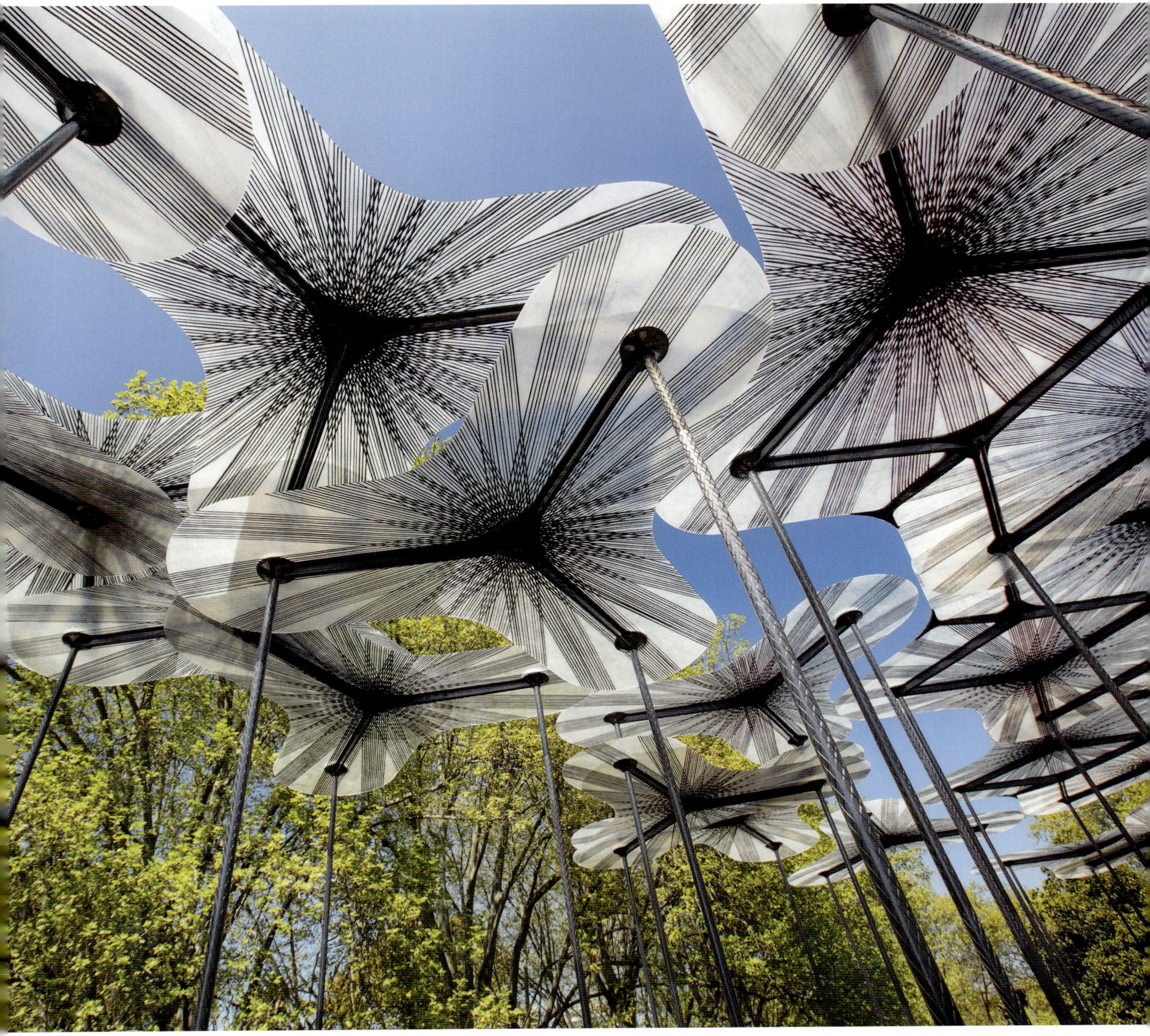

2

3

The twin natures of the seemingly ephemeral pavilion necessitated both swift construction and deconstruction methodologies while in its temporary home before becoming an embodiment of durability in order to persist in its permanent Docklands location thereafter.

Challenging the notion of public space

MPavilion was designed to challenge the notion of what a public space can and should be, a structure defined by an absence of walls and with physical and metaphorical connections to the surrounding landscape. This approach drove the design and thus the context for the research.

The sensation of a forest canopy was created by 44 seemingly fragile, translucent petals supported by 97 slender columns up to 4.2m high that sway gently in the breeze. At the centre, the petals should tightly cluster and be layered to produce a continuous shingled surface. As the pavilion fades out into the parkland, the size and number of spaces increase until the trees themselves take over the pavilion's role and the structure dissolves.

Moreover, the context was indelibly shaped by the personnel. The MPavilion recipe has ingredients from around the world and draws upon manufacturing experience derived from a number of different industries – but just like the design itself, the build solution is quite unique.

AL_A worked with the specialist fabricator mouldCAM (now ShapeShift) and engineers Arup to employ the boundary-pushing technology of composite materials to create the translucent petals.

As ShapeShift explains, "MPavilion is a great example of collaboration drawing together inspiring design, 3D technology, advanced materials and engineering and the all-important ingredients of practical experience and construction management."

AL_A have a long history of working with boatbuilders, of which Australia has some of the finest. Initial inspiration was provided by the innovative materials typically used in aerospace and in the surfboard industry and the latest technology used in nautical engineering – in particular, the large sails utilised in high-performance yachts that afforded a sense of the possibilities in both aesthetics and material capabilities.

The overall design was optimised to keep the fabrication simple by using symmetry. Therefore the final design was limited to petals of just two different sizes, while still allowing for multiple configurations.

Establishing a framework

The ambition and contexts – conceptual, physical, material – established the framework for a series of questions that in their answering would define the success of the pavilion.

At its heart was the notion of how to dematerialise a structure, albeit a temporary one, and to make it feel and look less like a permanent building. The solution of a forest of petals surmounting impossibly thin columns in order to make it as transparent and as light as possible simply uncovered further questions as to its material composition and fabrication methodology.

Consequently, the challenge became one of achieving sufficient lightness and transparency or translucency in the form of the petals without compromising its

1. The petals disperse at the edges, blending into the parkland tree canopies. Image: John Gollings.

2 & 3. The making of a petal. Image: mouldCAM.

4. MPavilion at dusk. Image: John Gollings.

5. The public enjoying MPavilion. Image: John Betts.

4

5

6

structural efficacy. This would impact on the relative sizes of the petals and their modularity, as well as their cross-section, with a flatter petal with additional reinforcement and a more profiled three-dimensional petal both initially appearing viable.

Simultaneously, there was a balancing act between allowing the columns to visibly move and the petals to shimmer with a gentle breeze and making them strong enough to withstand hurricane conditions.

The desire for ephemerality extended into an ambition for no visible wiring and for minimal light fittings and speakers, which posed questions to be answered in the design and fabrication of the pavilion.

Moreover, the brief necessitated a 200m² weather-protected area at the centre of pavilion, while the vision was for an unconditioned space that would allow the elements to participate in the performance.

A balancing act

The success of MPavilion would be determined by the delicate equipoises of flexibility and strength, of translucency and solidity. A solution to this balancing act could only be reached by a programme of comprehensive trials, of testing and prototyping composite materials.

It was determined early on in the design process that the larger petals would be hexagram-shaped in plan and 5m in diameter, assembled from three smaller components to aid fabrication, while the smaller petals would be trefoil-shaped in plan and 3m in diameter.

7

The larger petals would have columns positioned on their perimeters 2m apart and in the centre would be a 4 x 4m column-free space.

Nevertheless, the final choice of material was selected after due consideration of its high strength-to-weight ratio, very high tensile strength and mouldability that enabled it to be employed to generate exceptionally thin petals.

Each petal, measuring only 5mm to 7mm in thickness, is formed by a carbon fibre weave, interlacing structure and aesthetic together to form the pattern. Reinforcement is embedded into the surface of the borderless petal rather than as an encircling frame.

The greatest challenge was finding the perfect balance between achieving the thinnest possible petals, with the strength to support their weight over spans of 5m, and the desired level of transparency.

This necessitated a programme of testing utilising reinforcement in the petals to produce rigidity. These trials noted a direct correlation between reinforcement and translucency, whereby greater quantities of fibre produced an adverse impact on the desirable levels of transparency. At its extremes, the process fabricated petals that were clear but structurally flawed or petals that were sufficiently rigid but cloudy to the point of almost complete opacity.

The final solution was achieved thanks to a series of trial and error experiments conducted by the fabricator. Resin curing times and the correct placement and layering of the fibreglass and carbon fibre reinforcements were the keys to achieving the right balance between flexibility and rigidity of the petals.

After further trials of reinforcement fibres, the optimum balance between clarity and strength was ascertained and used to manufacture the petals for MPavilion. The team developed a cost-effective method of incorporating the carbon that still delivered maximum impact, using a custom-built tension loom manufactured by a specialist subcontractor.

Placement was not only optimised for structural performance, whereby the lines of fibre are never folded to allow for maximum efficiency, but also for the creation of a beautiful radial pattern that became the defining graphic of the entire project.

Rejecting the idea of a profiled three-dimensional petal to provide the necessary rigidity, the design and fabrication team opted for a flat petal with a carbon fibre 'backbone' reinforcement. This decision was partly influenced by the requirements for drainage and the need to cascade water from one petal to another. In turn, this backbone allowed the incorporation of the columns' capitals into the body of the petal without additional fixings.

The backbones were formed by injection moulding, which proved to be a more efficient solution with improved structural performance and a more elegant form. The moulds were CNC-cut from the 3D CAD to ensure accuracy and repeatability.

The opportunity to capitalise on the inherent strength of a shape is a singular advantage of the use of composite materials, as MPavilion proved.

Most significantly, it was decided to add external reinforcement by affixing multiple columns to each petal. This allowed the pavilion petals to be significantly reduced in thickness from 12mm down to 5-7mm, allowing for a great level of transparency.

MPavilion was a constant technological battle and necessitated the development of bespoke solutions in order to advance the potential offered by composite materials. There was undoubtedly a necessary balance made between the ideal, yet unrealisable, scenario of almost perfect transparency and the structural integrity of the pavilion.

Similarly, the balancing act of allowing the columns to move in the wind yet not be broken by gale force conditions was resolved by structural simulations of the 3D computer model before fabrication, testing different combinations of the diameter and thicknesses of the section of the columns.

The thin, high-strength columns used in the final pavilion were 45mm in diameter with 4mm wall thicknesses. Like the petals, they are the product of a process of research and development undertaken during their industrial manufacturing, which in this case saw the tubes initially developed for camera tripods.

In order to amplify the perceived movement, clusters of one, three and five small petals were created. This combination of the number of columns and petals created a different mass per column ratio, allowing them to sway gently in the breeze.

In turn, these were counterbalanced by a cluster of larger petals in the middle of the structure. This also assisted with making the cascading effect from the centre to the edges simpler, as well as allowing a wider column-free space for events.

The ambition to dematerialise the structure and blur the threshold between pavilion and park was achieved by material innovation working in parallel with the overall design. Once the 3D computer model was complete, a new fixing method was created for the vertically stacked petals, tying one into another. This opened up apertures in the pavilion and overlaps closer to the edge, allowing visitors to see glimpses of the sky and the surrounding tree canopies.

Each petal was mounted on slender carbon fibre columns that were designed to conceal the wiring of lights and speakers to augment the dematerialised ephemerality of the pavilion. This is heightened still further by a halo-like effect created by an LED strip forming the capital to the column, while pioneering technology turns the petals themselves into amplifiers. From the surrounding high-rises, the pavilion appears to have a glowing aura and a particular presence in the otherwise darkened garden at night.

6. Melbourne's skyline as a backdrop to MPavilion.
Image: Rory Gardiner.

7. MPavilion's canopy.
Image: Timothy Burgess.

8. MPavilion lights up at night.
Image: John Gollings.

An eye-catching architectural attraction

The commission and associated programme is quickly becoming one of Australia's leading design and architecture events and has become one of Melbourne's leading summer attractions.

By February 2016, AL_A's MPavilion had attracted more than 64,000 visitors over 126 days to 419 free events through collaborating with more than 260 cultural institutions, architects, artists, musicians, dancers, choreographers, scientists and designers.

In December 2015, *Wallpaper** named MPavilion 2015 as one of 15 installations that capture the global imagination. In January 2016, *designboom* named MPavilion 2015 in the 'Top 10 Temporary Structures of 2015'.

One of the unique features of the MPavilion project is that it is gifted to the city and the people of Melbourne. After its four-month programme, the pavilion was disassembled and moved to its new permanent site in Melbourne's parklands. This creates a permanent legacy that will become part of the cultural heritage and public amenities of Melbourne, attracting tourism, industry development and civic pride.

The 2015 pavilion opened to the public in its new permanent location in the Docklands public park in August 2016.

The pavilion's lasting legacy is a tribute to the ambition and collaboration that commissioned, conceived, developed and fabricated it. The willingness of the team to extend the boundaries of the possible in taking ordinary materials to new levels is testament to this spirit of innovation shared by all and to a mutual confidence in each member's expertise. MPavilion is a beautiful example of how taking materials and technology beyond their everyday applications can deliver extraordinary and unique results.

Project credits

Architect: AL_A.
Project directors: Amanda Levete, Maximiliano Arrocet, Ho-Yin Ng.
Project architect: Alex Bulygin.
Team: Alice Dietsch, Song Jie Lim, Filippo Previtali, Giulio Pellizzon.
Engineer: Arup.
Fabricator: mouldCAM (now ShapeShift).
Main contractor: Kane Constructions.
Lighting and sound design: Bluebottle and Sam Redston.

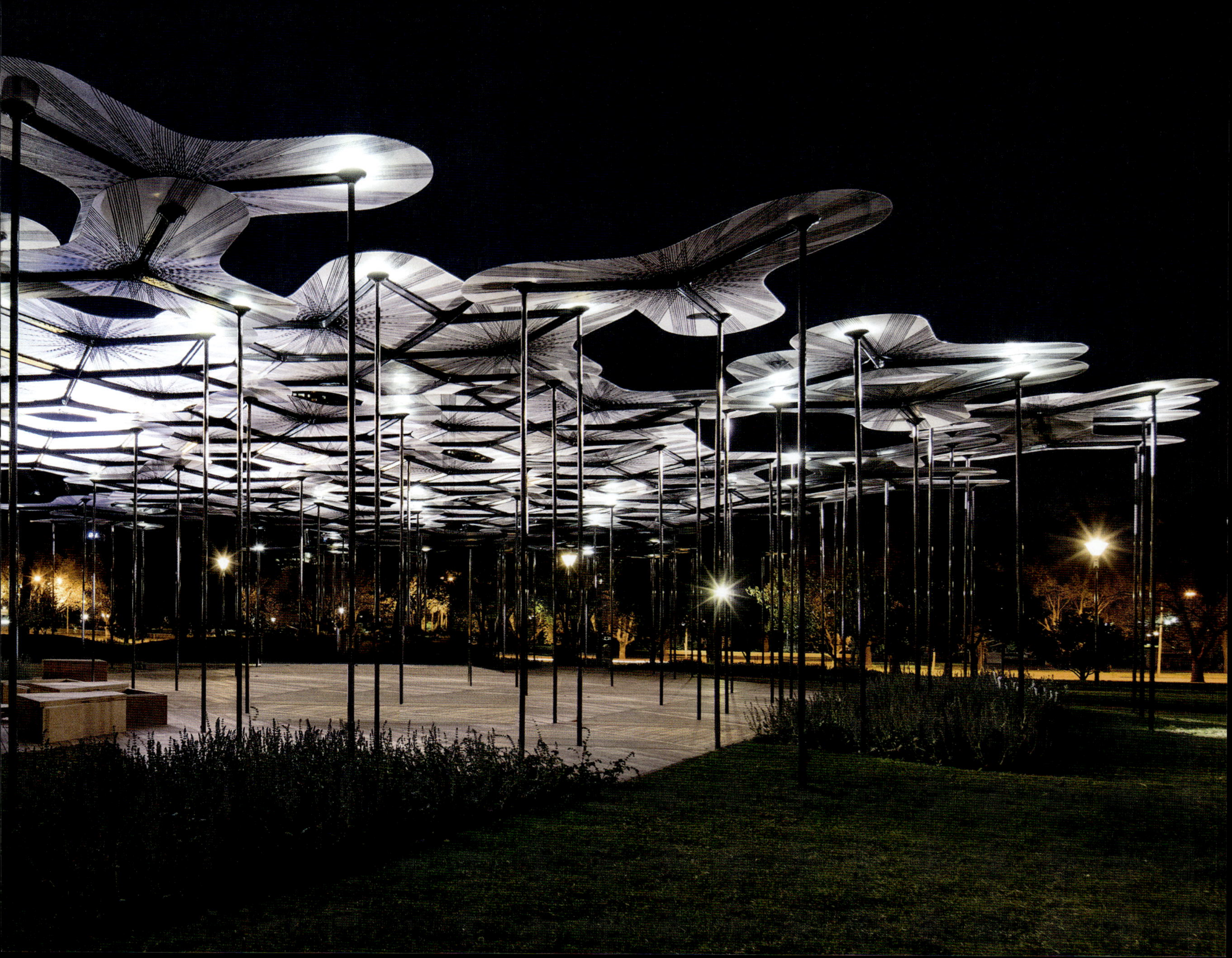

MULTI-PERFORMATIVE SKINS

EDOARDO TIBUZZI / DEYAN MARZEV
AKT II

In recent years, technology and digital innovation have provided a series of new design tools for the architectural world, which have dramatically morphed the massing of modern buildings. The envelopes which were traditionally constrained to a relatively planar or curved setting-out became complex forms, moulded in the digital environment – shapes that push the traditional boundaries of engineering, fabrication and performance.

This shift has generated a dichotomy, in which two solutions have become apparent: one where the envelope has its own supporting system, often quite complex, which is then dropped onto the main structural skeleton, and one where the skin is both the supporting system and the envelope at the same time.

In both cases, the challenges posed by these complex geometries required a series of digital tools and workflows to be developed, tools that interweave geometrical form-finding, structural optimisation and fabrication output. At AKT II, the work developed on new digital design-to-fabrication techniques has allowed the use of a building technology that integrates architectural form, structural armature and environmental enclosure in single multi-performative skins. This technology, which has been successfully tested on a number of built projects, has produced great savings in cost, use of material, energy and labour, by offering multiple functionalities within fewer building components.

An extension of monocoque construction, commonly used in aeronautical application, this technology is based on the prefabrication of large components in the factory, simplifying assembly in the field. Components are designed to be bolted as a kit of parts and then welded to form a smooth, waterproof enclosure.

Interrogatives

This applied research was propelled by two fundamental questions:

One: can the envelope, with its aesthetic and environmental functions, also provide a main structural function?

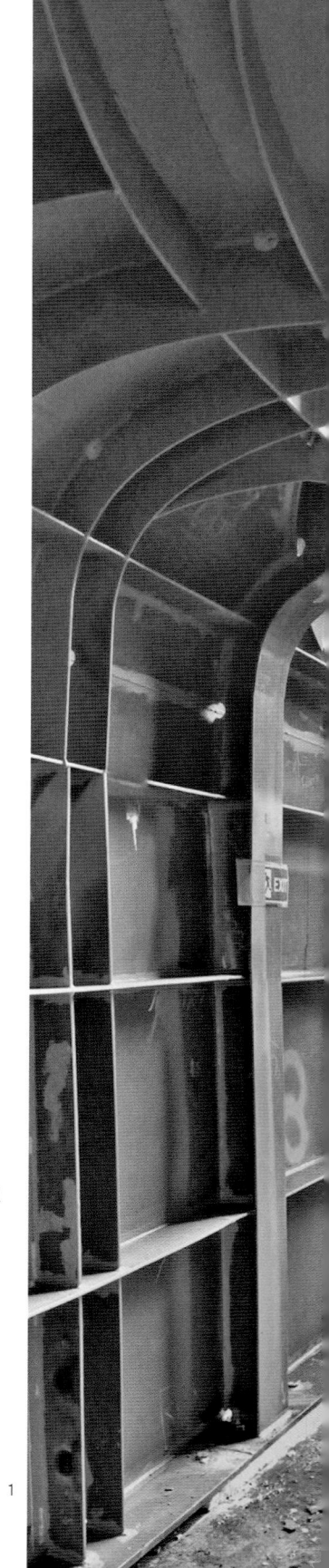

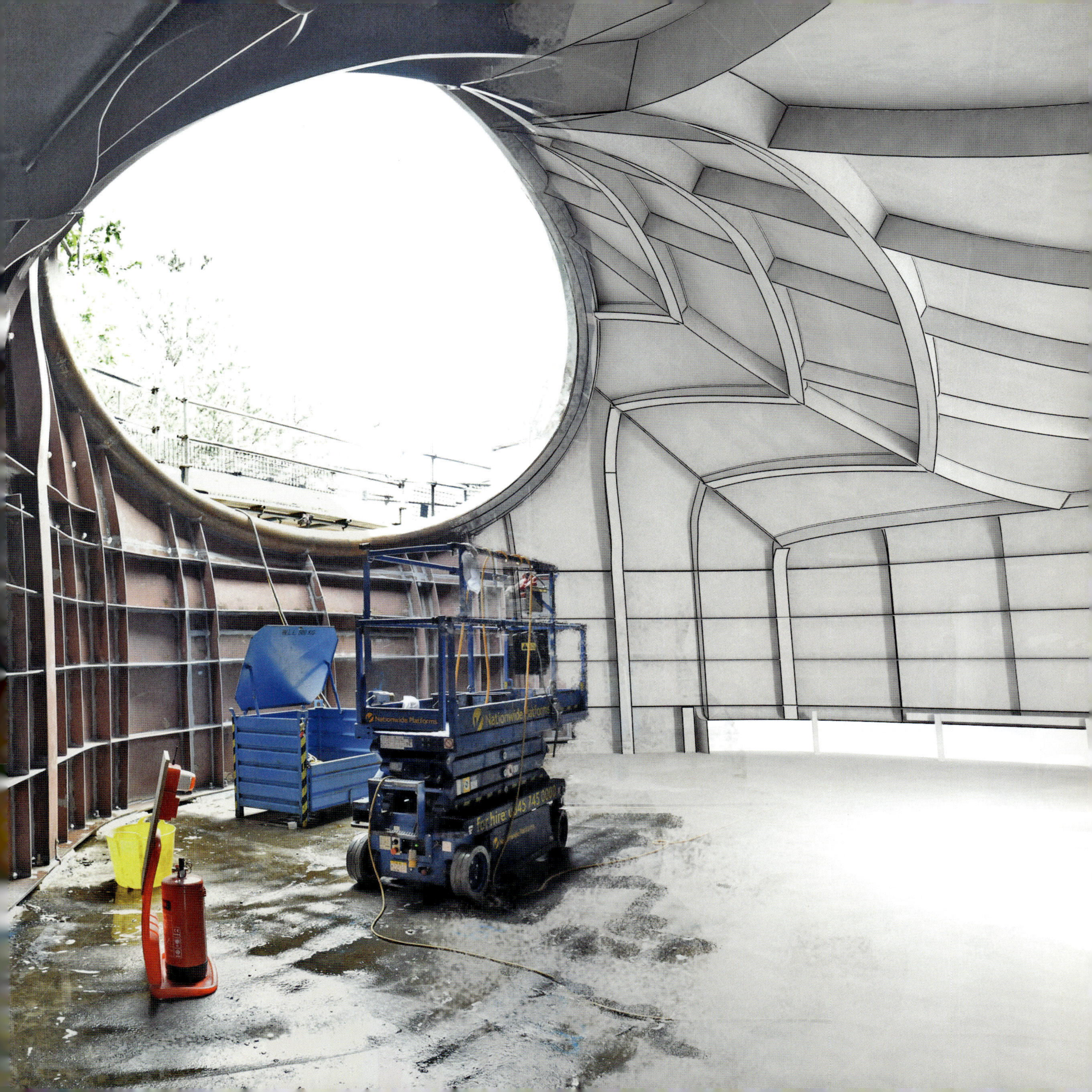
Nationwide Platforms
Nationwide Platforms

Two: which tools are needed to develop such a system, and what kind of workflow needs to be put in place to maximise efficiency and connect the architectural intent to the structural design while respecting the fabrication limits and tolerances?

Drawing Studio, Bournemouth

The Drawing Studio for Arts University Bournemouth by Crab Studio will be used as the first example to showcase the design process of a multi-performative skin. It is designed to create different conditions of light within the same space through openings within the organically shaped building. Constructed by specialist steel fabricators very much like the hull of a ship, the structure consists of an 8mm doubly-curved external plate, stiffened by thin internal welded steel rib-plates, creating a 16m-span column-free space. This was factory-prefabricated in large panels, then bolted and welded onsite to produce a smooth structural enclosure. The structural skin is internally insulated, and the internal finish fitted to the inside flange of the internal stiffeners. No external cladding or secondary framing was required, as the insulated structural skin provides complete climate control.

The organic shape proposed by the architects consisted of a large continuous massing, with one large opening for the main window and one wavy opening in the middle section for a secondary window, two entrances on each side of the building and some ground level openings to let light in at floor level.

Structure

The structural system envisaged in this case consisted of a skin manufactured using simple flat metal sheets which were laser-cut and formed into the final curvature, and welded together in the workshop using flat ribs welded orthogonally to the main plate to prevent buckling and provide stiffness. The waterproofing of the enclosure was guaranteed by the welding of the contiguous metal parts, and corrosion was prevented by paint.

An early stage detailed investigation was carried out to determine the factors influencing the construction and build-up of the steel semi-monocoque. The semi-monocoque is made up of a continuous top structural skin and ribs with and without a bottom flange.

The aim of the analysis was to minimise the plate thickness and maximise the rib spacing. The minimisation of the plate thickness was intended to reduce the overall weight of the structure and to reduce the amount of resources used in the construction. However, a lower limit of 8mm was taken to ensure weldability, because challenges can be encountered when welding thicknesses lower than this, due to excessive plate distortion.

From the analysis, it was found that the thickness of the top sheet of steel is controlled by plate buckling, due to the compression force that arises from the bending action. This was also affected by the spacing of the ribs, but to a lesser degree. Where large rib spacing was used, the stress at which the plate buckled reduced, making the section less structurally efficient.

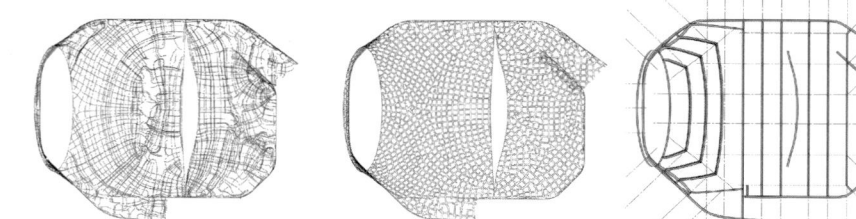

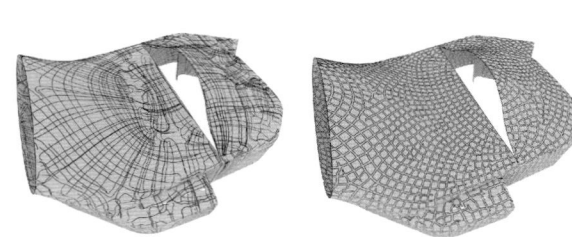

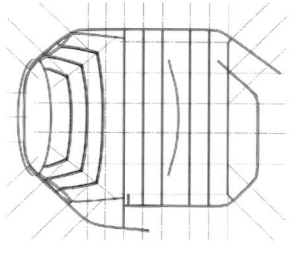

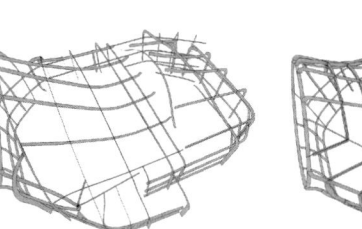

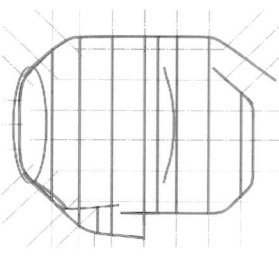

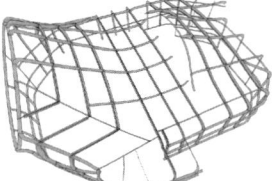

2

1. From digital to fabrication.
Image: © Cook Robotham Architectural Bureau / © AKT II.

2. Ribbing patterns.
Image: © AKT II.

3. Fabrication.
Image: © Cook Robotham Architectural Bureau / © CIG Architecture / © AKT II.

The overall structural depth is controlled by the deflection criteria. The semi-monocoque structure allows much smaller depths to be used than could be achieved through the use of traditional steel beams.

Workflow, optimisation, Re.AKT

The project used a complete workflow, from the modelling of the smooth external form of the building, achieved through bespoke smoothing algorithms, to the automated analysis of the structure, enabled by real-time interoperable models which interface with analysis and optimisation tools.

This workflow extends all the way through to patterning tools which set out plates and stiffeners, interfacing with automated fabrication methods designed to achieve high levels of precision in the final fabricated form. The three fundamental steps followed were:

1. Rationalisation of the provided surface to remove impurities in terms of curvature tangencies and incomplete boundaries.
2. Definition of the optimal pattern for the internal stiffening plates.
3. Definition of the subsequent patches of structural elements for fabrication.

The rationalisation of the surface was an essential step to ensure the elimination of folds which would have defined a clear interruption of the metal sheets. This process was implemented using a smoothing algorithm that was initially created for the movie animation industry. The Catmull-Clark algorithm takes an initial crude mesh and recursively subdivides it, averaging the faces' vertices. This algorithm was embedded in AKT II's internal toolkit (Re.AKT) and enhanced, introducing between other functionalities the option to assign constraint points, curves and surfaces. This allows the user to sculpt an interpolated smooth surface while still maintaining the original constraints. The differences between the original surface and the rationalised one were assessed by mapping the distortions as a coloured gradient on the surface.

By removing all the initial creases, it was possible to obtain one single smooth surface, and this therefore allowed the study of various patterns of stiffening ribs.

A parametric model was built to control the patterning of the ribs and, using an internal toolkit (Re.AKT), the initial geometry was plugged into the FEA structural solver. This step allowed the quick generation of various options for the ribs, with the aim of finding a solution that could balance the need for strength and stiffness to comply with the loading conditions with the need to limit the self-weight of the structure.

An initial proposal was to use the principal stress curves to define the rib patterns, which also generated a sub-option where a specific quadrangular module was mapped on the stress distribution. This option, although quite efficient, would not have been compatible with the fabrication splicing. The system was therefore simplified into a square grid subdivision first, and then, in order to reduce the spacing and weight, was optimised in the final configuration: a customised pattern, spaced circa 1.2m, that provided the best performance and the lightest configuration (Fig. 2).

Fabrication

The fabricator (CGI International) was able to use AKT II's optimised surface model to inform the subdivision of patches that could be cut from single metal sheets. This process was necessary to obtain strips that could be easily fabricated and transported to the site. The splices were coordinated with the stiffeners' locations in order to create a simple connection detail for the erection onsite. Once this information was added to the digital model, the fabricator started production and pre-assembly in its warehouse. The stiffeners were laser-cut and propped in place, then welded to form a skeletal network which would provide both a base support and a reference for the setting-out of the skin layers.

The flat metal sheets defining the skin were then welded onto the skeleton of ribs and locally adjusted to remove any distortion generated by welding and the imperfection generated in the fabrication process, to maintain tangency along the splits. To achieve tight curvatures around the openings, prefabricated metal tubes were used.

Installation

Once the building was fully pre-assembled, the parts were carefully dismounted and loaded onto trucks to be delivered to their final location. The vertical side walls were the first to be craned in and welded together, forming the boundary perimeter where the horizontal enclosure could then be supported and welded on. To make sure the structure was not going to distort in its temporary unconnected condition, props were used while the patches were craned into place. After every patch was placed and the local adjustments were made, an onsite welding process took place to seal all the edges and create a skin which could act as a singular structural element, at the same time providing waterproofing to the building. To complete the installation, several layers of paint were laid onto the structure to preserve the metal from corrosion and to give its final look (Figs. 1 and 3).

Library Walk, Manchester

The Library Walk Cloud pavilion is a link between the Manchester Town Hall and the adjacent Central Library by Ian Simpson Architects (Fig. 7). The 175m² pavilion uses frameless structural glass panels to support a 30-tonne, stainless steel roof structure. The distinctive shape of the roof was form-found using mathematical algorithms designed to create a smooth, organic but 'rational' undulation in the soffit, based on spherical distortions of a flat surface.

Different story, a shared path

Differing from the previous example, this case used digital tools to generate the final curved surface of the roof instead of optimising it for fabrication. In addition to this, the digital tools were set in such a way that the locations of the spheres distorting the flat surface could be altered in {xyz}; their radii of influence were also a parameter. By manipulating those values, the designer was able to position and alter the weight distribution of the overall roof, having total control of the design. The structure can be divided into two layers:

4. Shaping steel.
Image: © CIG Architecture.

5. *Cloud* stiffening ribs.
Image: © CIG Architecture.

6. *Cloud* installation.
Image: Courtesy of CIG Architecture.

7. Manchester Library Walk.
Image: © Valerie Bennet.

8. Two-layered structure.
Image: © AKT II.

1. The façade, which is a frameless set of 7.4m-high structural glass panels supporting the roof and providing lateral stability.
2. The roof, which consists of a polished stainless steel monocoque construction, allowing it to span 15m across a column-free space (Fig. 8). These external, exposed surfaces are welded to an internal armature of stiffeners, creating a rigid structure.

The simplicity and purity of this building is achieved by the simple combination of the two structural elements of the roof and the vertical glass façade, which are rigid in virtue of their form.

Another interesting difference of this installation when compared with the previous project is that the internal stiffening ribs follow a simple planar grid, and their distribution is regular due to the smaller number of patches required for the installation. The stainless steel is also welded on top of the stiffeners, following a similar procedure to the Drawing Studio (Fig. 4). The main difference in this case is in the external finish. The library entrance was envisaged to be a reflective surface from the beginning. To achieve this, the fabricator first ground the welding line until it disappeared, and the surfaces were then sand-blasted to further reduce the imperfections generated by the welds. The entire surface was then polished to create the mirror finish and protected with a robust film for transportation. Finally, the prefabricated sections of the upper Cloud were transported to the site and erected onto the pre-installed glass perimeter.

In a world where craft and science are merging, fusing such different expertise, there is a need for a deep and interactive collaborative process between disciplines. This union has ignited the development of bespoke digital tools for design, optimisation and fabrication that are pushing designers to think deeply about integration of purpose and systems. Multi-performative skins are one example that, with their integrated technology, can address many of the economic and environmental challenges our industry faces.

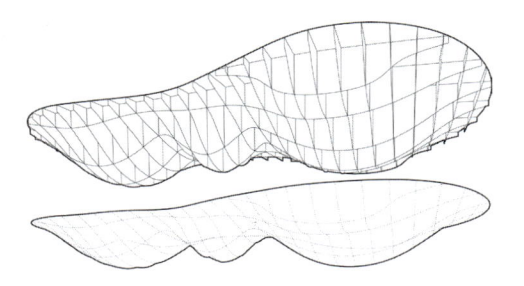

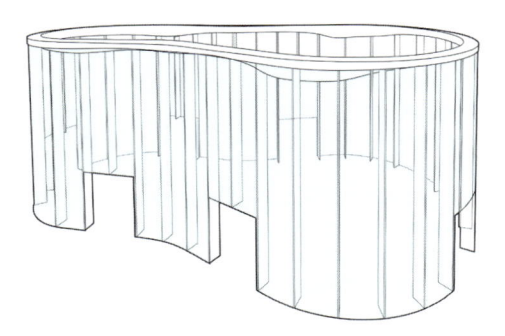

THE ARMADILLO VAULT
BALANCING COMPUTATION AND TRADITIONAL CRAFT

PHILIPPE BLOCK / MATTHIAS RIPPMANN / TOM VAN MELE
ETH Zurich – Block Research Group
DAVID ESCOBEDO
The Escobedo Group

This paper describes the development and fabrication of the Armadillo Vault, an unreinforced, freeform, cut-stone vault, which embodies the beauty of compression made possible through geometry. Specifically, the paper provides insights on how a highly interdisciplinary team managed to bridge the difficult gap between digital modelling and realisation by learning from historic precedent and by extending traditional craft with computation.

The vault is the centrepiece of *Beyond Bending*, a contribution to the 15th International Architecture Exhibition – La Biennale di Venezia 2016, curated by Alejandro Aravena (Fig. 2). Wrapping around the columns of the Corderie dell'Arsenale, the shell's shape comes from the same structural and constructional principles as stone cathedrals of the past, but is enhanced by computation and digital fabrication. Comprising 399 individually cut limestone voussoirs with a total weight of approximately 24 tonnes, the vault stands in pure compression, unreinforced and without mortar between the blocks. It spans more than 15m in multiple directions, covers an area of 75m² and has a minimum thickness of

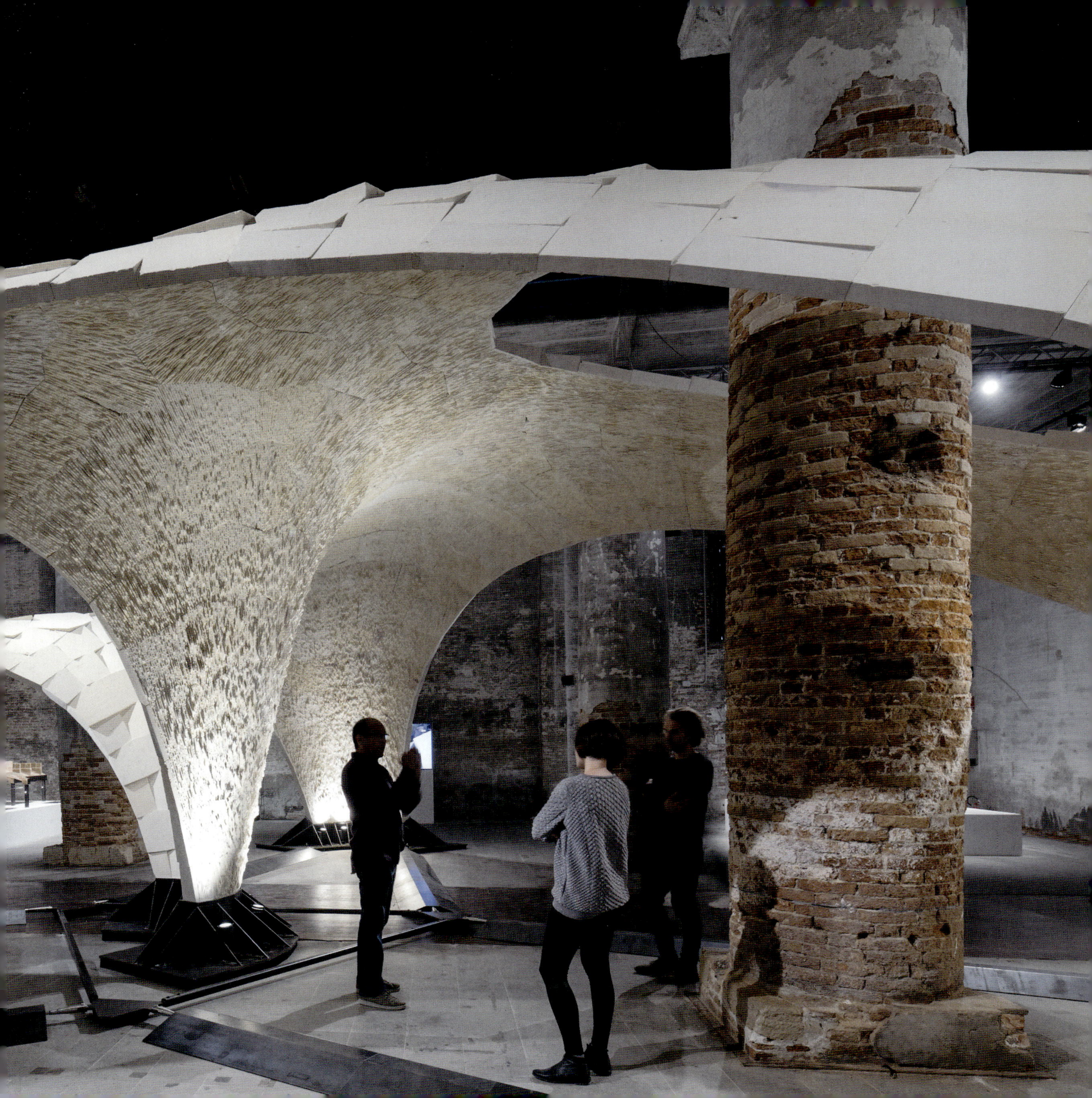

only 5cm (Fig. 1). Each stone is informed by structural logic, by the need for precise fabrication and assembly, by the hard constraints of a historically protected setting and by tight limitations on time, budget and construction. On the one hand, digital tools were developed for the form-finding process of the shell's funicular geometry, the discretisation of the thrust surface, the computational modelling and optimisation of the block geometry and the CNC machining process. On the other hand, together with master stonemasons, traditional strategies of stereotomy were investigated, analysed and revisited to develop appropriate and efficient stone-cutting and processing techniques and approaches to sequencing and assembly.

The lessons learned from historical precedent, combined with traditional craft enhanced by digital computation, allowed a collaborative team of engineers, designers and skilled masons to deal with the hard constraints of the project. Although such an interdisciplinary strategy reflects the holistic approach to design and construction of master builders in Gothic times that contrasts with today's more linear building processes, the presented work is not a romantic attempt to revive the Gothic. Rather, it is a direct critique of the current practice of planning and constructing freeform architecture. It is also a demonstration of how material and fabrication constraints are not equivalent to limited design possibilities, but can be the starting point for expressive and efficient structures.

The challenges of working in a historic setting

The Corderie dell'Arsenale is a historically protected building. Therefore nothing could be attached or anchored to the walls, columns or floor. Additionally, the average stress on the floor could not exceed 600kg/m^2, which corresponds to that caused by a tightly packed crowd of people. This also meant that no heavy equipment, such as a mobile crane, could be used for the assembly. Thus alternative methods for the manual setting of stones had to be developed. Furthermore, only five months were available for the entire project. This includes time needed for the design, engineering, fabrication and construction of the vault. The challenge was effectively to convert a 'perfect world' digital design into a 'real world' fabrication and construction process in an extremely short period of time for a constructional/material system without obvious mechanisms to compensate for tolerances.

Digital process

For this project, a smooth digital pipeline/workflow was developed to realise a structurally optimised and fabrication-driven generation of geometry.

Structural design and analysis
The vault's funicular geometry, which allows it to stand like an intricate, three-dimensional puzzle

1. The Armadillo Vault spans more than 15m with a maximum height of 4.3m and a minimum thickness of only 5cm. A system of tension ties balances the thrusts of the compression shell. Image: Iwan Baan.

2. The Armadillo Vault in the Corderie dell'Arsenale at the 15th International Architecture Exhibition – La Biennale di Venezia, 2016. Image: Anna Maragkoudaki/Block Research Group, ETH Zurich.

3. The local shell thickness ranges from just 5cm at the midspan to 12cm at the internal touch-down and point springing. Image: Anna Maragkoudaki/Block Research Group, ETH Zurich.

4. The overall tessellation design is defined by stone courses aligned perpendicular to the local force flow. Image: Aman Johnson.

in pure compression, results from a form-finding and optimisation process based on thrust network analysis (Block & Ochsendorf, 2007, Van Mele et al., 2014). These novel computational methods offer a more controlled, force-driven exploration of (inverted) hanging models.

The dominant self-weight of the vault was taken as a design load to define the middle surface of the structure, which was then offset according to assigned local thicknesses based on experience and weight constraints (Van Mele et al., 2016). The resulting intrados and extrados define a local shell thickness ranging from 5cm at the midspan and only 8cm along the line supports to 12cm at the internal touch-down and point springing (Fig. 3).

Based on the designed force flow, the stone envelope was discretised into courses and the courses into voussoirs. Staggering of the voussoirs, and alignment of the courses to the force flow and the boundary, guaranteed proper interlocking of all stones in the surface of the discrete shell (Fig. 4). To speed up the fabrication process, the voussoirs were made as large as possible with an approximate range of 45 to 135kg, so that they could still be handled by hand or with a lightweight jib crane. The stability of the unreinforced, dry-set assembly under various load conditions, including concentrated loads, settlements of the supports and earthquake loads, was confirmed using discrete element analysis (Van Mele et al., 2016).

Architectural geometry and fabrication

Due to the limited timeframe and large number of voussoirs, the main goal for the fabrication process was to reduce the average cutting time for each stone. Additionally, since there is no mortar between the voussoirs, which could have compensated for tolerances, the interfaces between stones had to be flush and therefore precisely cut and set.

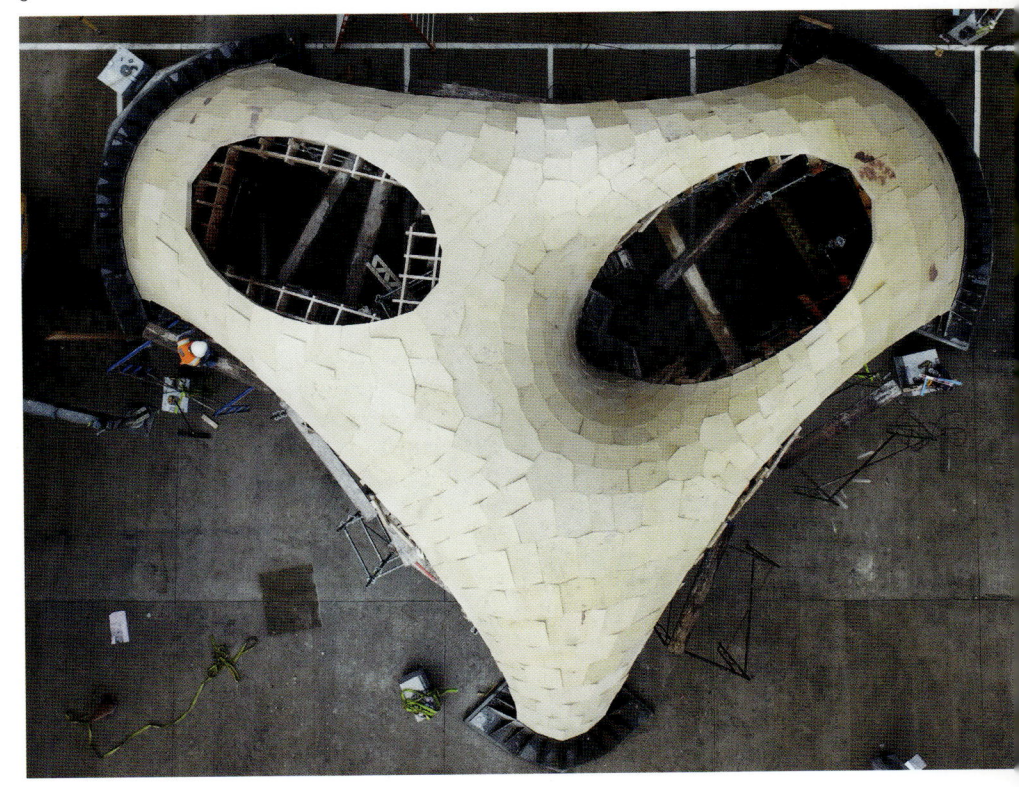

To optimise the fabrication process, the voussoirs were designed to have a convex cutting geometry along the interfaces, such that they could be cut efficiently with a circular saw (Rippmann et al., 2016). However, the vault has several areas with negative Gaussian curvature. Since it is geometrically impossible to discretise such a surface with a convex, planar mesh (Li, Liu & Wang, 2015), the faces of the extrados were allowed to disconnect and create a stepped, scale-like exterior. This visually emphasised the discrete nature of the shell and allowed the flat extrados faces of the voussoirs to be used as a base for the machining process. As a result, the cubic blanks no longer needed to be flipped and re-referenced, reducing fabrication time of the voussoirs significantly. The curved intrados faces were formed by side-by-side cuts with a circular blade, spaced such that thin stone fins remained. Rather than milling these away, the fins were hammered off manually to create a rough but precisely curved surface. The side surfaces perpendicular to the force flow were processed with custom profiling tools that create ruled surfaces with male/female registration grooves. These grooves are primarily used as reference geometry to assist assembly, but also prevent local sliding failure. The other side surfaces of each voussoir were created with simple planar cuts.

From digital to realisation

The vault was test-assembled offsite to allow a team of expert stonemasons to become familiar with the process. During the test assembly (and also during onsite assembly), each voussoir was fully supported by a falsework consisting of a standard scaffolding system with a custom-made wooden grid on top (Fig. 5). The voussoirs were placed manually, starting from the courses at the supports and converging towards the 'keystone' courses at the top. To gradually decentre the vault as evenly as possible, a specific sequence for lowering the falsework was determined, cycling through the independent scaffolding towers in several rounds.

Using imprecise formwork

In traditional cut-stone or stereotomic stone vaulting, voussoirs are never placed directly on falsework. Instead they are positioned using shims. This insight was used as a pragmatic formwork strategy that provided a way to deal with the rough interior surfaces of the stones. The wooden falsework was offset inward/downward by 3cm. As a result, large wooden shims could be placed in between the rough, knocked-off fins to support the stones on the falsework and precisely control their position. Additionally, this meant that precise positioning of the falsework sections was less critical. This resulted in significant time-saving and reduced logistical challenges. As an added bonus, the shims served as visual guides during decentring. Once they started falling on the ground, the shell was standing by itself.

Not building the designed geometry

Due to unavoidable machining tolerances, each of the voussoirs could only be within +/-0.4mm of the designed digital geometry. Since the vault was designed to have a high degree of structural redundancy and indeterminacy by introducing locally high degrees of double curvature, these small imprecisions had little or no effect on the structural integrity and behaviour of the overall

5. The falsework consists of a plywood waffle structure on top of standard scaffolding towers.
Image: Anna Maragkoudaki/Block Research Group, ETH Zurich.

6. The stone courses were built up starting from the supports. Voussoirs of the edge arches were positioned before closing the subsequent course to better control the global positioning of the stones.
Image: Anna Maragkoudaki/Block Research Group, ETH Zurich.

7. The different articulation of the intrados and extrados of the stone shell results from a combination of fabrication constraints, machining efficiency and aesthetic considerations. Image: Iwan Baan.

292 / 293

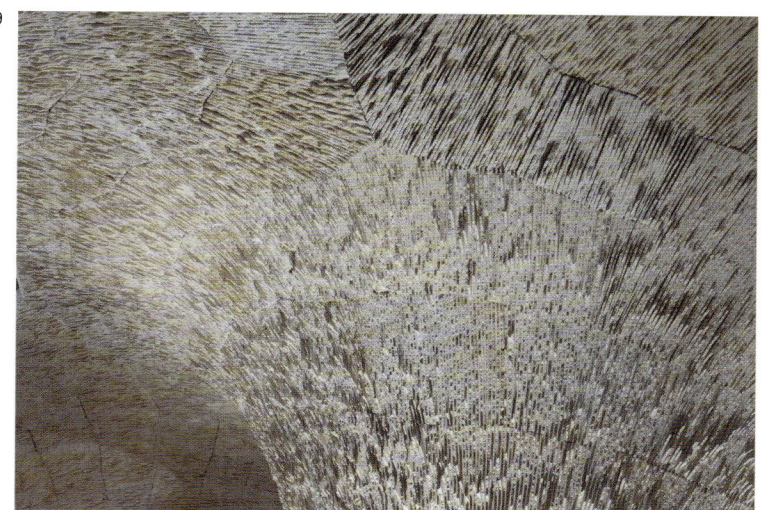

8

9

10

structure. However, over 14 courses of stones, these tolerances can quickly add up to large geometric discrepancies in the 'keystone' rows. Therefore the voussoirs in these rows were cut last, after construction had already begun, based on measurements of the partially assembled structure. Before decentring the test assembly, the 'as-built' geometry of the structure was recorded and the position of each stone relative to its neighbours was marked on the interfaces.

Since slight deviations of a fraction of a degree in placement angle at the base (or in fact anywhere along any row) cause significant deviations higher up, several strategies had to be developed. The masons would build a few rows, finish some of the edge arches and check that everything closed. If not, they would take down the rows, adjust, reposition and realign, repeating the entire process as needed (Fig. 6).

For structural reasons, it was much more important to have contacts that were as tight as possible between stones so that, after decentring, no uncontrollable and unpredictable settling of the assembly would occur. Using the above-mentioned shimming, the masons 'jiggled' every stone until all interfaces were tight. Where necessary, the interfaces were sanded off to improve the fit. The level of precision reached through manually trimming a stone depends on its initial geometry. Flat surfaces can easily be processed with simple templates and tools. Therefore the geometry of all interfaces was constrained to planar and ruled surfaces depending on their local alignment to the courses being perpendicular or parallel respectively.

A successful marriage of precision engineering and craft experience

The Armadillo Vault represents the close collaboration of engineers, designers and skilled stonemasons and builders. It is the culmination of over 10 years of joint research in stone construction, demonstrating that, with advanced, non-standard engineering approaches and novel equilibrium design methods, expressive geometries can be safely developed and – through combining optimised digital fabrication processes and experienced craft – successfully constructed. Proportionally only half as thick as an eggshell and standing without steel reinforcement, the expressively flowing stone surface challenges the conception that complex geometry need go hand-in-hand with inefficient use of material (Figs. 7–10). While the vault's architectural geometry was optimised in order to achieve all structural and fabrication constraints, and although a smooth digital pipeline with advanced data structures was developed to eliminate any possibility of human error in the handling and logistics, in the end it was the experienced human hand that locally controlled precision.

Acknowledgements

Structural design and architectural geometry: the Block Research Group, ETH Zurich – Philippe Block, Tom Van Mele, Matthias Rippmann, Edyta Augustynowicz, Cristián Calvo Barentin, Tomás Méndez Echenagucia, Mariana Popescu, Andrew Liew, Anna Maragkoudaki, Ursula Frick.

Structural engineering: Ochsendorf DeJong and Block (ODB Engineering) – Matthew DeJong, John Ochsendorf, Philippe Block, Anjali Mehrotra.

Fabrication and construction: the Escobedo Group – David Escobedo, Matthew Escobedo, Salvador Crisanto, John Curry, Francisco Tovar Yebra, Joyce I-Chin Chen, Adam Bath, Hector Betancourt, Luis Rivera, Antonio Rivera, Carlos Rivera, Carlos Zuniga Rivera, Samuel Rivera, Jairo Rivera, Humberto Rivera, Jesus Rosales, Dario Rivera.

Supported by Kathy and David Escobedo, ETH Zurich Department of Architecture, MIT School of Architecture + Planning, NCCR Digital Fabrication, Swiss Arts Council – Pro Helvetia, Artemide.

Thanks to Noelle Paulson for assistance with writing.

References

Block, P. and Ochsendorf, J., 2007, 'Thrust Network Analysis: a New Methodology for Three-dimensional Equilibrium' in *Journal of the International Association for Shell and Spatial Structures*, 48(3), p.1-8.

Li, Y., Liu, Y. and Wang, W., 2015, 'Planar Hexagonal Meshing for Architecture' in *IEEE Transactions on Visualization and Computer Graphics*, 21(1), p.95-106.

Rippmann, M., Van Mele, T., Popescu, M., Augustynowicz, E., Méndez Echenagucia, T., Calvo Barentin, C., Frick, U. and Block, P., 2016, 'The Armadillo Vault: Computational Design and Digital Fabrication of a Freeform Stone Shell' in *Advances in Architectural Geometry 2016*, p.344-363.

Van Mele, T., Panozzo, D., Sorkine-Hornung, O. and Block, P., 2014, 'Best-fit Thrust Network Analysis – Rationalization of Freeform Meshes' in *Shell Structures for Architecture: Form Finding and Optimization*, Routledge, London.

Van Mele, T., Mehrotra, A., Méndez Echenagucia, T., Frick, U., Augustynowicz, E., Ochsendorf, J., DeJong, M. and Block, P., 2016, 'Form Finding and Structural Analysis of a Freeform Stone Vault' in *Proceedings of the International Association for Shell and Spatial Structures (IASS) Symposium 2016*.

8. The finished cut-stone vault wraps around an existing column in the Corderie dell'Arsenale. Image: Iwan Baan.

9. The pattern on the intrados is carefully controlled to globally align with the flow of forces within the stone structure. Image: Anna Maragkoudaki/Block Research Group, ETH Zurich.

10. The intrados pattern makes the structure's geometry more legible. Image: Philippe Block/Block Research Group, ETH Zurich.

END

EDITORS BIOGRAPHIES

Achim Menges

Achim Menges, born in 1975, is a registered architect and Professor at the University of Stuttgart, where he has been the Founding Director of the Institute for Computational Design since 2008. He has also been Visiting Professor in Architecture at Harvard University's Graduate School of Design since 2009.

He graduated with honours from the AA School of Architecture in London, where he subsequently taught as Studio Master of the Emergent Technologies and Design Graduate Programme from 2002-09, as Visiting Professor from 2009-12 and as Unit Master of Diploma Unit 4 from 2003-06. From 2005-08, he was Professor of Form Generation and Materialisation at the HfG Offenbach University for Art and Design in Germany. In addition, he has held visiting professorships in Europe and the United States.

His practice and research focuses on the development of integral design processes at the intersection of morphogenetic design computation, biomimetic engineering and computer aided manufacturing that enables a highly articulated, performative built environment. His work is based on an interdisciplinary approach in collaboration with structural engineers, computer scientists, material scientists and biologists. He has published several books on this research and its related fields, and is the author/co-author of numerous articles and scientific papers. His projects and design research have received many international awards, have been published and exhibited worldwide and are exhibited in several renowned museum collections, including the permanent collection of the Centre Pompidou in Paris. Achim Menges is a member of several international research evaluation boards and a member of numerous scientific committees of leading peer-reviewed international journals and conferences.

Bob Sheil

Bob Sheil is co-founder and co-Chair of the FABRICATE Conference Series, Professor of Design through Production and Director of The Bartlett School of Architecture, UCL. Among numerous articles and papers, he is the editor of *Manufacturing the Bespoke* (Wiley, 2012) and has edited three issues of *AD*: *High Definition: Zero Tolerance in Design and Production* (Wiley, 2014), *Protoarchitecture: Analogue and Digital Hybrids* (Wiley, 2008) *and Design through Making* (Wiley, 2005). His built work includes the award-winning 55/02 Shelter at Kielder Water and Forest Park, Northumberland, UK (2008), designed and fabricated in collaboration with Stahlbau GmbH.

Ruairi Glynn

Ruairi Glynn is co-founder and co-Chair of the FABRICATE Conference Series. He is Director of the Interactive Architecture Lab at The Bartlett School of Architecture, UCL. Alongside his teaching and research, he practises as an installation artist, recently exhibiting at the Centre Pompidou in Paris, the National Art Museum of China in Beijing and the Tate Modern in London. He has worked with cultural and research institutions including the Royal Academy of the Arts, the Medical Research Council and the BBC, and has built public works for Twitter, Nike, Arup, BuroHappold and Bank of America Merrill Lynch. He co-edited *Digital Architecture, Passages Through Hinterlands* (2009), *Fabricate. Making Digital Architecture* (2011) and *Fabricate, Rethinking Design and Construction* (2017), alongside curating over a dozen international conferences, competitions and exhibitions. He has taught internationally, including at ETH Zurich, TU Delft, The Royal Danish Academy of Fine Arts, Copenhagen, and the Angewandte, Vienna. He is Programme Director of The Bartlett's new MArch Design for Performance and Interaction Masters, beginning in October 2017.

Marilena Skavara

Marilena Skavara has been an integral part of FABRICATE since its inaugural event in 2011. She holds a degree from the National Technical School of Architecture in Athens and completed the MSc in Adaptive Architecture and Computation at The Bartlett School of Architecture, both with distinction. Her Masters thesis project 'Adaptive Fa[CA]de' was exhibited during London Design Week in 2009 and has since been published in several publications and conferences. She has tutored and run Arduino/Processing workshops at The Bartlett, the Architectural Association, the École Speciale d'Architecture, University of Lund and the Centre for Information Technology and Architecture (CITA) in Copenhagen. She is Director at Codica, a London-based digital product studio, where she explores new interactions through digital services, wearables and prototyping.

CONTRIBUTORS BIOGRAPHIES

Mania Aghaei Meibodi is a practising architect. She co-founded meonia, a Stockholm and Toronto-based architecture and design practice. Alongside this, she has been a part-time postdoctoral scholar at Digital Building Technologies (DBT) at the Department of Architecture, ETH Zurich. Her research focuses on ways of creating geometrically complex architecture based on the close interplay between computational design, digital fabrication and new materials. At DBT, she exploits the potential of additive manufacturing (AM) for designing and realising geometrically complex forms. In this context, she uses AM to fabricate moulds with complex inner and outer features, which are then used to form cast concrete, aluminium and glass parts with geometric complexity and surface detailing.

Sean Ahlquist is an Assistant Professor of Architecture at the Taubman College of Architecture and Urban Planning. He is part of the Cluster in Computational Media and Interactive Systems, which connects architecture with the fields of material science, computer science, art, design and music. Research is centred on material computation, developing articulated material structures and modes of design which enable the study of spatial behaviours and human interaction. This is explored through the design and fabrication of pre-stressed lightweight structures, innovations in textile-reinforced composite materials and tactile sensorial environments as interfaces for physical interaction.

Yousef Alqaryouti is a PhD candidate at the University of Queensland with a particular interest in digital fabrication, modular housing and lightweight structures. Prior to enrolling at the University of Queensland, he worked for six years as a structural design engineer for different global consulting and engineering companies. He received an MSc in Civil Engineering/Structures with distinction from the University of Jordan (2014) and a BSc in Civil Engineering with distinction from the Hashemite University (2010).

Felix Amtsberg is a SUTD-MIT Postdoctoral Fellow at Singapore University of Technology and Design. He studied architecture at the University of Kassel, and received his diploma degree in 2008 and his Master of Science in 2010. From 2011, he worked as a Scientific Assistant at the Institute for Structural Design (Graz, University of Technology), where he was the coordinator of the ABB-Robotic Research and Teaching Laboratory and established the research focus on digital fabrication of load-carrying structures and sensory informed production. He received his PhD in 2016 from Stefan Peters and co-supervisor Prof. Martin Bechthold (Harvard GSD).

Inés Ariza is an architect and researcher at the Digital Structures research group at the Department of Architecture at MIT. Through her current research work, Ariza is developing digital workflows for physics-, fabrication- and assembly-based construction details for robotic assembly. Ariza holds a diploma in Architecture from the University of Buenos Aires and an MS in Design and Computation from MIT, and has served as a 2014-16 Fulbright Scholar and 2016 Quarra Matter Fellow.

Kim Baber is a Fellow in Civil Engineering and Architecture at the University of Queensland. He is also a registered architect and Principal of Baber Studio, and in 2015 his practice was awarded the Queensland Emerging Architect Prize by the Australian Institute of Architects. In 2016, Baber received the Gottstein Trust Fellowship to research timber fabrication technologies being developed in Europe and Japan. Through his practice and research, Baber works with architecture and civil engineering students to design, develop and construct prototypes that test innovations in the structural use of timber.

Ehsan Baharlou is a doctoral candidate at the Institute for Computational Design and Construction (ICD) at University of Stuttgart. He holds a Master of Science in Architecture with distinction from the Islamic Azad University of Tehran. Along with pursuing his doctoral research, he has taught seminars on computational design techniques and design thinking at the ICD since 2010. His research is currently focused on the integration of fabrication and construction characteristics into computational design for form generation via agent-based modelling and simulation.

Ball-Nogues Studio, established in 2004, is a collaboration between **Benjamin Ball** and **Gaston Nogues**. Their work is informed by the exploration of craft. Essential to each project is the design of the production process itself, with the aim of creating environments that enhance sensation, generate spectacle and invite physical engagement. While both partners were trained as architects at the Southern California Institute of Architecture, they have produced a diverse body of work that explores the intersection of public art, architecture and fabrication.

Ricardo Baptista joined AKT II in 2005 after completing a Master's degree at Imperial College London, balancing his previous academic experience gained from the Technical Superior Institute in Lisbon. As a director at AKT II, he has enjoyed the design-led approach and emphasis on creativity, playing a lead role in a number of ground-breaking projects in the UK and abroad, including in the US, Scandinavia and the Middle East. Baptista has collaborated with a host of high-profile architectural practices, producing buildings that range from small-scale schemes in the private and public sectors to large museums, airports and stadia.

Michael Bergin is Principal Research Scientist, Product Designer and Team Leader for the User Experience team in the Design Research group, part of the Office of the CTO and Autodesk Research at Autodesk in San Francisco, California. Bergin contributes to the emerging field of generative design systems in domains ranging from automotive and aerospace to manufacturing and the built environment.

Mathias Bernhard studied architecture at ETH Lausanne and ETH Zurich. After specialising in digital fabrication while working at the Rapid Architectural Prototyping Laboratory and gaining a Master of Advanced Studies in Computer-Aided Architectural Design (MAS CAAD), he joined the CAAD group at ETH Zurich as a research and teaching assistant. His work mainly focuses on the intersection between architecture and information technology, combining machine learning and artificial intelligence with computational geometry. He joined the group for Digital Building Technologies in autumn 2015, investigating large-scale additive manufacturing in architecture.

Phil Bernstein is an architect and an Autodesk Fellow. He teaches Professional Practice at Yale, where he received both his BA and his MArch. He is co-editor of *Building (In) The Future: Recasting Labor in Architecture and BIM In Academia*, and he is a Senior Fellow of the Design Futures Council. Formerly, Bernstein was Chair of the AIA National Contract Documents Committee and a Principal with Pelli Clarke Pelli Architects.

Shajay Bhooshan is an Associate at Zaha Hadid Architects, where he heads the research activities of the Computation and Design (CoDe) group. He also works as a studio master at the AA DRL Masters programme in London. He pursues his research in structure and fabrication-aware architectural geometry as a Research Fellow at the Block Research Group, ETH Zurich. Previously, he worked at Populous, and completed an MPhil at the University of Bath and an MArch at the AA. He has taught and presented work at various professional conferences events and institutions, including AAG 2016, simAud 2015, 2014 and 2010, Siggraph 2008, ETH Zurich, Yale University and University of Applied Arts, Vienna, CITA Copenhagen, IAAC Barcelona and Innsbruck University.

Philippe Block is Associate Professor at the Institute of Technology in Architecture, ETH Zurich, where he co-directs the Block Research Group (BRG) and is Deputy Director of the Swiss National Centre of Competence in Research – Digital Fabrication. BRG's research focuses on structural form-finding as well as the optimisation and fabrication of shell structures. Block studied architecture and structural engineering at the VUB in Belgium and at MIT in the US, where he obtained his PhD in 2009. As a Partner at Ochsendorf DeJong & Block, he provides structural assessment of historic monuments and design and engineering of novel compression structures.

Hua Chai holds a Bachelors degree in Architecture from Tongji University, Shanghai. Currently, he is a postgraduate student at Tongji University. His research focuses on the digital design and robotic fabrication of wood tectonics based on structural performance.

Chin-Yi Cheng is a designer, engineer and programmer, as well as a cook. He loves to switch between these roles and even play more than one of them at the same time. He has a dual Masters degree in Computational Design and Computer Science at MIT, and majored in Architecture and Mechanical Engineering during his undergraduate studies. As part of a multidisciplinary research team at MIT Media Lab, he explores new ideas to combine design, materials, interaction, digital fabrication, biology and molecular gastronomy. As an independent researcher, he focuses on helping designers and engineers interact with artificial intelligence.

Esben Clausen Nørgaard is a civil engineer. He graduated from Aalborg University in Architectural Design in 2014, and joined CITA after graduation. His primary research and interest lies in prototyping, fabrication and rationalisation. Since joining CITA, his primary focus has been on fabrication with industrial robots and how these can be used to create relationships between traditional craftsmanship and digital environments.

Brandon Clifford is an Assistant Professor at MIT and Principal at Matter Design. Clifford received his Master of Architecture from Princeton University in 2011 and his Bachelor of Science in Architecture from Georgia Tech in 2006. He worked as project manager at Office dA from 2006-09, was a LeFevre Fellow at OSU from 2011-12 and was a Belluschi Lecturer at MIT from 2012-16. Clifford has been awarded the Design Biennial Boston Award, the Architectural League Prize and the prestigious SOM Prize. Clifford's translation of past knowledge into contemporary practice continues to provoke new directions for digital design.

David Correa is an Assistant Professor at the University of Waterloo and a Doctoral Candidate at the Institute for Computational Design (ICD) at the University of Stuttgart. At the ICD, Correa initiated and led the research field of bio-inspired 3D printed programmable material systems. His doctoral research investigates the reciprocal relationship between material design and fabrication from a multi-scalar perspective. With a focus on climate-responsive materials for the built environment, the research integrates computational tools, simulation and digital fabrication with bio-inspired design strategies for material architectures. As a designer in architecture, product design and commercial digital media, Correa's professional work engages multiple disciplines and environments, from dense urban settings to remote northern regions.

Graham Cranston is a Structural Engineer with Simpson Gumpertz & Heger (SGH). He specialises in the analysis and design of non-standard structures, as well as investigations of underperforming and failed structures. Cranston uses high fidelity structural models, finite element analysis and computational tools to solve diverse structural problems. He is currently developing tools and methods to enable the use of 3D printed structural components in buildings. Cranston is a licensed structural engineer in Vermont and Hawaii.

Xavier De Kestelier is a Principal at HASSELL, where he is the Head of Design Technology and Innovation. He was previously a Partner at Foster + Partners, where he jointly headed its Specialist Modelling Group, a project-driven R&D group that specialises in complex geometrical problems, computation and building physics. He has also been involved with several funded research projects on the application of additive manufacturing on an architectural scale with both NASA and ESA. He has been a Visiting Professor at the University of Ghent, an Adjunct Professor at Syracuse University and a Teaching Fellow at The Bartlett School of Architecture, UCL. In 2010, he became a Director at Smartgeometry, a non-profit organisation that promotes advances in digital design in architectural research and practice.

Benjamin Dillenburger is an architect and Assistant Professor for Digital Building Technologies at the Institute of Technology in Architecture (ITA) at ETH Zurich. His research focuses on the development of building technologies based on the close interplay of computational design methods, digital fabrication and new materials. In this context, he searches for ways to exploit the potential of additive manufacturing for building construction and investigates the specific performance and tectonic possibilities of high-resolution additive printing processes. The aim is to improve the quality of planning and construction, as well as to open up new paths towards a richer and more engaging architecture.

Moritz Dörstelmann is a Research Associate at the Institute for Computational Design and Construction (ICD) at Stuttgart University and Guest Professor for Emerging Technologies at the Technical University of Munich. He studied architecture at the RWTH Aachen University and the University of Applied Arts in Vienna. Dörstelmann's research focuses on integrated computational design and fabrication strategies for novel fibre composite building systems. Through interdisciplinary cooperation, he has realised a series of fibrous lightweight structures which are material-efficient and explore a novel architectural design repertoire. His research achieves the simultaneous advancement of building culture and technology.

Gershon Dublon is an artist, electrical engineer and PhD student at the MIT Media Lab. In his research, he develops new interfaces to sense data and explores how sensor networks might become prosthetic through attention-aware devices and environments. His artwork explores similar themes of transduction, perception and attention. His work has appeared at conferences including IEEE Sensors, NIME and CHI, in media such as the *NYT*, *Wired* and *Scientific American* and in museums and festivals worldwide. Dublon has a Masters from MIT and a BSEE from Yale.

Jeg Dudley is a computational designer currently working within the Parametric Applied Research Team (P.art) at AKT II. He received architecture degrees from the University of Bath before working for companies across the design spectrum, including RoboFold, Heatherwick Studio and Atmos. Dudley has extensive industry experience in the design, optimisation and construction of complex geometries. He has run workshops and lectured at several academic institutions, including the Royal College of Art, UCL and the University of Westminster, where he is a Visiting Lecturer.

David Escobedo is founding owner of the Escobedo Group, where he leads a six-division vertically integrated general contracting firm. Escobedo's primary focus is utilising digital fabrication and computation in the steel, stone and millwork divisions. Additionally, Escobedo has developed a panelised construction process that delivers fully systems-integrated prefabricated wall, floor and ceiling panels that mitigate the customary high-end residential construction issues of weather, labour force scarcity and site risks. Escobedo is known within the industry for his technical design capabilities, his skills in fabricating complex and challenging structures and his problem-solving techniques.

Jelle Feringa is co-founder of Odico Formwork Robotics and EZCT Architecture and Design Research. While developing his PhD thesis at Hyperbody TU Delft, Feringa established a robotics lab at the RDM innovation dock in the Rotterdam harbour in 2011. This research formed the technological foundation for Odico Formwork Robotics. Feringa is CTO at Odico and developer of the offline robotics programming platform PyRAPID that lies at the heart of Odico's operation. Since 2014, Odico has rolled out its fabrication technology and as such is involved in the realisation of several high-profile construction projects.

Ida Katrine Friis Tinning is an architect and Kaospilot member, with a focus on computational design and textile architecture. At the Royal Academy of Fine Arts School of Architecture, Tinning graduated from CITA Studio in Computation in Architecture. Since 2014, Tinning has worked for CITA on the Complex Modelling Projects Tower and Lace Wall and as an assistant teacher. Besides the research projects at CITA, she has worked for and helped to set up File under Pop, and worked as a project leader and architect for architectural practices.

Augusto Gandía studied architecture at the University of Mendoza, Argentina, and completed a Masters in Media Architecture at the Bauhaus University Weimar, Germany. He has worked for architectural offices in Madrid and Stuttgart. In June 2015, he joined Gramazio Kohler Research at the National Centre of Competence in Research (NCCR) Digital Fabrication at ETH Zurich as a PhD researcher. His research focuses on the relationship between computational design and the simulation of architectural fabrication processes.

Joseph M. Gattas is a civil engineer and Lecturer at the University of Queensland. He leads the Folded Structures Lab, a research group working in origami-inspired engineering and more broadly on parametric geometry, sheet material fabrication methods and applications developed at the intersection of the two. He is also a member of the Queensland Centre for Future Timber Structures and is interested in all aspects of design, making and coding.

Hélori Geglo is Project Manager (International) at Hess Timber. He holds a technology degree in Mechanical Engineering from the University of Rennes, France, as well as an engineering diploma in Wood Construction and Technology from the École Supérieure du Bois in Nantes, France. At Hess Timber, he has managed projects such as the H16 aircraft maintenance hall in Cannes airport, France, the wooden sculpture Mobiversum by architect J. Mayer H. in Wolfsburg, Germany, the construction of wooden lighting high masts in Busan, South Korea, and the Seine Musicale by Shigeru Ban in Paris, France.

Fabio Gramazio is an architect with multidisciplinary interests, ranging from computational design and robotic fabrication to material innovation. In 2000, he founded the architecture practice Gramazio Kohler Architects in conjunction with his partner Matthias Kohler, where numerous award-winning designs have been realised. Also responsible for the world's first robotic architectural laboratory at ETH Zurich, their research has been formative in the field of digital architecture, merging computational design and additive fabrication through the customised use of industrial robots. Gramazio's work has been widely published and internationally exhibited, and is comprehensively documented in the book *The Robotic Touch – How Robots Change Architecture*.

Anthony Hauck has been involved in architecture, engineering, construction and technology for more than 30 years. After a career in architecture and construction, he joined the Autodesk Revit team in 2007, holding a succession of Product Management positions in the group until becoming Director of Product Strategy for Autodesk AEC Generative Design in 2015, where he is responsible for helping define the next generation of building software products and services for the AEC industry.

Darron Haylock joined Foster + Partners in 1995, working initially on a research and development laboratory and office tower for Daewoo Electronics in Seoul. He has worked in a number of academy schools, including Thomas Deacon Academy and West London Academy. He was Partner-In-Charge of a series of hospitals for Circle Healthcare at sites in Bath, Southampton and Manchester. More recently, he has led the design team for the Edmund and Lily Safra Brain Sciences Building at the Hebrew University in Jerusalem and the new Maggie's at the Robert Parfett Building in Manchester. He was made a Partner in 2004.

Volker Helm is a Professor at the University of Applied Sciences and Arts of Dortmund. He completed his studies in architecture at the University of Siegen, Germany, before specialising in CAAD as a Master of Advanced Studies under the Chair of Professor Ludger Hovestadt at ETH Zurich. His studies focused on computer-aided architectural design and automated production. He also worked for six years at Herzog & de Meuron in Basle, with a focus on the development, programming and realisation of complex geometries. From 2010-17, he was a researcher for the Chair of Architecture and Digital Fabrication at ETH Zurich. In 2015, he completed his doctoral thesis, with a focus on robot-based construction processes onsite.

In a career spanning three decades, **Ralph Helmick** has created over forty major public sculptures across the US, generating experimental civic designs linked more by underlying aesthetic principles than a signature style. Optical consolidation is a consistent theme in his artwork, manifesting itself in sculptures that knit classic visual phenomena (anamorphosis, pointillism) with modern fabrication technologies (3D scanning of handmade forms, 3D manufacturing). He is currently working on his first international commission.

Anders Holden Deleuran is a PhD Fellow at CITA (KADK) in Copenhagen. His research extends the interdisciplinary Complex Modelling project, specifically investigating integrative development in non-linear, dynamic and multi-stage computational design modelling. He previously worked for the international architectural practice Aedas as a computational design researcher within their R&D group. Prior to this, he was a research assistant at CITA, working on numerous projects including collaborations with Mark Burry and Philip Beesley. Deleuran has extensive tutoring experience from CITA and other institutes, including RMIT, KTH Stockholm, TU Delft, UDK Berlin, ETH Zurich, Architectural Association, Aalborg University and Smartgeometry.

Christopher Hutchinson is a Professor in Materials with an expertise in the processing, structure and properties of metals and alloys. His work spans fundamental aspects of both metal alloy design and structure, through to applied projects involving new steels for automotive applications or aluminium alloys for aerospace. Hutchinson makes use of both advanced experimentation, including characterisation of an alloy's micro and atomic structure, and simulation models for the relationship between the structure and its properties.

Hiroshi Ishii is the Jerome B. Wiesner Professor of Media Arts and Sciences at the MIT Media Lab, where he currently directs the Tangible Media Group. He joined the MIT Media Lab in October 1995. Ishii and his team have presented their visions of Tangible Bits and Radical Atoms at a variety of academic, design and artistic venues (including ACM SIGCHI, ACM SIGGRAPH, the Industrial Design Society of America, AIGA, Ars Electronica, ICC, Centre Pompidou, the Victoria and Albert Museum and Milan Design Week), emphasising that the development of tangible interfaces requires the rigour of both scientific and artistic review.

Manuel Jiménez García is the co-founder and Principal of madMdesign, a computational design practice based in London. His work has been exhibited worldwide in venues such as Centre Pompidou (Paris), Canada's Design Museum (Toronto), the Royal Academy of Arts (London), Zaha Hadid Design Gallery (London), Clerkenwell Design Week (London) and X Spanish Architectural Biennale (Madrid). Alongside his practice, Manuel is currently a Lecturer at The Bartlett School of Architecture, UCL, where he directs the Architectural Design Research Cluster 4 and Unit 19 on the MArch programme; in addition, he curates Plexus, a multidisciplinary lecture series based on computational design.

Andrei Jipa is a doctoral student at the Digital Building Technologies Chair, ITA, D-Arch, ETH Zurich. He studied architecture at Ion Mincu University in Bucharest, the University of Sheffield and the University of Westminster in London. After his diploma, he founded jamD, a digital fabrication and parametric design studio based in London. Jipa taught Computational Design to MArch, MSc and MAS students at the University of Westminster, Oxford Brookes University, The Bartlett and ETH Zurich. His current research focuses on 3D printing formwork for concrete building components.

Paul Kassabian is a structural engineer and Associate Principal at Simpson Gumpertz & Heger (SGH). He works on a wide range of structural systems such as buildings, bridges and sculptures. At SGH, Kassabian develops innovative methods of design and construction using SGH's in-house materials lab, computational design approaches and digital fabrication techniques. He taught graduate structural engineering students at MIT for ten years and currently teaches the Structures course at Harvard GSD. He is researching construction techniques informed by termite collective construction and robotic swarm behaviour with Harvard's Wyss Institute. Kassabian is a licensed structural engineer in 14 US states and one Canadian province, and is a chartered engineer in the UK.

James Kingman studied civil and structural engineering at the University of Leeds. He is currently a Design Engineer at AKT II, where he has worked on a wide range of projects involving advanced analysis and simulation, including commercial developments, long span tensile structures and innovative bridges. Kingman has published a number of academic works on applications of topology optimisation in structural engineering and also contributed to the AKT II publication *Design Engineering Re-Focused*. He has lectured at UCL and the University of Bristol.

Since 2011, **Harald Kloft** has been a Professor in Structural Design and the Head of the Institute of Structural Design (ITE) at Braunschweig University of Technology in Germany. His research is focused on designing resource-efficient non-standard structures in the context of digital fabrication processes. Before Braunschweig, he taught as a full Professor for Structural Design at the University of Kaiserslautern (2000-11) and at Graz University of Technology (2007-09). From 2000-08, he was also Visiting Professor for Structural Design in the architectural class of the Städelschule in Frankfurt. Besides his academic background, Kloft is co-founder and Partner of the international engineering practice osd – office for structural design in Frankfurt.

Jan Knippers specialises in lightweight structures and innovative materials. He is Partner and co-founder of Knippers Helbig Advanced Engineering. The focus of their work is on efficient structural design for international architecturally demanding projects. Since 2000, Knippers has been Head of the Institute for Building Structures and Structural Design (ITKE) at the University of Stuttgart and is involved in many research projects on fibre-based materials and biomimetics in architecture. As such, he is a speaker at the Collaborative Research Centre 'Biological Design and Integrative Structures'. Knippers completed his PhD in civil engineering at the Technical University of Berlin in 1992.

Matthias Kohler is an architect with multidisciplinary interests ranging from computational design and robotic fabrication to material innovation. In 2000, he founded the architecture practice Gramazio Kohler Architects in conjunction with his partner Fabio Gramazio, where numerous award-winning designs have been realised. Also responsible for the world's first robotic architectural laboratory at ETH Zurich, their research has been formative in the field of digital architecture, merging computational design and additive fabrication through the customised use of industrial robots. Kohler's work has been widely published and internationally exhibited, and is comprehensively documented in the book *The Robotic Touch – How Robots Change Architecture*. Since 2014, Kohler has also been the Director of the National Centre of Competence in Research (NCCR) Digital Fabrication.

Radovan Kovacevic is Professor of Mechanical Engineering and holds the Herman Brown Chair in Engineering at Southern Methodist University, Dallas, Texas. He is Director of the Research Center for Advanced Manufacturing and the Center for Laser-Aided Manufacturing at SMU. He has over 40 years of research and teaching experience in the field of mechanical engineering. He holds seven US patents and has authored or co-authored over 600 technical papers and six books. He is a Fellow of the American Society of Mechanical Engineers, the Society of Manufacturing Engineers and the American Welding Society. He has held fellowships from the Fulbright Foundation, the Carl Duisberg Foundation and the Alexander von Humboldt Foundation.

Benjamin S. Koren was born in Frankfurt and grew up in Miami, Florida. He studied architecture, film and music at NYU, the University of Miami, the Architectural Association in London and the Angewandte in Vienna. He went on to work for the Advanced Geometry Unit at Arup in London and for Herzog & de Meuron in Basel. He is the founder and Managing Director of ONE TO ONE (www.onetoone.net), a computational geometry and digital fabrication consultancy set up in 2009, with offices in Frankfurt and New York. He lives and works in New York.

William Kreysler is founder and CEO of Kreysler & Associates (K&A), a custom moulder of fibre-reinforced products located in Napa County, California. K&A has won awards for excellence in the manufacture of fibre-reinforced polymer (FRP) architectural products, industrial products and large-scale sculptures. The firm has customers throughout North America, Europe and Asia. Kreysler is also a founding member and President of the Digital Fabrication Network, an international network of cross-industry professionals working together to improve processes and accelerate the adoption of digital fabrication tools and techniques.

Riccardo La Magna is a structural engineer and PhD candidate at the Institute of Building Structures and Structural Design (ITKE) at the University of Stuttgart. In his research, he focuses on simulation technology, innovative structural systems and new materials for building applications.

Ting-Uei (Jeff) Lee is a PhD candidate at the University of Queensland, where he explores the intersection of curved-crease folded geometries, parametric modelling and the manufacture of composite structures.

Juhun Lee is a computational designer at Simpson Gumpertz & Heger. His architectural design background includes both his undergraduate work and two years of professional experience in Korea. Lee studied computational design related to building structures and energy simulation in his graduate program at Harvard University. Currently, he is working closely with structural design and building science groups to provide Simpson Gumpertz & Heger and its clients with innovative design solutions.

Scott Leinweber was a visiting Fulbright researcher at CITA during 2015-16, where he investigated how computational tools are changing traditional Danish design culture. Now back in New York, Leinweber is a creative technologist and designer who explores the digital-physical dialogue of craftsmanship today. He works with artists, product designers and architects to realise projects as diverse as interactive art and sculpture, product design, video, information mapping and architectural spaces.

AL_A (Amanda Levete Architects) is the award-winning architecture and design studio founded in 2009 by the RIBA Stirling Prize-winning architect Amanda Levete with Directors Ho-Yin Ng, Alice Dietsch and Maximiliano Arrocet. AL_A's approach to design balances the intuitive with the strategic, drawing on a foundation of rigorous research, innovation, collaboration and painstaking attention to detail. In every project, however modest in scale, AL_A try to advance debate, be it analytical response, social purpose, manufacturing technique or material innovation.

Hendrik Lindemann works as a research assistant at the Institute of Structural Design in Braunschweig and is the Design by Technology Chair at Folkwang University of the Arts in Essen. Lindemann studied architecture at the Technical University Braunschweig in Germany and the Technical University in Delft. In 2015, he graduated from Braunschweig with high honours. He has led several workshops on digital design and fabrication techniques. The focus of his current research is the influence of digital fabrication methods on the design workflow and its physical results in industrial products and architectural design.

Henry David Louth is a Senior Designer at Zaha Hadid Architects as part of the Computation and Design (CoDe) group. His research examines ruled and developable constraint-based solving, curve-crease folding (CCF) applications and digital simulation of analogue material behaviour. He completed his MArch from the Architectural Association (AA) Design Research Lab (DRL) in 2014 and has been a registered architect and LEED AP in the US since 2011. He has taught and presented at the AA London, the AA Visiting School in India, AIA Louisiana and USGBC, and is a contributor to eCAADe and simAUD.

Richard Maddock is an architect, builder and software engineer. He holds a degree in Computer Systems Engineering from the University of Tasmania and for more than a decade wrote software for companies such as Porsche, IBM and ANZ Banking Group. After working in construction for several years, he turned his attention to architecture, receiving a Master of Architecture from The University of Melbourne. In 2014, he joined Foster + Partners as a member of the Specialist Modelling Group.

Deyan Marzev is a 3D designer based in London, UK, with a Masters degree in Structural Engineering from the University of Architecture, Civil Engineering and Geodesy in Sofia, Bulgaria. In 2007, he joined Adams Kara Taylor (now AKT II) as a structural CAD technician, and in 2013 became a member of the computational research team (p.art®). His work at AKT II includes projects such as Knight Architects' Merchant Square Footbridge (2014), AHMM's 240 Blackfriars (2012-14), Foster + Partners' Bloomberg Headquarters, London (2012-present) and multiple Zaha Hadid Architects projects, including Grand Theatre de Rabat (2014-present).

Heath May AIA is Associate Principal and Director of HKS Laboratory for INtensive Exploration. Holding a Master of Architecture degree from Texas Tech University CoA, May leads a design team responsible for projects including future:GSA, a net-zero renovation solution that earned the 2012 WAN Commercial Building of the Year Award, and Sustainable Urban Living, a winner of the 2010 Chicago Athenaeum Green Good Design Award. May currently serves on the Advisory Board of the PACCAR Technology Institute at the University of North Texas and is an Advanced Graduate Design Studio Lecturer at University of Texas Arlington, CAPPA, School of Architecture.

Wes McGee is an Assistant Professor in Architecture and the Director of the FABLab at the University of Michigan Taubman College of Architecture and Urban Planning. His work revolves around the interrogation of machinic craft and material performance, with a research and teaching agenda focused on developing new connections between design, engineering, materials and process through the creation of customised software and hardware tools. As a founding Partner and Senior Designer at the studio Matter Design, his work spans a broad range of scales and materials, always dedicated to reimagining the role of the designer in the digital era.

Rich Merlino is the President of Addaero Manufacturing. For most of his career, he served in various sales and operations capacities for Pratt & Whitney, a division of United Technologies. He has a BS in Mechanical Engineering from UMass Amherst and an MS in Supply Management and Finance from Indiana University.

Ammar Mirjan is an architect and researcher with a background in automation engineering. He received a BA in Architecture from the Bern University of Applied Sciences and an MArch from The Bartlett School of Architecture, UCL. In 2011, he took up the Professorship for Architecture and Digital Fabrication (Prof. Fabio Gramazio, Prof. Matthias Kohler) at ETH Zurich and completed his PhD on architectural fabrication processes with flying robots in 2016. At Gramazio Kohler Research, he oversees and is involved in a variety of research projects related to robotic fabrication in architecture.

Caitlin Mueller is a researcher, designer and educator working at the interface of architecture and structural engineering. She is currently an Assistant Professor at the Massachusetts Institute of Technology's Department of Architecture and Department of Civil and Environmental Engineering in the Building Technology Programme, where she leads the Digital Structures research group. Mueller earned a PhD in Building Technology from MIT, an SM in Computation for Design and Optimisation from MIT, an MS in Structural Engineering from Stanford University and a BS in Architecture from MIT, and has practised at several architecture and engineering firms across the US.

Tobias Müller, from Munich, studied civil engineering at the Augsburg University of Applied Sciences. While working as a project manager at Peuckert GmbH, he was responsible for the development, planning and implementation of special constructions on a variety of different building projects, such as the ceiling system of the Stuttgart and Nuremberg airports, several subway stations in Cologne and Düsseldorf, the RTL TV station in Cologne and the concert hall of the Elbphilharmonie Hamburg. Since 2012, he has been part of the management at Peuckert GmbH.

Stefan Neudecker holds the Design by Technology Chair at Folkwang University of the Arts in Essen. Within the field of industrial design, he teaches experimental fabrication methods, interdisciplinary design processes and mass customisation strategies. He initially studied electrical engineering at the University of Erlangen and later studied architecture at the Technical University at Braunschweig. Besides freelance activities at architectural offices including realities:united, Graftlab and GMP, he has worked as a research assistant at the IMD (Institute for Media and Design) and the ITE (Institute of Structural Design) at the Technical University at Braunschweig. He developed and managed the DFG Research facility DBFL at the ITE.

Paul Nicholas holds a PhD in Architecture from RMIT University, Melbourne, Australia. Having previously practised with Arup Consulting Engineers from 2005 and AECOM/Edaw from 2009, Paul joined the Centre for Information Technology and Architecture (CITA), Copenhagen, Denmark, in 2011. He currently leads the CITAstudio international Masters programme. Nicholas' particular interest is in the development of innovative computational approaches that establish new bridges between design, structure and materiality. His recent research explores sensor-enabled robotic fabrication, multi-scale modelling and the idea that designed materials such as composites necessitate new relationships between material, representation, simulation and making.

Jifei Ou (欧冀飞) is a designer, researcher and PhD candidate at the MIT Media Lab, where he focuses on designing and fabricating transformable materials across scales (from μm to m). Physical materials are usually considered as static, passive and permanent. Ou is interested in redesigning physical materials with the characteristics of digital information. As much as his work is informed by digital technology, he is inspired in equal measure by the natural world around him. He has been leading projects that study biomimicry and bio-derived materials to design shape-changing packaging, garments and furniture.

Ştefana Parascho is a PhD researcher at Gramazio Kohler Research, as part of the National Centre of Competence in Research (NCCR) Digital Fabrication at the ETH Zurich. She studied architecture at the University of Bucharest and the University of Stuttgart. She has worked for Knippers Helbig Advanced Engineering and Design-to-Production Stuttgart and has taught and researched at the University of Stuttgart Institute of Building Structures and Structural Design. Her current research focus lies in the development of multi-robotic assembly processes for architectural applications and integrative design methods for the design of robotically fabricated structures.

Gernot Parmann has been Assistant Professor at the Institute for Structural Design at Graz University of Technology, where he studied architecture, since 2012. Before his studies, he was an accomplished cast mechanic and machine tool operator, working in this profession for 10 years. Since his diploma thesis, 'Modular Standard', in 2012, he has worked in the field of applied research for manufacturing processes, specifically in concrete construction. At present, he is writing his PhD thesis about the fabrication of UHPC shell structures and works as a Project Manager for product development at Max Bögl Bauservice GmbH.

Stefan Peters is Professor at the Institute for Structural Design at Graz University of Technology, and since 2013 has been the Dean of the Faculty of Architecture. He received his doctorate from Jan Knippers on the subject of the structural application of GFK and glass. In 2007, together with Stephan Engelsmann, he founded the civil engineering company Engelsmann Peters Beratende Ingenieure, with branches in Stuttgart and Graz. His most important projects are breathe.austria, Austrian Pavilion Expo Milano 2015; the glass dome on the Mansueto Library in Chicago; the ZOB Schwäbisch Hall; the ZOB Pforzheim (awarded the Staatspreis Baukultur Baden Württemberg 2016); and the pedestrian and cyclists' bridge in Stuttgart-Vaihingen (awarded the Ingenieurpreis des Deutschen Stahlbaus). His publications include *Faustformel Tragwerksentwurf* (2013) and *Pre-Fabricated Non-Standard Shell Structures Made of UHPC* (2015).

Jörg Petri
Before taking up the post of Director of Innovation at NOWlab@BigRep in 2016, and establishing NOWlab in 2014, Petri was Assistant Professor and PhD fellow at the ITE, TU Braunschweig. His research focuses on robotic and additive fabrication methods in construction, design and architecture. In an academic context, he has published, taught and lectured at IASS Amsterdam and Tokyo, AvB Amsterdam, University of Kentucky, TU Cottbus, TU Coburg and TU Kassel. From 2006-13, Petri was an Associate Architect and Project Leader at UNStudio, van Berkel & Bos in Amsterdam, and also led the Knowledge Platform Architectural Sustainability ASP. Petri is a registered architect at Architektenkammer Berlin.

Marshall Prado is a Research Associate at the Institute for Computational Design at the University of Stuttgart. He holds a Bachelor of Architecture degree from North Carolina State University, and a Master of Architecture and a Master of Design Studies in Technology from the Harvard University Graduate School of Design. Prado has taught at the University of Hawaii and has been an invited studio critic at the University of Pennsylvania, Carnegie Mellon University, the University of Michigan and the Wentworth Institute of Technology. He has led several workshops on digital design and fabrication techniques. His current research interests include the integration of computation and fabrication techniques into material systems and spatial design strategies.

David Reeves is a designer, programmer and researcher currently working at Zaha Hadid Architects as a member of the Computation and Design (CoDe) group. Here, his research focuses on geometry processing and novel applications of numerical methods within architectural design modelling. Reeves has taught and presented work at various institutions, including the Architectural Association, The Bartlett, CITA and Yale University. His work has also been published and presented at international conferences and events such as Acadia, SimAud, IASS and Smartgeometry.

Gilles Retsin is the founder of Gilles Retsin Architecture, a young award-winning London-based architecture and design practice. The practice has developed numerous provocative proposals for international competitions, and most recently was a finalist in the international competition for the New National Gallery in Budapest. Currently, the practice is working on a range of projects, among them a 10,000m^2 museum in China. Retsin graduated from the Architectural Association in London. Prior to founding his own practice, he worked in Switzerland as a project architect with Christian Kerez, and in London with Kokkugia. Alongside his practice, Gilles directs a research cluster at The Bartlett School of Architecture, UCL, and is also a Senior Lecturer at the University of East London. His work has been acquired by the Centre Pompidou in Paris and he has exhibited internationally in museums including the Museum of Art and Design in New York. His visionary designs have been featured in publications such as *Wired*, *The Guardian* and *Dezeen*, as well as in academic conferences such as ECAADE, ACADIA and FABRICATE.

Roger Ridsdill Smith leads Foster + Partners' Structural Engineering team. He gained his degree in structural engineering from Cambridge University and began his professional career in Paris. In 1994, he joined Ove Arup and Partners, becoming a director of the firm in 2003 and subsequently running a multidisciplinary engineering group in Arup's London office. His other projects include the Millennium Bridge in London, Chateau Margaux winery in France and Tocumen International Airport in Panama. In 2010, he won the Royal Academy of Engineering Silver Medal.

Matthias Rippmann has been a member of the Block Research Group at ETH Zurich since 2010, obtaining his doctorate in 2016. In 2015, he joined the Swiss National Centre of Competence in Research Digital Fabrication as a Postdoctoral Fellow. He conducts research in structurally informed design and digital fabrication, and is the developer of the form-finding software RhinoVAULT. He studied architecture at the University of Stuttgart and the University of Melbourne. He worked in Stuttgart at Behnisch Architekten, LAVA, the Institute for Lightweight Structures and Conceptual Design and Werner Sobek Engineers. In 2010, he co-founded the consultancy firm Rippmann Oesterle Knauss GmbH (ROK).

Christopher Robeller is an architect and postdoctoral researcher at the Swiss National Centre for Competence in Research (NCCR) Digital Fabrication. Christopher received his architecture diploma with distinction from London Metropolitan University in 2008 and worked at ICD Stuttgart from 2008-10, where he developed integral timber plate joints for the award-winning ICD/ITKE Research Pavilion 2010. Since 2011, he has worked at IBOIS and received a doctoral degree from EPFL in 2015 for his thesis 'Integral Mechanical Attachment for Timber Folded Plate Structures'. His research has been published in journals and at conferences such as *The International Journal of Space Structures*, *Bauingenieur*, ACADIA, RobArch and AAG, where he received the Best Paper Award in 2014.

Sarah Rodrigo is an artist and arts manager with over twenty years' experience working on major public art projects. She is currently overseeing Helmick Sculpture's first international commissions.

Jean Roulier trained as a joiner, carpenter and wood building engineer. Having accumulated extensive experience in CAD in practice, he co-founded the company Lignocam SA in 2006 in order to develop a CAM software for the wood industry. Since then, Lignocam has become the leading CAM software for interpreting BTL files. Its objective is the promotion of wood in construction – even in the most daring projects – as well as the realisation of a smooth digital chain in the construction and fabrication process.

Fabian Scheurer is co-founder of Design-to-Production and leads the company's office in Zurich. He graduated from the Technical University of Munich with a diploma in computer science and architecture and gained professional experience as a CAD-trainer, software developer and new media consultant. In 2002, he joined Ludger Hovestadt's CAAD group at ETH Zurich, where he co-founded Design-to-Production as a research group to explore the connections between digital design and fabrication. At the end of 2006, Design-to-Production teamed up with architect Arnold Walz and became a commercial consulting practice, supporting architects, engineers and fabricators in the digital production of complex design.

Simon Schleicher is an Assistant Professor in the Department of Architecture at the University of California, Berkeley. His transdisciplinary work draws from architecture, engineering and biology. By cross-disciplinary pooling of knowledge, he aims to transfer bending and folding mechanisms found in nature to lightweight and responsive systems in architecture.

Patrik Schumacher has been Principal at Zaha Hadid Architects since its RIBA Gold Medal and Pritzker prize-winning founder's passing in April 2016. He has been a co-author on multiple projects since he joined the firm in 1988, and since then, alongside Hadid, has built the practice into the 400-strong global architecture and design brand it is today. In 1996, he founded the Design Research Laboratory at the Architectural Association, where he continues to teach. He lectures worldwide and recently held the John Portman Chair in Architecture at Harvard's GSD. Over the last 20 years, he has contributed over 100 articles to architectural journals and anthologies, including his manifestos on Parametricism, a two-volume theoretical magnum opus *The Autopoiesis of Architecture* and the magazine *AD – Parametricism 2.0*.

Tobias Schwinn is a Research Associate and doctoral candidate at the Institute for Computational Design (ICD) at the University of Stuttgart under the supervision of Professor Achim Menges. In his research, he focuses on behaviour-based approaches for computational design integration in the context of robotic fabrication of segmented shell structures. Prior to joining the ICD in 2011, he worked as a Senior Designer for Skidmore, Owings and Merrill in New York and London. Schwinn studied architecture at the Bauhaus-University in Weimar, Germany, and at the University of Pennsylvania in Philadelphia as part of the US-EU Joint Consortium for Higher Education. He received his diploma in engineering in architecture in 2005.

Martin Self is Director of Hooke Park, the Architectural Association's woodland campus, and is co-Director of the Design + Make Masters programme, which he founded in 2009. He holds degrees in aerospace engineering and architecture theory, and worked as a consultant engineer at Ove Arup & Partners between 1996 and 2007, where he was a founding member of its Advanced Geometry Group. He has also consulted within practices including Zaha Hadid Architects and Antony Gormley Studio. He has taught at the Architectural Association since 2005, including as tutor for Intermediate Unit 2's series of summer pavilions, and has directed Hooke Park since 2010, where he is overseeing the production of a series of student-designed experimental buildings.

Timothy Shan Sutherland received a Master of Fine Arts degree, with a concentration on metalworking, from the Cranbrook Academy of Art in 2005. He is a Master of Architecture and Master of Science in Digital Fabrication candidate at the Taubman College of Architecture, University of Michigan. He lectures in Architectural Tectonics and Visual Communication at Lawrence Technical University. He has executed several private architectural commissions and many public sculptural installations in the US and abroad. His work continues to explore the combination of industrial and pre-industrial processes with digital design and fabrication methods.

Yuliya Sinke Baranovskaya is currently a research assistant at CITA, at The Royal Danish Academy of Fine Arts School of Architecture, Design and Conservation. She joined the Centre in early 2016 and became part of the Complex Modelling Project, which focuses on the integration of material performances with digital design cultures. Prior to this, Baranovskaya graduated from the ITECH MSc Programme (2015) at the University of Stuttgart and was highly involved in design development and robotic fabrication for the ICD/ITKE Research Pavilions 2013 and 2014-15. Her interest lies in the realms of fibrous morphologies and innovative material applications for architecture and large-scale installations.

Vicente Soler is an architect who consults and lectures as a specialist in computational design and digital fabrication, participating in multiple internationally recognised projects. In academia, he has worked as a researcher in robotics applied to architecture and has taught in several postgraduate architectural programs, both in Spain and the UK. Currently, he works at The Bartlett School of Architecture, co-tutoring a unit in the Architectural Design postgraduate programme and offering support for computational design and robotics. He has developed software for the programming and control of industrial robots that is actively used in multiple architecture schools around the world.

James Solly is a Research Associate at the Institute for Building Structures and Structural Design (ITKE) at University of Stuttgart and a member of the Innochain ETN network, where he is working on virtual prototyping strategies for fibre-reinforced polymers. He is a chartered engineer (CEng MICE) who previously worked at Ramboll UK, BuroHappold Engineering and Format Engineers, delivering projects from experimental sculptures to international museums, before moving to his current research and teaching role.

Asbjørn Søndergaard is co-founder and Chief Development Officer at Odico, where he heads its industrial research and development. This entails several high-profile research efforts to develop novel fabrication technologies within architectural construction, such as robotic hot blade cutting, augmented reality interfaces for robotic production programming and automation of non-repetitive robotic manufacturing. A trained architect and Research Fellow at Aarhus School of Architecture, Søndergaard's research synthesises an interest in topology optimisation and robotic fabrication as constitutive instruments of architectural design, an interest that is being investigated over several collaborations with, among others, Gramazio Kohler Research, ETH Zurich and the Israel Institute of Technology, Haifa.

David Stasiuk is the Director of Applied Research at Proving Ground, a technology consultancy for architects, engineers and manufacturers that focuses on the development of advanced computational tools that facilitate data-driven design and project collaboration. His academic research exists within the larger framework of CITA's Complex Modelling project, which investigates the digital infrastructures of design models, examining concerns of feedback and scale across the expanded digital design chain. His work discusses adaptive reparameterisation, focusing on the dynamic activation of data structures that allow for model networks to operate holistically as representational engines in the realisation of complex material assemblies.

Hanno Stehling is Partner and Head of Software Development at the digital fabrication consultancy Design-to-Production in Zurich. He graduated with a diploma in architecture from University of Kassel, Germany, where he studied under Prof. Manfred Grohmann (Bollinger + Grohmann) and Prof. Frank Stepper (Coop Himelb(l)au). He has a strong background in computer programming and has gradually focused his studies onto the intersection between architecture and computer science. He worked as a freelance programmer and computational designer for renowned architects like Bernhard Franken before joining Design-to-Production in 2009. Stehling is co-founder of the online platform RhinoScript.org and gives modelling and scripting classes to both academic and professional audiences.

Martin Tamke is Associate Professor at the Centre for Information Technology and Architecture (CITA) in Copenhagen. He joined the newly founded research centre in 2006 and shaped its design-based research practice. Projects on new design and fabrication tools for wood and composite materials led to work which focuses on material-aware design strategies and complex modelling strategies. This is explored through demonstrators that investigate an architectural practice engaged with bespoke materials and behaviour. Currently, he is involved in the Danish-funded four-year Complex Modelling research project and the EU-funded adapt-r and InnoChain PhD research networks.

Andreas Thoma is a Research Assistant at Gramazio Kohler Research, ETH Zurich and the Swiss National Centre for Competence in Research (NCCR) Digital Fabrication. He graduated with a BSc in Architecture from Bauhaus-University Weimar in Germany in 2009, and in 2014 with a Master of Science in Architecture from ETH Zurich. From 2010-12, he worked for Herzog & de Meuron in Basel, Switzerland. In 2012, he joined Gramazio Kohler Research, where he co-led the projects Iridescence Print and Rock Print. Since May 2016, he has led the CTI project Spatial Timber Units.

Mette Ramsgaard Thomsen is Professor and Head of the Centre for Information Technology and Architecture (CITA) in Copenhagen. Her research examines how computation is changing the material cultures of architecture. In projects such as Complex Modelling and Innochain, she explores the infrastructures of computational modelling, including open topologies and adaptive parametrisation. She is pursuing design-led research in the interface and implications of computational design and its materialisation. Recent projects focus on advanced modelling concepts with highly interdependent materials systems and computational design models with integrated simulation of material behaviour.

Geoffrey Thün is Associate Dean of Research and Creative Practice and Associate Professor at the University of Michigan Taubman College of Architecture and Urban Planning. He is a founding Partner of the research-based practice RVTR, which serves as a platform for exploration and experimentation in the agency of architecture and urban design within the context of dynamic ecological systems, infrastructures, materially and technologically mediated environments and emerging social organisations. The work ranges in scale from regional territories and urbanities to full-scale installation-based prototypes that explore responsive and kinetic envelopes that mediate energy, atmosphere and interaction.

Edoardo Tibuzzi graduated from the Civil Engineering Faculty of Rome University in 2005 and joined AKT II in 2007, where he now leads their Computational Design Unit. He has a profound interest in parametric design, interoperable modelling, multi-scale modelling of structures and multidisciplinary design. Since joining AKT II, he has completed various projects, including the UK Pavilion at the Shanghai 2010 Expo, the BMW Pavilion and Coca-Cola BeatBox Pavilion at the London 2012 Olympics, Birmingham New Street Station re-development and Bloomberg's London Headquarters. Tibuzzi has taught Architectural Technology at KTH in Stockholm and led various workshops at Tor Vergata University in Rome, at UCL and at Colgate University. He is a co-author of *Design Engineering Re-Focused*, recently published in the AD Smart03 series.

Daniel Tish is a Research Associate at RVTR. He received his MArch with distinction from the University of Michigan in 2015. He was a 2014 Dow Sustainability Fellow, and his thesis was awarded an honorable mention in the 2015 Jacques Rougerie Innovation and Architecture for Space international competition. His interests lie in the computational investigations of complex systems and geometries, especially those pertaining to responsive architectural elements. He is working towards the development of new metrics and methodologies for designing architectural interactions with the transient nature of the occupants and environments around them.

Andreas Trummer is Associate Professor for Structural Design and Robotic Fabrication at Graz University of Technology. He studied structural engineering at Graz University of Technology and earned his PhD at the University of Natural Resources and Life Sciences, Vienna, in 2002. Since 2009, he has been establishing the 'Roboter Design Labor' and set his research focus on questions of digital fabrication of load-carrying building elements. These range from prefabricated concrete shell elements to the Ceramic Shell Project in collaboration with the GSD Design Robotics Group and an ongoing project about the future of concrete 3D printing processes.

Tom Van Mele is co-Director of the Block Research Group (BRG) at ETH Zurich, where he has led research and development since 2010. In 2008, he received his PhD from the Department of Architectural Engineering at the Vrije Universiteit Brussel in Belgium. His current research projects include the analysis of three-dimensional collapse mechanisms of masonry structures, the development of flexible formwork systems for concrete shells and the development of graphical design and analysis methods such as three-dimensional graphic statics. He is the developer of the online interactive learning platform eQUILIBRIUM, and of compAS, the computational research framework for architecture and structures.

Kathy Velikov is Associate Professor at the University of Michigan Taubman College of Architecture and Urban Planning. She is a licensed architect and a founding Partner of the research-based practice RVTR, which serves as a platform for exploration and experimentation in the agency of architecture and urban design within the context of dynamic ecological systems, infrastructures, materially and technologically mediated environments and emerging social organisations. The work ranges in scale from regional territories and urbanities to full-scale installation-based prototypes that explore responsive and kinetic envelopes that mediate energy, atmosphere and interaction. Velikov currently serves on the Board of Directors for ACADIA.

Emmanuel Vercruysse holds a set of leading roles at the Architectural Association. He is co-Director of the Postgraduate Design + Make course based at the AA's satellite campus out in Hooke Park, positioning the campus at the forefront of architectural research through prototyping and large-scale fabrication. The roles of Director of the Robotic Fabrication Visiting School and Curator of Robotic Development allow for developing research areas within the cutting-edge field of robotic fabrication, both in Bedford Square and at Hooke Park.

James Warton is a computational designer and applications developer with HKS LINE. He received a Master's degree in Architecture and Urbanism from the Architectural Association's Design Research Laboratory (DRL). He is currently enrolled as a PhD student in SMU's Mechanical Engineering Department, where he holds a Research Assistantship through the Research Center for Advanced Manufacturing (RCAM). At HKS, his responsibilities encompass a range of activities from preliminary analysis and conceptual design through to the design development and documentation of complex architectural systems.

Karl D.D. Willis is a member of the Integrated Additive Manufacturing team at Autodesk. He holds a PhD in Computational Design from Carnegie Mellon University and previously worked in the Interaction Group at Disney Research. He has over 20 publications and 10 patents issued and pending, and his work has been covered extensively in the media, including the BBC, NBC, *Wired* and *New Scientist*.

Maria Yablonina is a researcher and a PhD candidate at the Institute for Computational Design (ICD) at the University of Stuttgart. With a strong interest in mobile robotics and distributed collaborative systems, she is currently focusing on exploring potential fabrication techniques enabled through the introduction of architecture-specific custom robotic tools for construction and fabrication. Her work includes the development of hardware and software tools, as well as complementing material systems. Prior to joining the ICD in 2016, she completed her Masters at the ITECH programme in 2015 and was an Artist-in-Residence at Autodesk Pier 9 in 2016.

Philip F. Yuan is a Professor and PhD Advisor in the College of Architecture and Planning (CAUP) at Tongji University and a council member of the Architectural Society of China (ASC). He is the co-founder of the Digital Architectural Design Association (DADA) of the ASC and founding Partner of Shanghai Archi-Union Architects and Fab-Union Intelligent Engineering Co. Ltd. He was a Visiting Scholar at MIT during 2008–09. He established the Digital Design Research Center (DDRC) in 2010 and has organised the DigitalFUTURE Shanghai summer school programme for the past six years. His research focuses on computational design and digital fabrication.

Mateusz Zwierzycki is a Research Assistant at the Centre for Information Technology and Architecture (CITA) in Copenhagen, as well as an architect, developer, computational design populariser and tutor for many international coding and design workshops. His focus is on parametric and generative design, which he considers to be a natural way of design thinking. Currently, he is working on the application of machine learning to generative design within academia and practice. He has created a variety of plug-ins and libraries for Grasshopper3d, such as Anemone, Starling, Volvox, Squid and, most recently, Owl.

Fabricate 2017
Co-organisers

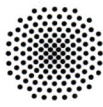

University of Stuttgart

ICD Institute for Computational Design and Construction

UCL

The Bartlett School of Architecture

Diamond sponsor

AUTODESK

Platinum sponsors

ARUP   KUKA

CIG OSTSEE STAAL   bigrep

Gold sponsors

FARO   design to production   Trimble

With the support of

DFG

Fabricate 2017

Editors:
Bob Sheil, Achim Menges, Ruairi Glynn and Marilena Skavara
Project Editor:
Eli Lee

First published in 2017 by
UCL Press
University College London
Gower Street
London WC1E 6BT

Available to download free: www.ucl.ac.uk/ucl-press

Text © Bartlett School of Architecture and the authors
Images © Bartlett School of Architecture and the authors

A CIP catalogue record for this book is available from
The British Library.

This book is published under a Creative Commons
Attribution Non-commercial Non-derivative 4.0 International
license (CC BY-NC-ND 4.0). This license allows you to share,
copy, distribute and transmit the work for personal and
non-commercial use providing author and publisher
attribution is clearly stated. Attribution should include
the following information:

Bob Sheil, Achim Menges, Ruairi Glynn and Marilena Skavara
(eds.), *Fabricate*. London, UCL Press, 2017. https://doi.
org/10.14324/111. 9781787350014

Further details about CC BY licenses are available at
creativecommons.org/licenses.

ISBN: 978-1-78735-000-7 (Hbk.)
ISBN: 978-1-78735-001-4 (PDF)
DOI: https://doi.org/10.14324/111. 9781787350014

Design:
Patrick Morrissey / Unlimited

Printing:
Albe de Coker, Antwerp, Belgium

UCLPRESS

fabricate.org

We are pleased to announce that we have secured the
domain name **fabricate.org** and are looking for partners to
support us in building this space as a world-leading resource
for design and making.

Our existing domains (fabricate2011.org, fabricate2014.org and
fabricate2017.org) will migrate to the new timeless address soon.

If you are interested in finding out more about **fabricate.org**
and future FABRICATE events, please email **partners@fabricate.org**